The IMA Volumes in Mathematics and its Applications

T0143136

For other titles published in this series, go to
www.springer.com/series/811

Institute for Mathematics and its Applications (IMA)

The **Institute for Mathematics and its Applications** was established by a grant from the National Science Foundation to the University of Minnesota in 1982. The primary mission of the IMA is to foster research of a truly interdisciplinary nature, establishing links between mathematics of the highest caliber and important scientific and technological problems from other disciplines and industries. To this end, the IMA organizes a wide variety of programs, ranging from short intense workshops in areas of exceptional interest and opportunity to extensive thematic programs lasting a year. IMA Volumes are used to communicate results of these programs that we believe are of particular value to the broader scientific community.

The full list of IMA books can be found at the Web site of the Institute for Mathematics and its Applications:

http://www.ima.umn.edu/springer/volumes.html.

Presentation materials from the IMA talks are available at

http://www.ima.umn.edu/talks/.

Video library is at http://www.ima.umn.edu/videos/.

Fadil Santosa, Director of the IMA

* * * * * * * * * *

IMA ANNUAL PROGRAMS

1982–1983	Statistical and Continuum Approaches to Phase Transition
1983–1984	Mathematical Models for the Economics of Decentralized Resource Allocation
1984–1985	Continuum Physics and Partial Differential Equations
1985–1986	Stochastic Differential Equations and Their Applications
1986–1987	Scientific Computation
1987–1988	Applied Combinatorics
1988–1989	Nonlinear Waves
1989–1990	Dynamical Systems and Their Applications
1990–1991	Phase Transitions and Free Boundaries
1991–1992	Applied Linear Algebra
1992–1993	Control Theory and its Applications
1993–1994	Emerging Applications of Probability
1994–1995	Waves and Scattering
1995–1996	Mathematical Methods in Material Science
1996–1997	Mathematics of High Performance Computing

Continued at the back

John C. Baez · J. Peter May
Editors

Towards Higher Categories

 Springer

Editors
John C. Baez
Department of Mathematics
University of California, Riverside
USA
http://www.math.ucr.edu/home/baez

J. Peter May
Department of Mathematics
University of Chicago
USA
http://www.math.uchicago.edu/~may/

ISBN 978-1-4614-2463-5   e-ISBN 978-1-4419-1524-5
DOI 10.1007/978-1-4419-1524-5
Springer New York Dordrecht Heidelberg London

Mathematics Subject Classification (2000): 18A99, 18D99, 18F99, 18G99

Springer is part of Springer Science+Business Media (www.springer.com)

FOREWORD

This IMA Volume in Mathematics and its Applications

TOWARDS HIGHER CATEGORIES

contains expository and research papers based on a highly successful IMA Summer Program on n-Categories: Foundations and Applications. We are grateful to all the participants for making this occasion a very productive and stimulating one.

We would like to thank John C. Baez (Department of Mathematics, University of California Riverside) and J. Peter May (Department of Mathematics, University of Chicago) for their superb role as summer program organizers and editors of this volume. We take this opportunity to thank the National Science Foundation for its support of the IMA.

Series Editors

Fadil Santosa, Director of the IMA

Markus Keel, Deputy Director of the IMA

PREFACE

DEDICATED TO **MAX KELLY**, JUNE 5 1930 TO JANUARY 26 2007.

This is *not* a proceedings of the 2004 conference "n-Categories: Foundations and Applications" that we organized and ran at the IMA during the two weeks June 7–18, 2004! We thank all the participants for helping make that a vibrant and inspiring occasion. We also thank the IMA staff for a magnificent job. There has been a great deal of work in higher category theory since then, but we still feel that it is not yet time to offer a volume devoted to the main topic of the conference. At that time, we felt that we were at the beginnings of a large new area of mathematics, but one with many different natural approaches in desperate need of integration into a cohesive field. We feel that way still.

So, instead of an introduction to higher category theory, we have decided to publish a series of papers that provide useful *background* for this subject. This volume is aimed towards the wider mathematical community, rather than those knowledgable in category theory and especially higher category theory. We are particularly sensitive to the paucity of young Americans knowledgeable enough in the subject to be potential readers.

The focus of the conference was on comparing the many approaches to higher category theory. These approaches, as they existed at the time of the conference, have been summarized by Leinster [29] and by Cheng and Lauda [19]. The earliest was based on filtered simplicial sets. It is due to Street [48, 50] and is being developed in detail by Verity [54, 55, 56], with a contribution by Gurski [25]. Another approach, called "opetopic" since operads were used to describe the shapes of diagrams used in the theory, is due to Baez and Dolan [1] and has been developed by Leinster [30], Cheng [13, 14, 15, 16], and Makkai and his collaborators [26, 36]. There is also a topologically motivated approach using operads due to Trimble [53], which has been studied and generalized by Cheng and Gurski [17, 18]. There is a quite different and more extensively developed operadic approach to weak ∞-categories due to Batanin [3, 49], with a variant due to Leinster [31]. Penon [42] gave a related, very compact definition of ∞-category, which was later improved and corrected by Batanin [4] and Cheng and Makkai [20]. Another highly developed approach, based on n-fold simplicial sets and inspired by work of Segal in infinite loop space theory, is due to Tamsamani and Simpson [43, 44, 45, 46, 51]. Yet another theory, due to Joyal [8, 27], is based on a presheaf category called the category of "theta-sets". Also, Lurie [35] has begun making extensive use of Barwick's ideas on (∞, n)-categories, which are roughly weak ∞-categories where all morphisms above dimension n are invertible [2].

The theme of the 2004 conference was comparisons among all these approaches to higher categories — or at least, those that existed at the time. While there are papers that tackle aspects of this immense unification project [9, 15, 17, 47], it is still quite unclear how a unified theory of higher categories will evolve. In proposing the conference, we wrote as follows: "It is *not* to be expected that a single all embracing definition that is equally suited for all purposes will emerge. It is not a question as to whether or not a good definition exists. Not one, but many, good definitions already do exist, although they have been worked out to varying degrees. There is growing general agreement on the basic desiderata of a good definition of n-category, but there does not yet exist an axiomatization, and there are grounds for believing that only a partial axiomatization may be in the cards."

One can make analogies with many other areas where a number of interrelated definitions exist. In algebraic topology, there are various symmetric monoidal categories of spectra, and they are related by a web of Quillen equivalences of model categories. In this context, all theories are in some sense "the same", but the applications require use of different models: many things that can be proven in one model cannot easily be proven in another [38]. In algebraic geometry, there are many different cohomology theories, definitely not all the same, but connected by various comparison functors. Motivic theory is in part a search for a universal source of such comparisons.

In the case of weak n-categories, it is unclear whether there is a useful sense in which all known theories are the same. We do not have a complete web of comparison maps relating different theories. Nor are we sure what it *means* for two theories to be "the same", despite important insights by Grothendieck [24] and Makkai [37]. The terms in which comparisons should be made are not yet clear. Quillen model category theory should capture some comparisons, but it seems likely to be too coarse to give the complete story.

A smaller related theme of the conference was that there should be a "baby" comparison project, for which model category theory would in fact be sufficient. Precisely, the idea was that there should be a web of Quillen equivalences among the various notions of $(\infty, 1)$ category. These include topologically or simplicially enriched categories, Segal categories, complete Segal spaces, and quasi-categories. In the years since the conference, this comparison project has been largely completed by Bergner [10, 11] and Joyal and Tierney [28].

Another smaller related theme was the higher categorical modelling of homotopy n-types of topological spaces. It was a dream of Grothendieck [24] that weak n-groupoids should model n-types. Brown has shown that strict n-groupoids are easy to compute with [12]; unfortunately, they capture only part of the information in an n-type. However, Loday [32] and others have found other strict algebraic structures that can fully model n-types. More recently we are seeing work that implements Grothendieck's original idea in various approaches to weak n-categories and that compares these approaches to the algebraic approaches [5, 6, 21, 40, 41, 52].

As mentioned, the goal of this volume is merely to prepare the reader for more detailed study of these fast-moving topics. So, we begin with a light-hearted paper that treats Grothendieck's dream as a starting-point for speculations on the relationship between n-categories and cohomology. It is based on notes that Michael Shulman took of John Baez's 2007 Namboodiri Lectures at Chicago. Higher category theory has largely developed from a series of analogies with and potential applications to other subjects, including algebraic topology, algebraic geometry, mathematical physics, computer science, logic, and, of course, category theory. This paper illustrates this, and raises the challenge of formalizing the patterns that become visible thereby.

The second paper, by Julie Bergner, is a survey of $(\infty, 1)$-categories. She begins by describing four approaches: simplicial categories, complete Segal spaces, Segal categories and quasi-categories. Then she describes the network of Quillen equivalences relating these: the "baby comparison project" mentioned above.

The third paper, by Simona Paoli, focuses on ways of modelling homotopy n-types of topological spaces in terms of strict algebraic structures. Her focus is on "internal" structures, meaning structures that live in categories other than the category of sets. She surveys the known comparisons among such algebraic models for n-types, and she also compares them to Tamsamani's weak n-groupoids.

Logically, we might next delve into various approaches to weak n-categories and full-fledged ∞-categories. But this seems premature. So instead, the rest of the volume goes back to the beginnings of the subject. By now every well-educated young mathematician can be expected to be familiar with categories, as introduced by Eilenberg and Mac Lane in 1945. It has taken longer to understand that what they introduced was a 2-category: Cat, with categories as objects, functors as morphisms, and natural transformations as 2-morphisms. Ehresmann [22] introduced strict n-categories sometime in the 1960's, and Eilenberg and Kelly discussed them in 1965 [23], with Cat as a key example of a strict 2-category. Bénabou introduced the more general weak 2-categories or "bicategories" the following year [7]. But even today, the mathematics of 2-categories is considered somewhat recondite, even by many mathematicians who implicitly use these structures all the time. There is a great deal of basic 2-category theory that can illuminate everyday mathematics. The second editor rediscovered a chunk of this while writing a book on parametrized homotopy theory [39], and he was chastened to see how little he knew of something that was so very basic to his own work.

For this reason, the next paper is the longest in this volume: a thorough introduction to the theory of 2-categories, by Steve Lack. This paper gives a solid grounding for anyone who wants some idea of what lies beyond mere categories and how to work with higher categorical notions. Anybody interested in higher category theory must learn something of the richness of 2-categories.

Lawrence Breen's paper, on gerbes and 2-gerbes, gives an idea of how naturally 2-categorical algebra arises in the study of algebraic and differential geometry. His paper also illustrates the need for "enriched" higher category theory, in which one deals with hom objects that have more structure than is seen in merely set-based categories. Many more such applications could be cited.

Steve Lack is an Australian, and it is noteworthy that the premier world center for category theory has long been Sydney. We have dedicated this volume to Max Kelly, the founder of the Australian school of category theory, who died in 2007. Kelly visited Chicago in 1970-71, just before becoming Head of the Department of Pure Mathematics at the University of Sydney. That was long before e-mail, and Max was considering how best to build up a department that would necessarily suffer from a significant degree of isolation. He decided to focus in large part on the development of his own subject, and he succeeded admirably. The final paper in this volume, by Kelly's student Ross Street, gives a fascinating mathematical and personal account of the development of higher category theory in Australia.

We had very much hoped to include a survey by André Joyal of his important work on quasi-categories. As shown in the work of Lurie [33, 34], these give a very tractable model of $(\infty, 1)$-categories. Joyal's work showing that one "can do category theory" in quasi-categories is an essential precursor to Lurie's work and is unquestionably one of the most important recent developments in higher category theory. However, Joyal's survey is not yet complete: it has grown to hefty proportions and is still growing. So, it will appear separately.

REFERENCES

[1] J. Baez and J. Dolan, Higher-dimensional algebra III: n-categories and the algebra of opetopes, *Adv. Math.* **135** (1998), 145–206. Also available as arXiv:q-alg/9702014.

[2] C. Barwick, *Weakly Enriched* **M**-*Categories*, work in progress.

[3] M. Batanin, Monoidal globular categories as natural environment for the theory of weak n-categories, *Adv. Math.* **136** (1998), 39–103.

[4] M. Batanin, On Penon method of weakening algebraic structures, *Jour. Pure Appl. Algebra* **172** (2002), 1–23.

[5] M. Batanin, The symmetrisation of n-operads and compactification of real configuration spaces, *Adv. Math.* **211** (2007), 684–725. Also available as arXiv:math/0301221.

[6] M. Batanin, The Eckmann–Hilton argument and higher operads, *Adv. Math.* **217** (2008), 334–385. Also available as arXiv:math/0207281.

[7] J. Bénabou, Introduction to bicategories, in *Reports of the Midwest Category Seminar*, Springer, Berlin, 1967, pp. 1–77.

[8] C. Berger, Iterated wreath product of the simplex category and iterated loop spaces, *Adv. Math.* **213** (2007), 230–270. Also available as arXiv:math/0512575.

[9] C. Berger, A cellular nerve for higher categories, *Adv. Math.* **169** (2002), 118–175. Also available at ⟨http://math.ucr.edu.fr/∼cberger/⟩.

[10] J. Bergner, A model category structure on the category of simplicial categories, *Trans. Amer. Math. Soc.* **359** (2007), 2043-2058. Also available as arXiv:math/0406507.

[11] J. Bergner, Three models of the homotopy theory of homotopy theory, *Topology* **46** (2006), 1925–1955. Also available as arXiv:math/0504334.
[12] R. Brown, P. Higgins and R. Sivera, *Nonabelian Algebraic Topology: Higher Homotopy Groupoids of Filtered Spaces*. Available at ⟨http://www.bangor.ac.uk/~mas010/nonab-a-t.html⟩.
[13] E. Cheng, The category of opetopes and the category of opetopic sets, *Th. Appl. Cat.* **11** (2003), 353–374. Also available as arXiv:math/0304284.
[14] E. Cheng, Weak *n*-categories: opetopic and multitopic foundations, *Jour. Pure Appl. Alg.* **186** (2004), 109–137. Also available as arXiv:math/0304277.
[15] E. Cheng, Weak *n*-categories: comparing opetopic foundations, *Jour. Pure Appl. Alg.* **186** (2004), 219–231. Also available as arXiv:math/0304279.
[16] E. Cheng, Opetopic bicategories: comparison with the classical theory, available as arXiv:math/0304285.
[17] E. Cheng, Comparing operadic theories of *n*-category, available as arXiv:0809.2070.
[18] E. Cheng and N. Gurski, Toward an *n*-category of cobordisms, *Th. Appl. Cat.* **18** (2007), 274–302. Available at ⟨http://www.tac.mta.ca/tac/volumes/18/10/18-10abs.html⟩.
[19] E. Cheng and A. Lauda, *Higher-Dimensional Categories: an Illustrated Guidebook*. Available at ⟨http://www.dpmms.cam.ac.uk/~elgc2/guidebook/⟩.
[20] E. Cheng and M. Makkai, A note on the Penon definition of *n*-category, to appear in *Cah. Top. Géom. Diff.*
[21] D.-C. Cisinski, Batanin higher groupoids and homotopy types, in *Categories in Algebra, Geometry and Mathematical Physics*, eds. A. Davydov *et al*, *Contemp. Math.* **431**, AMS, Providence, Rhode Island, 2007, pp. 171–186. Also available as arXiv:math/0604442.
[22] C. Ehresmann, *Catégories et Structures*, Dunod, Paris, 1965.
[23] S. Eilenberg and G. M. Kelly, Closed categories, in *Proceedings of the Conference on Categorical Algebra (La Jolla, California, 1965)*, eds. S. Eilenberg *et al*, Springer, New York, 1966.
[24] A. Grothendieck, *Pursuing Stacks*, letter to D. Quillen, 1983. To be published, eds. G. Maltsiniotis, M. Künzer and B. Toen, *Documents Mathématiques*, Soc. Math. France, Paris, France.
[25] M. Gurski, Nerves of bicategories as stratified simplicial sets. To appear in *Jour. Pure Appl. Alg.*.
[26] C. Hermida, M. Makkai, and J. Power: On weak higher-dimensional categories I, II. *Jour. Pure Appl. Alg.* **157** (2001), 221–277.
[27] A. Joyal, Disks, duality and θ-categories, preprint, 1997.
[28] A. Joyal and M. Tierney, Quasi-categories vs Segal spaces, available as arXiv:math/0607820.
[29] T. Leinster, A survey of definitions of *n*-category, *Th. Appl. Cat.* **10** (2002), 1–70. Also available as arXiv:math/0107188.
[30] T. Leinster, Structures in higher-dimensional category theory. Available as arXiv:math/0109021.
[31] T. Leinster, *Higher Operads, Higher Categories*, Cambridge U. Press, Cambridge, 2003. Also available as arXiv:math/0305049.
[32] J. L. Loday, Spaces with finitely many non-trivial homotopy groups, *Jour. Pure Appl. Alg.* **24** (1982), 179–202.
[33] J. Lurie, *Higher Topos Theory*, available as arXiv:math/0608040.
[34] J. Lurie, Stable infinity categories, available as arXiv:math/0608228.
[35] J. Lurie, On the classification of topological field theories, available as arXiv:0905.0465.
[36] M. Makkai, The multitopic ω-category of all multitopic ω-categories. Available at ⟨http://www.math.mcgill.ca/makkai⟩.
[37] M. Makkai, On comparing definitions of "weak *n*-category". Available at ⟨http://www.math.mcgill.ca/makkai⟩.
[38] J. P. May, What precisely are E_∞ ring spaces and E_∞ ring spectra, *Geometry and*

Topology Monographs **16** (2009), 215–284.

[39] J. P. May and J. Sigurdsson, *Parametrized Homotopy Theory*, AMS, Providence, Rhode Island, 2006.

[40] S. Paoli, Semistrict models of connected 3-types and Tamsamani's weak 3-groupoids. Available as arXiv:0607330.

[41] S. Paoli, Semistrict Tamsamani n-groupoids and connected n-types. Available as arXiv:0701655.

[42] J. Penon, Approche polygraphique des ∞-catégories non strictes, *Cah. Top. Géom. Diff.* **40** (1999), 31–80. Also available at ⟨http://www.numdam.org/item?id=CTGDC_1999_40_1_31_0⟩

[43] C. Simpson, A closed model structure for n-categories, internal Hom, n-stacks and generalized Seifert–Van Kampen. Available as arXiv:alg-geom/9704006.

[44] C. Simpson, Limits in n-categories, available as arXiv:alg-geom/9708010.

[45] C. Simpson, Calculating maps between n-categories, available as arXiv:math/0009107.

[46] C. Simpson, On the Breen–Baez–Dolan stabilization hypothesis for Tamsamani's weak n-categories, available as arXiv:math/9810058.

[47] C. Simpson, Some properties of the theory of n-categories, available as arXiv:math/0110273.

[48] R. Street, The algebra of oriented simplexes, *Jour. Pure Appl. Alg.* **49** (1987), 283–335.

[49] R. Street, The role of Michael Batanin's monoidal globular categories, in *Higher Category Theory*, Contemp. Math. **230**, AMS, Providence, Rhode Island, 1998, pp. 99–116. Also available at ⟨http://www.math.mq.edu.au/~street⟩.

[50] R. Street, Weak omega-categories, in *Diagrammatic Morphisms and Applications*, Contemp. Math. 318, AMS, Providence, RI, 2003, pp. 207–213. Also available at ⟨http://www.math.mq.edu.au/~street⟩.

[51] Z. Tamsamani, Sur des notions de n-catégorie et n-groupoide non-strictes via des ensembles multi-simpliciaux, *K-Theory* **16** (1999), 51–99. Also available as arXiv:alg-geom/9512006.

[52] Z. Tamsamani, Equivalence de la théorie homotopique des n-groupïodes et celle des espaces topologiques n-tronqués. Also available as arXiv:alg-geom/9607010.

[53] Trimble n-category, nLab entry available at ⟨http://ncatlab.org/nlab/show/Trimble+n-category⟩.

[54] D. Verity, *Complicial Sets: Characterising the Simplicial Nerves of Strict ω-Categories*, Memoirs AMS **905**, 2005. Also available as arXiv:math/0410412.

[55] D. Verity, Weak complicial sets, a simplicial weak ω-category theory. Part I: basic homotopy theory. Available as arXiv:math/0604414.

[56] D. Verity, Weak complicial sets, a simplicial weak ω-category theory. Part II: nerves of complicial Gray-categories. Available as arXiv:math/0604416.

Editors:

John C. Baez
Department of Mathematics
University of California, Riverside
http://www.math.ucr.edu/home/baez

J. Peter May
Department of Mathematics
University of Chicago
http://www.math.uchicago.edu/~may/

CONTENTS

LECTURES ON N-CATEGORIES AND COHOMOLOGY

JOHN C. BAEZ* AND MICHAEL SHULMAN†

Abstract. This is an explanation of how cohomology is seen through the lens of n-category theory. Special topics include nonabelian cohomology, Postnikov towers, the theory of 'n-stuff', and n-categories for $n = -1$ and -2. A lengthy appendix clarifies certain puzzles and ventures into deeper waters such as higher topos theory. An annotated bibliography provides directions for further study.

Key words. n-category, cohomology.

AMS(MOS) subject classifications. 18G50 (Primary), 18D05, 18B25, 20JXX, 55S45 (Secondary).

Preface. The goal of these talks was to explain how cohomology and other tools of algebraic topology are seen through the lens of n-category theory. The talks were extremely informal, glossing over the difficulties involved in making certain things precise, just trying to sketch the big picture in an elementary way. A lot of the material is hard to find spelled out anywhere, but nothing new is due to me: anything not already known by experts was invented by James Dolan, Toby Bartels or Mike Shulman (who took notes, fixed lots of mistakes, and wrote the Appendix).

The first talk was one of the 2006 Namboodiri Lectures in Topology at the University of Chicago. It's a quick introduction to the relation between Galois theory, covering spaces, cohomology, and higher categories. The remaining talks, given in the category theory seminar at Chicago, were more advanced. Topics include nonabelian cohomology, Postnikov towers, the theory of 'n-stuff', and n-categories for $n = -1$ and -2. Some questions from the audience have been included.

Mike Shulman's extensive Appendix (§5) clarifies many puzzles raised in the talks. It also ventures into deeper waters, such as the role of posets and fibrations in higher category theory, alternate versions of the periodic table of n-categories, and the theory of higher topoi. For readers who want more details, we have added an annotated bibliography. — JB

1. The basic principle of Galois theory.

1.1. Galois theory. Around 1832, Galois discovered a basic principle:

> **We can study the ways a little thing k can sit in a bigger thing K:**

$$k \hookrightarrow K$$

---

*Department of Mathematics, University of California, Riverside, CA 92521, USA (baez@math.ucr.edu).

†Department of Mathematics, University of Chicago, 5734 S. University Avenue, Chicago, IL 60637, USA (shulman@math.uchicago.edu).

J.C. Baez, J.P. May (eds.), *Towards Higher Categories*, The IMA Volumes in Mathematics and its Applications 152, DOI 10.1007/978-1-4419-1524-5_1, © Springer Science+Business Media, LLC 2010

**by keeping track of the symmetries of K that fix
k. These form a subgroup of the symmetries of K:**

$$\text{Gal}(K|k) \subseteq \text{Aut}(K).$$

For example, a point k of a set K is completely determined by the subgroup of permutations of K that fix k. More generally, we can recover any subset k of a set K from the subgroup of permutations of K that fix k.

However, Galois applied his principle in a trickier example, namely commutative algebra. He took K to be a field and k to be a subfield. He studied this situation by looking at the subgroup $\text{Gal}(K|k)$ of automorphisms of K that fix k. Here this subgroup does not determine k unless we make a further technical assumption, namely that K is a 'Galois extension' of k. In general, we just have a map sending each subfield of K to the subgroup of $\text{Aut}(K)$ that fixes it, and a map sending each subgroup of $\text{Aut}(K)$ to the subfield it fixes. These maps are not inverses; instead, they satisfy some properties making them into what is called a 'Galois connection'.

When we seem forced to choose between extra technical assumptions or less than optimal results, it often means we haven't fully understood the general principle we're trying to formalize. But, it can be very hard to take a big idea like the basic principle of Galois theory and express it precisely without losing some of its power. That is not my goal here. Instead, I'll start by considering a weak but precise version of Galois' principle as applied to a specific subject: not commutative algebra, but *topology*.

Topology isn't really separate from commutative algebra. Indeed, in the mid-1800s, Dedekind, Kummer and Riemann realized that commutative algebra is a lot like topology, only backwards. Any space X has a commutative algebra $\mathcal{O}(X)$ consisting of functions on it. Any map

$$f \colon X \to Y$$

gives a map

$$f^* \colon \mathcal{O}(Y) \to \mathcal{O}(X).$$

If we're clever we can think of any commutative ring as functions on some space — or on some 'affine scheme':

$$[\text{Affine Schemes}] = [\text{Commutative Rings}]^{\text{op}}.$$

Note how it's backwards: the *inclusion* of commutative rings

$$p^* \colon \mathbb{C}[z] \hookrightarrow \mathbb{C}[\sqrt{z}]$$

corresponds to the *branched cover* of the complex plane by the Riemann surface for $\sqrt{z}$:

$$\begin{array}{ccc} p \colon \mathbb{C} & \to & \mathbb{C} \\ z & \mapsto & z^2. \end{array}$$

So: classifying how a little commutative algebra can *sit inside* a big one amounts to classifying how a big space can *cover* a little one. Now the Galois group gets renamed the group of **deck transformations**: in the above example it's $\mathbb{Z}/2$:

$$\sqrt{z} \mapsto -\sqrt{z}.$$

The theme of 'branched covers' became very important in later work on number theory, where number fields are studied by analogy to function fields (fields of functions on Riemann surfaces). However, it's the simpler case of unbranched covers where the basic principle of Galois theory takes a specially simple and pretty form, thanks in part to Poincaré. This is the version we'll talk about now. Later we'll generalize it from unbranched covers to 'fibrations' of various sorts — fibrations of spaces, but also fibrations of n-categories. Classifying fibrations using the basic principle of Galois theory will eventually lead us to cohomology.

1.2. The fundamental group. Around 1883, Poincaré discovered that any nice connected space B has a connected covering space that covers all others: its **universal cover**. This has the biggest deck transformation group of all: the **fundamental group** $\pi_1(B)$.

The idea behind Galois theory — turned backwards! — then says that:

Connected covering spaces of B are classified by subgroups

$$H \subseteq \pi_1(B).$$

This is the version we all learn in grad school. To remove the 'connectedness' assumption, we can start by rephrasing it like this:

Connected covering spaces of B with fiber F are classified by transitive actions of $\pi_1(B)$ on F.

This amounts to the same thing, since transitive group actions are basically the same as subgroups: given a subgroup $H \subseteq \pi_1(B)$ we can define F to be $\pi_1(B)/H$, and given a transitive action of $\pi_1(B)$ on F we can define H to be the stabilizer group of a point. The advantage of this formulation is that we can generalize it to handle covering spaces where the total space isn't connected:

Covering spaces of B with fiber F are classified by actions of $\pi_1(B)$ on F:

$$\pi_1(B) \to \mathrm{Aut}(F).$$

Here F is any set and $\mathrm{Aut}(F)$ is the group of permutations of this set:

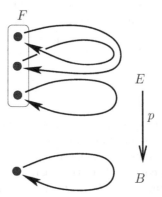

You can see how a loop in the base space gives a permutation of the fiber. The basic principle of Galois theory has become 'visible'!

1.3. The fundamental groupoid. So far the base space B has been connected. What if B is not connected? For this, we should replace $\pi_1(B)$ by $\Pi_1(B)$: the **fundamental groupoid** of B. This is the category where:
- objects are points of B: $\bullet x$
- morphisms are homotopy classes of paths in B:

The basic principle of Galois theory then says this:

> **Covering spaces $F \hookrightarrow E \to B$ are classified by actions of $\Pi_1(B)$ on F: that is, functors**
>
> $$\Pi_1(B) \to \operatorname{Aut}(F).$$

Even better, we can let the fiber F be different over different components of the base B:

> **Covering spaces $E \to B$ are classified by functors**
>
> $$\Pi_1(B) \to \mathbf{Set}.$$

What does this mean? It says a lot in a very terse way. Given a covering space $p\colon E \to B$, we can uniquely lift any path in the base space to a path in E, given a lift of the path's starting point. Moreover, this lift depends only on the homotopy class of the path. So, our covering space assigns a set $p^{-1}(b)$ to each point $b \in B$, and a map between these sets for any homotopy class of paths in B. Since composition of paths gets sent to composition of maps, this gives a *functor* from $\Pi_1(B)$ to **Set**.

Conversely, given any functor $F\colon \Pi_1(B) \to \mathbf{Set}$, we can use it to cook up a covering space of B, by letting the fiber over b be $F(b)$, and so on. So, with some work, we get a one-to-one correspondence between isomorphism classes of covering spaces $E \to B$ and natural isomorphism classes of functors $\Pi_1(B) \to \mathbf{Set}$.

But we actually get more: we get an *equivalence of categories*. The category of covering spaces of B is equivalent to the category where the objects are functors $\Pi_1(B) \to \mathbf{Set}$ and the morphisms are natural transformations between these guys. This is what I've really meant all along by saying "X's are classified by Y's." I mean there's a category of X's, a category of Y's, and these categories are equivalent.

1.4. Eilenberg–Mac Lane spaces. In 1945, Eilenberg and Mac Lane published their famous paper about categories. They *also* published a paper showing that any group G has a 'best' space with G as its fundamental group: the **Eilenberg-Mac Lane space** $K(G, 1)$.

In fact their idea is easiest to understand if we describe it a bit more generally, not just for groups but for groupoids. For any groupoid G we can build a space $K(G, 1)$ by taking a vertex for each object of G:

an edge for each morphism of G:

a triangle for each composable pair of morphisms:

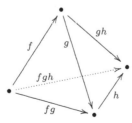

a tetrahedron for each composable triple:

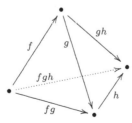

and so on. This space has G as its fundamental groupoid, and it's a **homotopy 1-type**: all its homotopy groups above the 1st vanish. These facts characterize it.

Using this idea, one can show a portion of topology is just groupoid theory: homotopy 1-types are the same as groupoids! To make this precise requires a bit of work. It's not true that the category of homotopy 1-types and maps between them is equivalent to the category of groupoids and functors between them. But, they form Quillen equivalent model categories. Or, if you prefer, they form 2-equivalent 2-categories.

1.5. Grothendieck's dream. Since the classification of covering spaces

$$E \to B$$

only involves the fundamental groupoid of B, we might as well assume B is a homotopy 1-type. Then E will be one too.

So, we might as well say E and B are *groupoids!* The analogue of a covering space for groupoids is a **discrete fibration**: a functor $p: E \to B$ such that for any morphism $f: x \to y$ in B and object $\tilde{x} \in E$ lifting x, there's a unique morphism $\tilde{f}: \tilde{x} \to \tilde{y}$ lifting f:

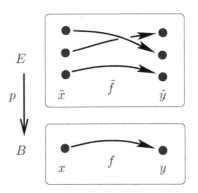

The basic principle of Galois theory then becomes:
Discrete fibrations $E \to B$ are classified by functors

$$B \to \mathbf{Set}.$$

This is true even when E and B are categories, though then people use the term 'opfibrations'. This — and much more — goes back to Grothendieck's 1971 book *Étale Coverings and the Fundamental Group*, usually known as SGA1.

Grothendieck dreamt of a much bigger generalization of Galois theory in his 593-page letter to Quillen, *Pursuing Stacks*. Say a space is a **homotopy n-type** if its homotopy groups above the nth all vanish. Since homotopy 1-types are 'the same' as groupoids, maybe homotopy n-types are 'the same' as n-groupoids! It's certainly true if we use Kan's simplicial approach to n-groupoids — but we want it to emerge from a general theory of n-categories.

For n-groupoids, the basic principle of Galois theory should say something like this:

Fibrations $E \to B$ where E and B are n-groupoids are classified by weak $(n+1)$-functors

$$B \to n\mathbf{Gpd}.$$

Now when we say 'classified by' we mean there's an equivalence of $(n+1)$-categories. 'Weak' $(n+1)$-functors are those where everything is preserved *up to equivalence*. I include the adjective 'weak' only for emphasis: we need all n-categories and n-functors to be weak for Grothendieck's dream to have any chance of coming true, so for us, everything is weak by default.

Grothendieck made the above statement precise and proved it for $n = 1$; later Hermida did it for $n = 2$. Let's see what it amounts to when $n = 1$. To keep things really simple, suppose E, B are just *groups*, and fix the fiber F, also a group. With a fixed fiber our classifying 2-functor will land not in all of **Gpd**, but in $\mathrm{AUT}(F)$, which is the 'automorphism 2-group' of F — I'll say exactly what that is in a minute. In this simple case fibrations are just *extensions*, so we get a statement like this:

Extensions of the group B by the group F, that is, short exact sequences

$$1 \to F \to E \to B \to 1,$$

are classified by weak 2-functors

$$B \to \mathrm{AUT}(F).$$

This is called **Schreier theory**, since a version of this result was first shown by Schreier around 1926. The more familiar classifications of abelian or central group extensions using Ext or H^2 are just watered-down versions of this.

$\mathrm{AUT}(F)$ is the **automorphism 2-group** of F, a 2-category with:

- F as its only object: $\bullet F$
- automorphisms of F as its morphisms:

- elements $g \in F$ with $g\alpha(f)g^{-1} = \beta(f)$ as its 2-morphisms:

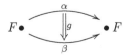

In other words, we get $\mathrm{AUT}(F)$ by taking **Gpd** and forming the sub-2-category with F as its only object, all morphisms from this to itself, and all 2-morphisms between these.

Given a short exact sequence of groups, we classify it by choosing a **set-theoretic section**:

$$1 \longrightarrow F \xrightarrow{\ i\ } E \underset{p}{\overset{s}{\rightleftarrows}} B \longrightarrow 1 \,,$$

meaning a function $s \colon B \to E$ with $p(s(b)) = b$ for all $b \in B$. This gives for any $b \in B$ an automorphism $\alpha(b)$ of F:

$$\alpha(b)(f) = s(b) f s(b)^{-1}.$$

Since s need not be a homomorphism, we may not have

$$\alpha(b)\,\alpha(b') = \alpha(bb')$$

but this holds *up to conjugation* by an element $\alpha(b, b') \in F$. That is,

$$\alpha(b, b')\,[\alpha(b)\,\alpha(b')f]\,\alpha(b, b')^{-1} = \alpha(bb')f$$

where

$$\alpha(b, b') = s(bb')\,(s(b)\,s(b'))^{-1}.$$

This turns out to yield a *weak* 2-functor

$$\alpha \colon B \to \mathrm{AUT}(F).$$

If we consider two weak 2-functors equivalent when there's a 'weak natural isomorphism' between them, different choices of s will give equivalent 2-functors. Isomorphic extensions of B by F also give equivalent 2-functors.

The set of equivalence classes of weak 2-functors $B \to \mathrm{AUT}(F)$ is often called the **nonabelian cohomology** $H(B, \mathrm{AUT}(F))$. So, we've described a map sending isomorphism classes of short short exact sequences

$$1 \to F \to E \to B \to 1$$

to elements of $H(B, \mathrm{AUT}(F))$. And, this map is one-to-one and onto.

This is *part* of what we mean by saying extensions of B by F are classified by weak 2-functors $B \to \mathrm{AUT}(F)$. But as usual, we really have something much better: an equivalence of 2-categories.

There's a well-known category of extensions of B by F, where the morphisms are commutative diagrams like this:

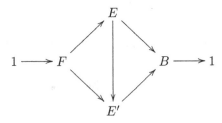

But actually, we get a 2-category of extensions using the fact that groups are special groupoids and **Gpd** is a 2-category. Similarly, $\hom(B, \mathrm{AUT}(F))$ is a 2-category, since $\mathrm{AUT}(F)$ is a 2-category and B is a group, hence a special sort of category, hence a special sort of 2-category. And, the main result of Schreier theory says the 2-category of extensions of B by F is equivalent to $\hom(B, \mathrm{AUT}(F))$. This implies the earlier result we stated, but it's much stronger.

In short, generalizing the fundamental principle of Galois theory to fibrations where everything is a *group* gives a beautiful classification of group extensions in terms of nonabelian cohomology. In the rest of these lectures, we'll explore how this generalizes as we go from groups to n-groupoids and n-categories.

2. The power of negative thinking.

2.1. Extending the periodic table. Now I want to dig a lot deeper into the relation between fibrations, cohomology and n-categories. At this point I'll suddenly assume that you have some idea of what n-categories are, or at least can fake it. The periodic table of n-categories shows what various degenerate versions of n-category look like. We can think of an $(n + k)$-category with just one j-morphism for $j < k$ as a special sort of n-category. They look like this:

THE PERIODIC TABLE

	$n = 0$	$n = 1$	$n = 2$
$k = 0$	sets	categories	2-categories
$k = 1$	monoids	monoidal categories	monoidal 2-categories
$k = 2$	commutative monoids	braided monoidal categories	braided monoidal 2-categories
$k = 3$	''	symmetric monoidal categories	sylleptic monoidal 2-categories
$k = 4$	''	''	symmetric monoidal 2-categories
$k = 5$	''	''	''
$k = 6$	''	''	''

For example, in the $n = 0$, $k = 1$ spot we have 1-categories with just one 0-morphism, or in normal language, categories with just one object — i.e., monoids. Indeed a 'monoid' is the perfect name for a one-object category, because 'monos' means 'one' — but that's not where the name comes from, of course. It's a good thing, too, or else Eilenberg and Mac Lane might have called categories 'polyoids'.

We'll be thinking about all these things in the most weak manner possible, so '2-category' means 'weak 2-category', aka 'bicategory', and so on. Everything I'm going to tell you *should* be true, once we really understand what is going on. Right now it's more in the nature of dreams and speculations, but I don't think we'll be able to prove the theorems until we dream enough.

Eckmann and Hilton algebraicized a topological argument going back to Hurewicz, which proves that strict monoidal categories with one object are commutative monoids. Eugenia Cheng and Nick Gurski have studied this carefully, and they've shown that things are a little more complicated when we consider *weak* monoidal categories, but I'm going to proceed in a robust spirit and leave such issues to smart young people like them.

Things get interesting in the second column, when we get *braided* monoidal categories. These are not the most obvious sort of 'commutative' monoidal categories that Mac Lane first wrote down, namely the symmetric ones. James Dolan and I were quite confused about why braided and symmetric both exist, until we started getting the hang of the periodic table.

Noticing that in the first column we stabilize after 2 steps at commutative monoids, and in the second after 3 steps at symmetric monoidal categories, we enunciated the **stabilization hypothesis**, which says that the nth column should stabilize at the $(n + 2)$nd row. We believed this because of the Freudenthal suspension theorem in homotopy theory, which says that if you keep suspending and looping a space, it gets nicer and nicer, and if it only has n nonvanishing homotopy groups, eventually it's as nice as it can get, and it stabilizes after $n + 2$ steps. This is related, because any space gives you an n-groupoid of points, paths, paths-of-paths, and so on.

Such a beautiful pattern takes the nebulous, scary subject of n-categories and imposes some structure on it. There are all sorts of operations that take you hopping between different squares of this chart.

We call this chart the 'Periodic Table' — not because it's periodic, but because we can use it to predict new phenomena, like Mendeleev used the periodic table to predict new elements.

After we came up with the periodic table I showed it to Chris Isham, who does quantum gravity at Imperial College. I was incredibly happy with it, but he said: "That's obviously not right — you didn't start the chart at the right place. First there should be a column with just one interesting row, then a column with two, and *then* one with three."

I thought he was crazy, but it kept nagging me. It's sort of weird to start counting at three, after all. But there are no (-1)-categories or (-2)-categories! Are there?

It turns out there are! Eventually Toby Bartels and James Dolan figured out what they are. And they realized that Isham was right. The periodic table really looks like this:

THE EXTENDED PERIODIC TABLE

	$n = -2$	$n = -1$	$n = 0$	$n = 1$	$n = 2$
$k = 0$	?	?	sets	categories	2-categories
$k = 1$	''	?	monoids	monoidal categories	monoidal 2-categories
$k = 2$	''	''	commutative monoids	braided monoidal categories	braided monoidal 2-categories
$k = 3$	''	''	''	symmetric monoidal categories	sylleptic monoidal 2-categories
$k = 4$	''	''	''	''	symmetric monoidal 2-categories

You should be dying to know what fills in those question marks. Just for fun, I'll tell you what two of them are now. You probably won't think these answers are obvious — but you will soon:

- (-1)-categories are just truth values: there are only two of them, True and False.
- (-2)-categores are just 'necessarily true' truth values: there is only one of them, which is True.

I know this sounds crazy, but it sheds lots of light on many things. Let's see why (-1)- and (-2)-categories really work this way.

2.2. The categorical approach. Before describing (-1)-categories and (-2)-categories, we need to understand a couple of facts about the n-categorical world.

The first is that in the n categorical universe, every n-category is secretly an $(n + 1)$-category with only identity $(n + 1)$-morphisms. It's common for people to talk about sets as discrete categories, for example. A way to think about it is that these identity $(n + 1)$-morphisms are really *equations*.

When you play the n-category game, there's a rule that you should never say things are equal, only isomorphic. This makes sense up until the top level, the level of n-morphisms, when you break down and allow yourself to say that n-morphisms are equal. But actually you aren't breaking the rule here, if you think of your n-category as an $(n + 1)$-category with only identity morphisms. Those equations are really isomorphisms: it's just that the only isomorphisms existing at the $(n + 1)$st level are identities. Thus when we assert equations, we're refusing to think about things still more categorically, and saying "all I can take today is an n-category" rather than an $(n + 1)$-category.

We can iterate this and go on forever, so every n-category is really an ∞-category with only identity j-morphisms for $j > n$.

The second thing is that big n-categories have lots of little n-categories inside them. For example, between two objects x, y in a 3-category, there's a little 2-category $\hom(x, y)$. James and I jokingly call this sort of thing a 'microcosm', since it's like a little world within a world:

A microcosm

In general, given objects x, y in an n-category, there is an $(n-1)$-category called $\hom(x, y)$ (because it's the 'thing' of morphisms from x to y) with the morphisms $f \colon x \to y$ as its objects, etc..

We can iterate this and look at microcosms of microcosms. A couple of objects in $\hom(x, y)$ give an $(n - 2)$-category, and so on. It's handy to say that two j-morphisms x and y are **parallel** if they look like this:

This makes sense if $j > 0$; if $j = 0$ we decree that all j-morphisms are parallel. The point is that it only makes sense to talk about something like $f \colon x \to y$ when x and y are parallel. Given parallel j-morphisms x and y, we get an $(n - j - 1)$-category $\hom(x, y)$ with $(j + 1)$-morphisms $f \colon x \to y$ as objects, and so on. This is a little microcosm.

In short: *given parallel j-morphisms x and y in an n-category,* $\hom(x, y)$ *is an* $(n - j - 1)$*-category.* Now take $j = n$. Then we get a (-1)-category! If x and y are parallel n-morphisms in an n-category, then $\hom(x, y)$ is a (-1)-category. What is it?

You might say "that's cheating: you're not allowed to go that high." But it isn't really cheating, since, as we said, every n-category is secretly an ∞-category. We just need to work out the answer, and that will tell us what a (-1)-category is.

The objects of $\hom(x, y)$ are $(n + 1)$-morphisms, which here are just identities. So, if $x = y$ there is one object in $\hom(x, y)$, otherwise there's none. So there are really just two possible (-1)-categories. There aren't any (-1)-categories that don't arise in this way, since in general any n-category can be stuck in between two objects to make an $(n+1)$-category.

Thus, there are just two (-1)-categories. You could think of them as the 1-element set and the empty set, although they're not exactly sets. We can also call them '$=$' and '$\neq$', or 'True' and 'False.' The main thing is, there are just two.

I hope I've convinced you this is right. You may think it's silly, but you should think it's right.

Now we should take $j = n+1$ to get the (-2)-categories. If we have two parallel $(n+1)$-morphisms in an n-category, they are both identities, so being parallel they must both be $1_z \colon z \to z$ for some z, so they're equal. At this level, they *have* to be equal, so there *is* an identity from one from to the other, necessarily. It's like the previous case, except we only have one choice.

So there's just one (-2)-category. When there's just one of something, you can call it anything you want, since you don't have to distinguish it from anything. But it might be good to call this guy the 1-element set, or '$=$', or 'True'. Or maybe we should call it 'Necessarily True', since there's no other choice.

So, we've worked out (-1)-categories and (-2)-categories. We could keep on going, but it stabilizes past this point: for $n > 2$, $(-n)$-categories are all just 'True.' I'll leave it as a puzzle for you to figure out what a monoidal (-1)-category is.

2.3. Homotopy n-types. I really want to talk about what this has to do with topology. We're going to study very-low-dimensional algebra and apply it to very-low-dimensional topology—in fact, so low-dimensional that they never told you about it in school. Remember Grothendieck's dream:

HYPOTHESIS 1 (Grothendieck's Dream, aka the Homotopy Hypothesis). *n-Groupoids are the same as homotopy n-types.*

Here 'n-groupoids' means *weak* n-groupoids, in which everything is invertible up to higher-level morphisms, in the weakest possible way. Similarly 'the same' is meant in the weakest possible way, which we might make precise using something called 'Quillen equivalence'. A **homotopy n-type** is a nice space (e.g. a CW-complex) with vanishing homotopy groups π_j for $j > n$. (If it's not connected, we have to take π_j at every basepoint.) We could have said 'all spaces' instead of 'nice spaces,' but then we'd need to talk about *weak* homotopy equivalence instead of homotopy equivalence.

People have made this precise and shown that it's true for various low values of n, and we're currently struggling with it for higher values. It's known (really well) for $n = 1$, (pretty darn well) for $n = 2$, (partly) for $n = 3$, and in a somewhat fuzzier way for higher n. Today we're going to do it for *lower* values of n, like $n = -1$ and $n = -2$.

Now, they never told you about the negative second homotopy group of a space—or 'homotopy thingy', since after all we know π_0 is only a set, not a group. In fact we won't define these negative homotopy thingies as thingies yet, but only what it means for them to vanish:

DEFINITION 2. *We say* $\pi_j(X)$ **vanishes for all basepoints** *if given any* $f\colon S^j \to X$ *there exists* $g\colon D^{j+1} \to X$ *extending* f. *We say* X *is a* **homotopy** n**-type** *if* $\pi_j(X)$ *vanishes for all basepoints whenever* $j > n$.

We'll use this to figure out what a homotopy 0-type is, then use it to figure out what a homotopy (-1)-type and a homotopy (-2)-type are.

X is a *homotopy 0-type* when all circles and higher-dimensional spheres mapped into X can be contracted. So, X is just a disjoint union of connected components, all of which are contractible. (The fact that being able to contract all spheres implies the space is contractible, for nice spaces, is Whitehead's theorem.) From the point of view of a homotopy theorist, such a space might as well just be a set of points, i.e. a discrete space, but the points could be 'fat'. This is what Grothendieck said should happen; all 0-categories are 0-groupoids, which are just sets.

Now let's figure out what a homotopy (-1)-type is. If you pay careful attention, you'll see the following argument is sort of the same as what we did before to figure out what (-1)-categories are.

By definition, a *homotopy (-1)-type* is a disjoint union of contractible spaces (i.e. a homotopy 0-type) with the extra property that maps from S^0 can be contracted. What can X be now? It can have just one contractible component (the easy case)—or it can have none (the sneaky case). So X is a disjoint union of 0 or 1 contractible components. From the point of view of homotopy theory, such a space might as well be an empty set or a 1-point set. This is the same answer that we got before: the absence or presence of an equation, 'False' or "True".

Finally, a *homotopy (-2)-type* is thus a space like this such that any map $S^{-1} \to X$ extends to D^0. Now we have to remember what S^{-1} is. The n-sphere is the unit sphere in $\mathbb{R}^{n+1}$, so S^{-1} is the unit sphere in $\mathbb{R}^0$. It consists of all unit vectors in this zero-dimensional vector space, i.e. it is the empty set, $S^{-1} = \emptyset$. Well, a map from the empty set into X is a really easy thing to be given: there's always just one. And D^0 is the unit disc in $\mathbb{R}^0$, so it's just the origin, $D^0 = \{0\}$. So this extension condition says that X has to have at least one point in it. Thus a homotopy (-2)-type is a disjoint union of precisely one contractible component. Up to homotopy, it is thus a one-point set, or 'True'. So Grothendieck's idea works here too.

Now I think you agree; I gave you *two* proofs!

What about (-3)-types? Even I get a little scared about S^{-2} and D^{-1}. But, they're both empty, so any homotopy (-2)-type is automatically a homotopy (-3)-type as well, and indeed a homotopy $(-n)$-type for all $n \geq 2$.

2.4. Stuff, structure, and properties. What's all this nonsense about? In math we're often interested in equipping things with extra structure, stuff, or properties, and people are often a little vague about what these mean. For example, a group is a set (*stuff*) with operations (*structure*) such that a bunch of equations hold (*properties*).

You can make these concepts very precise by thinking about forgetful functors. It always bugged me when reading books that no one ever defined 'forgetful functor'. Some functors are more forgetful than others. Consider a functor $p\colon E \to B$ (the notation reflects that later on, we're going to turn it into a fibration when we use Grothendieck's idea). There are various amounts of forgetfulness that p can have:

- p **forgets nothing** if it is an equivalence of categories, i.e. faithful, full, and essentially surjective. For example the identity functor **AbGp** → **AbGp** forgets nothing.
- p **forgets at most properties** if it is faithful and full. E.g. **AbGp** → **Gp**, which forgets the property of being abelian, but a homomorphism of abelian groups is just a homomorphism between groups that happen to be abelian.
- p **forgets at most structure** if it is faithful. E.g. the forgetful functor from groups to sets, **AbGp** → **Sets**, forgets the structure of being an abelian group, but it's still faithful.
- p **forgets at most stuff** if it is arbitrary. E.g. **Sets**^{2} → **Sets**, where we just throw out the second set, is not even faithful.

There are different ways of slicing this pie. For now, we are thinking of each level of forgetfulness as subsuming the previous ones, so 'forgetting at most structure' means forgetting structure and/or properties and/or nothing, but we can also try to make them completely disjoint concepts. Later I'll define a concept of 'forgetting *purely* structure' and so on.

What's going on here is that in every case, what you can do is take objects downstairs and look at their *fiber* or really *homotopy fiber* upstairs. An object in the homotopy fiber upstairs is an object together with a morphism from its image to the object downstairs, as follows.

Given $p\colon E \to B$ and $x \in B$, its **homotopy fiber** or **essential preimage**, which we write $p^{-1}(x)$, has:

- objects $e \in E$ equipped with isomorphism $p(e) \cong x$
- morphisms $f\colon e \to c'$ in E compatible with the given isomorphisms:

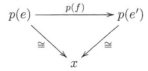

It turns out that the more forgetful the functor is, the bigger and badder the homotopy fibers can be. In other words, switching to the language of topology, they can have bigger **homotopical dimension**: they can have nonvanishing homotopy groups up to dimension d for bigger d.

FACT 3. *If E and B are groupoids, then:*

- *p forgets stuff if all $p^{-1}(x)$ are arbitrary (1-)groupoids. For example, there's a whole* groupoid *of ways to add an extra set to some set.*

- *p forgets structure if all $p^{-1}(x)$ are 0-groupoids, i.e. groupoids which are (equivalent to) sets. For example, there's just a* set *of ways of making a set into a group.*

- *p forgets properties if all $p^{-1}(x)$ are (-1)-groupoids. For example, there's just a* truth value *of ways of making a group into an abelian group: either you can or you can't (i.e. it is or it isn't).*

- *p forgets nothing if all $p^{-1}(x)$ are (-2)-groupoids. For example, there's just a 'necessarily true' truth value of ways of making an abelian group into an abelian group: you always can, in one way.*

(In the examples above, we are considering the groupoids *of sets, pairs of sets, groups and abelian groups. We'll consider the case of categories in §5.2.)*

We can thus study how forgetful a functor is by looking at what homotopy dimension its fibers have.

Note that to make this chart work, we really needed the negative dimensions. You should want to also go in the other direction, say if we had 2-groupoids or 3-groupoids; then we'll have something 'even more substantial than stuff.' James Dolan dubbed that *eka-stuff*, by analogy with how Mendeleev called elements which were missing in the periodic table 'eka-?', e.g. 'eka-silicon' for the missing element below silicon, which now we call germanium. He guessed that eka-silicon would be a lot like silicon, but heavier, and so on. Like Mendeleev, we can use the periodic table to guess things, and then go out and check them.

2.5. Questions and comments.

2.5.1. What should forgetful mean?

Peter May: *Only functors with left adjoints should really be called 'forgetful.' Should the free group functor be forgetful?*

MS: *A set is the same as a group with the property of being free and the structure of specified generators.*

JB: Right, you can look at it this way. Then the free functor $F \colon \mathbf{Set} \to \mathbf{Grp}$ 'forgets at most structure': it's faithful, but neither full nor essentially surjective. Whether you want to call it 'forgetful' is up to you, but this is how I'm using the terminology now.

2.5.2. Monoidal (-1)-categories. Puzzle: What's a monoidal (-1)-category?

Answer: A (-1)-category is a truth value, and the only monoidal (-1)-category is True.

To figure this out, note that a monoidal (-1)-category is what we get when we take a 0-category with just one object, say x, and look at $\hom(x, x)$. A 0-category with one object is just a one-element set $\{x\}$, and $\hom(x, x)$ is just the equality $x = x$, which is 'True.'

Note that monoidal (-1)-categories are stable, so we are just adding a property to the previous one, just like when we pass from monoids to commutative monoids, or braiding to symmetry.

In general, we have forgetful functors marching *up* the periodic table, which forget different amounts of things. We forget nothing until we get up to the end of the stable range, then we forget a property (symmetry, or commutativity), then a structure (monoid structure, or braiding), then stuff (a monoidal structure, of which there are a whole category of ways to add to a given category), then eka-stuff (ways to make a 2-category into a monoidal 2-category), and so on.

2.5.3. Maps of truth values. Here's a funny thing. Note that for categories, we have a composition *function*

$$\hom(x, y) \times \hom(y, z) \to \hom(x, z).$$

For a set, the substitute is transitivity:

$$(x = y) \,\&\, (y = z) \Rightarrow (x = z).$$

In other words, we can 'compose equations.' But here $\Rightarrow$ is acting as a map between truth values. What sort of morphisms of truth values do we have? We just have a 0-category of (-1)-categories, so there should be only identity morphisms. The implication $F \Rightarrow T$ doesn't show up in this story.

That's a bit sad. Ideally, a bunch of *propositional logic* would show up at the level of -1-categories. Toby Bartels has a strategy to fix this. In his approach, posets play a much bigger role in the periodic table, to include the notion of implication between truth values.

In fact, it seems that the periodic table is just a slice of a larger 3-dimensional table relating higher categories and logic... see §5.1 for more on this.

3. Cohomology: The layer-cake philosophy. We're going to continue heading in the direction of cohomology, but we'll get there by a perhaps unfamiliar route. Last time we led up to the concept of *n-stuff*, although we stopped right after discussing ordinary stuff and only mentioned eka-stuff briefly.

3.1. Factorizations. What does it mean to forget *just* stuff, or *just* properties?

DEFINITION 4. *An* ∞*-functor* $p\colon E \to B$ *is* n-**surjective** *(perhaps* **essentially** n-**surjective***) if given any parallel* $(n-1)$*-morphisms* e *and* e' *in* E*, and any* n*-morphism* $f\colon p(e) \to p(e')$*, there is an* n*-morphism* $\widetilde{f}\colon e \to e'$ *such that* $p(\widetilde{f})$ *is equivalent to* $f\colon p(\widetilde{f}) \simeq f$*.*

For example, suppose $p\colon E \to B$ is a function between *sets*. It is

- 0-surjective if it is *surjective* in the usual sense, since in this case equivalences are just equalities. (The presence of e and e' here is a bit confusing, unless you believe that all n-categories go arbitrarily far down as well.)
- 1-surjective if it is *injective*; since in this case all 1-morphisms are identities, 1-surjective means that if $p(e) = p(e')$, then $e = e'$.

That's a nice surprise: *injective means 'surjective on equations'!*

Now, another thing that you can do in the case of sets is to take any old function $p\colon E \to B$ and factor it as first a surjection and then an injection:

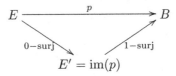

The interesting thing is that this keeps generalizing as we go up to higher categories.

To see how this works, first suppose E and B are categories. A functor $p\colon E \to B$ is:

- 0-surjective if it is essentially surjective;
- 1-surjective if it is full;
- 2-surjective if it is faithful;
- 3-surjective always, and so on.

Do you see why?

- Crudely speaking, 0-surjective means 'surjective on objects'. But you have to be a bit careful: it's sufficient that every object in B is *isomorphic* to $p(e)$ for some $e \in E$. So, 0-surjective really means *essentially* surjective.
- Crudely speaking, 1-surjective means 'surjective on arrows,' but you just have to be a little careful: it's not fair to ask that an arrow downstairs be the image of something unless we already know that its source and target are the images of something. So, it really means our functor is *full*
- Similarly, 2-surjective means 'surjective on equations between morphisms,' i.e. injective on hom-sets. So, it means our functor is *faithful*.

Note that the conjunction of all three of these means our functor is an equivalence, just as a surjective and injective function is an isomorphism of sets.

The notions of forgetting at most stuff, structure, or properties can also be defined using conjunctions of these conditions, namely:
- Forgets nothing: 0, 1, and 2-surjective
- Forgets at most properties: 1 and 2-surjective
- Forgets at most structure: 2-surjective
- Forgets stuff: arbitrary

As with functions between sets, we get a factorization result for functors between categories. Any functor factors like this:

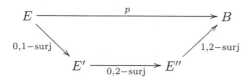

I think this is a well-known result. You build these other categories as 'hybrids' of E and B: E gradually turns into B from the top down. We start with E; then we throw in new 2-morphisms (equations between morphisms) that we get from B; then we throw in new 1-morphisms (morphisms), and finally new 0-morphisms (objects). It's like a horse transforming into a person from the head down. First it's a horse, then it's a centaur, then it's a faun-like thing that's horse from the legs down, and finally it's a person.

In more detail:
- The objects of E' are the same objects as E, but a morphism from e to e' is a morphism $p(e) \to p(e')$ in B which is in the image of p; and
- The objects of E'' are the objects of B in the (essential) image of p, with all morphisms between them.

This is the same thing as is happening for sets, but there it's happening so fast that you can't see it happening.

The next example will justify this terminology:
- A functor which is 0- and 1-surjective **forgets purely stuff**;
- A functor which is 0- and 2-surjective **forgets purely structure**;
- A functor which is 1- and 2-surjective **forgets purely properties**.

EXAMPLE 5. *Let's take the category of pairs of vector spaces and forget down to just the underlying set of the first vector space (so that we have an interesting process at every stage).*

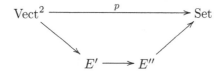

The objects of E' are again pairs of vector spaces, but its morphisms are just linear maps between the first ones. We write this as

$$[\textit{pairs of vector spaces, linear maps between first ones}].$$

In fact, this category is equivalent to Vect; the extra vector space doesn't participate in the morphisms, so it might as well not be there. Our factorization cleverly managed to forget the stuff (the second vector space) but still keep the structure on what remains.

The objects of E'' are sets with the property *that they can be made into vector spaces, and its morphisms are arbitrary functions between them. Here we forgot just the structure of being a vector space, but we cleverly didn't forget the* property *of being vector-space-izable. That's sort of cool.*

Having seen how these factorizations work for sets and categories, we can guess how they go for ∞-categories. Let's say an ∞-functor $p\colon E \to B$ **forgets purely j-stuff** if it's i-surjective for all $i \neq j + 1$. Note the funny '+1' in there: we need this to make things work smoothly. For ordinary categories, we have:

- A functor that 'forgets purely 1-stuff' forgets purely stuff;
- A functor that 'forgets purely 0-stuff' forgets purely structure;
- A functor that 'forgets purely (-1)-stuff' forgets purely properties;
- A functor that 'forgets purely (-2)-stuff' forgets nothing.

If we go up to 2-categories we get a new concept, '2-stuff'. This is what we called 'eka-stuff' last time.

Given the pattern we're seeing here, and using what they knew about Postnikov towers, James Dolan and Toby Bartels guessed a factorization result like this:

HYPOTHESIS 6. *Given a functor $p\colon E \to B$ between n-categories, it admits a factorization:*

$$E_n = E \xrightarrow{\quad\quad\quad p \quad\quad\quad} B = E_{-2}$$

with $E_n = E \xrightarrow{p_n} E_{n-1} \xrightarrow{p_{n-1}} \cdots \quad \cdots \xrightarrow{p_0} E_{-1} \xrightarrow{p_{-1}} B = E_{-2}$

where p_j forgets purely j-stuff.

Someone must have already made this precise and proved this for 2-categories. If not, someone should go home and do it tonight; it shouldn't be hard.

In fact, this result is already known for all n-groupoids, but only if you believe Grothendieck that they are the same as homotopy n-types. In this case, there's a topological result which says that every map between homotopy n-types factors like this. Such a factorization is called a **Moore–Postnikov tower**. When we factor a map from a space to a *point* this way, it's called a **Postnikov tower**. As we'll see, this lets us view a space as being made up out of layers, one for each homotopy group. And this lets us *classify homotopy types using cohomology* — at least in principle.

3.2. Cohomology and Postnikov towers.

In a minute we'll see that from the viewpoint of homotopy theory, a space is a kind of 'layer cake' with one layer for each dimension. I claim that cohomology is fundamentally the study of classifying 'layer cakes' like this. There are many other kinds of layer cakes, like chain complexes (which are watered-down versions of spaces), L_∞-algebras and A_∞-algebras (which are chain complexes with extra bells and whistles), and so on. But let's start with spaces.

How does it work? If E is a homotopy n-type, we study it as follows. We map it to something incredibly boring, namely a point, and then work out the Postnikov tower of this map:

Here we think of E as an n-groupoid and the point as the terminal n-groupoid, which has just one object, one morphism and so on. The Postnikov tower keeps crushing E down, so that E_j is really just a j-groupoid. This process is called **decategorification**. At the end of the day, E_{-2} is the only (-2)-groupoid there is: the point.

Of course, in the world of topology, they don't use our category-theoretic terminology to describe these maps. Morally, p_j forgets purely j-stuff — but topologists call this 'killing the jth homotopy group.' More precisely, the map

$$p_j \colon E_j \to E_{j-1}$$

induces isomorphisms

$$\pi_i(E_j) \to \pi_i(E_{j-1})$$

for all i *except* $i = j$, in which case it induces the zero map.

We won't get into how topologists actually construct Postnikov towers. Once Grothendieck's dream comes true, it will be a consequence of the result for n-categories.

It's fun to see how this Postnikov tower works in the shockingly low-dimensional cases $j = 0$ and $j = -1$, where the jth 'homotopy thingy' isn't a group — just a set, or truth value. When we get down to E_0, our space is just a set, at least up to homotopy equivalence. Killing its 0th 'homotopy set' then collapses all its points to the same point (if it has any to begin with). We're left with either the one-point set or the empty set. Killing the (-1)st 'homotopy truth value' then gives us the one-point set.

But enough of this negative-dimensional madness. Let's see how people use Postnikov towers to classify spaces up to homotopy equivalence. Consider any simplification step $p_j: E_j \to E_{j-1}$ in our Postnikov tower. By the wonders of homotopy theory, we can describe this as a fibration. One of the great things about homotopy theory is that even a map that doesn't look anything like a 'bundle' is always equivalent to a fibration, so we can think of it as some kind of bundle-type thing. Thus we can consider the homotopy fiber F_j of this map, which can either be constructed directly, the way we constructed the 'essential preimage' for a functor (here using paths in the base space) or by first converting the map into an actual fibration and then taking its literal fiber.

(We say 'the' fiber as if they were all the same, but if the space isn't connected they won't necessarily all be the same. Let's assume for now that E was connected, for simplicity.)

So, we get a fibration

$$F_j \to E_j \xrightarrow{p_j} E_{j-1}.$$

Since p_j doesn't mess with any homotopy groups except the jth, the long exact sequence of homotopy groups for a fibration

$$\ldots \longrightarrow \pi_i(F_j) \longrightarrow \pi_i(E_j) \longrightarrow \pi_i(E_{j-1}) \longrightarrow \pi_{i-1}(F_j) \longrightarrow \ldots$$

tells us that the homotopy fiber must have only one non-vanishing homotopy group: $\pi_i(F_j) = 0$ unless $i = j$. We killed the jth homotopy group, so where did that group go? It went up into the fiber.

Such an F_j, with only one non-vanishing homotopy group, is called an **Eilenberg–Mac Lane space**. The great thing is that a space with only its jth homotopy group nonzero is completely determined by that group — up to homotopy equivalence, that is. For fancier spaces, the homotopy groups aren't enough to determine the space: we also have to say how the homotopy groups talk to each other, which is what this Postnikov business is secretly doing. The Eilenberg–Mac Lane space with G as its jth homotopy group is called $K(G, j)$. (Of course we need G abelian if $j > 1$.)

So, we've got a way of building any homotopy n-type E as a 'layer cake' where the layers are Eilenberg–Mac Lane spaces, one for each dimension. At the jth stage of this procress, we get a space E_j as the total space of this fibration:

$$K(\pi_j(E), j) = F_j \to E_j \xrightarrow{p_j} E_{j-1}.$$

These spaces E_j become better and better approximations to our space as j increases, and $E_n = E$.

If we know the homotopy groups of the space E, the main task is to understand the fibrations p_j. The basic principle of Galois theory says how to classify fibrations:

Fibrations of n-groupoids

$$F \to E \xrightarrow{\ p\ } B$$

with a given base B and fiber F are classified by maps

$$k \colon B \to \mathrm{AUT}(F).$$

Here $\mathrm{AUT}(F)$ is the **automorphism $(n+1)$-group** of F, i.e. an $(n+1)$-groupoid with one object. For example, a set has an automorphism group, a category has an automorphism 2-group, and so on. So, the map k is an $(n+1)$-functor, of the weakest possible sort.

How do we get the map k from the fibration? Topologists have a trick that involves turning $\mathrm{AUT}(F)$ into a space called the 'classifying space' for F-bundles, at least when B and F are spaces.

But now I want you to think about it n-categorically. How does it work? Think of B as an ∞-category. Then a path in B (a 1-morphism) lifts to a path in E, which when we move along it, induces some automorphism of the fiber (a 1-morphism in the one-object $(n+1)$-groupoid $\mathrm{AUT}(F)$). Similarly, a path of paths induces a morphism between automorphisms. This continues all the way up, which is how we get a map $k \colon B \to \mathrm{AUT}(F)$.

This is a highbrow way of thinking about cohomology theory. We may call it 'nonabelian cohomology.' You've probably seen cohomology with coefficients in some *abelian* group, which is a special case that's easy to compute; here we are talking about a more general version that's supposed to explain what's really going on.

Specifically, we call the set of $(n+1)$-functors $k \colon B \to \mathrm{AUT}(F)$ modulo equivalence the **nonabelian cohomology** of B with coefficients in $\mathrm{AUT}(F)$. We denote it like this:

$$H(B, \mathrm{AUT}(F)).$$

We purposely leave off the little superscript i's that people usually put on cohomology; our point of view is more global. The element $[k]$ in the cohomology $H(B, \mathrm{AUT}(F))$ corresponding to a given fibration is called its **Postnikov invariant**.

So, to classify a space E, we think of it as an n-groupoid and break it down with its Postnikov tower, getting a whole *list* of guys

$$k_j \colon E_{j-1} \to \mathrm{AUT}(F_j)$$

and thus a list of Postnikov invariants

$$[k_j] \in H(E_{j-1}, \mathrm{AUT}(F_j)).$$

Together with the homotopy groups of E (which determine the fibers F_j), these Postnikov invariants classify the space E up to homotopy equivalence. Doing this in practice, of course, is terribly hard. But the idea is simple.

Next time we'll examine certain low-dimensional cases of this and see what it amounts to. In various watered-down cases we'll get various famous kinds of cohomology. The full-fledged n-categorical version is beyond what anyone knows how to handle except in low dimensions — even for n-groupoids, except by appealing to Grothendieck's dream. Street has a nice paper on cohomology with coefficients in an ∞-category; he probably knew this stuff I'm talking about way back when I was just a kid. I'm just trying to bring it to the masses.

3.3. Questions and comments.

3.3.1. Internalizing n-surjectivity.

Tom Fiore: *You can define epi and mono categorically and apply them in any category, not just sets. Can you do a similar thing and define analogues of 0-, 1-, and 2-surjectivity using diagrams in any 2-category, etc.?*

JB: I don't know. That's a great question.

Eugenia Cheng: *You can define a concept of 'essentially epic' in any 2-category, by weakening the usual definition of epimorphism. But in **Cat**, 'essentially epic' turns out to mean essentially surjective and full. I expect that in n-categories, it will give 'essentially surjective on j-morphisms below level n'. You can't isolate the action on objects from the action on morphisms, so you can't characterize a property that just refers to objects.*

(For a more thorough discussion see §5.5: there is a way to characterize essentially surjective functors 2-categorically, as the functors that are left orthogonal to full and faithful functors.)

3.3.2. How normal people think about this stuff.

Aaron Lauda: *What do mortals call this $(n+1)$-group $\mathrm{AUT}(F)$ that you get from an n-groupoid F?*

JB: Well, suppose we have any $(n+1)$-group, say G. The first thing to get straight is that there are *two ways* to think of this in terms of topology.

First, by Grothendieck's dream, we can think of G as a topological group, say $|G|$, that just happens to be a homotopy n-type. Why do the numbers go down one like this? It's just like when people see a category with one object: they call it a monoid. There's a level shift here: the *morphisms* of the 1-object category get called *elements* of the monoid.

Second, we can just admit that our $(n+1)$-group G is a special sort of $(n+1)$-groupoid. Following Grothendieck's dream, we can think of this as a homotopy $(n+1)$-type, called $B|G|$. But, we should always think of this as a 'connected pointed' homotopy $(n+1)$-type.

Why? Well, an $(n + 1)$-group is just an $(n + 1)$-groupoid with one object. More generally, an $(n + 1)$-groupoid is *equivalent* to an $(n + 1)$-group if it's **connected** — if all its objects are equivalent. But to actually turn a connected $(n + 1)$-groupoid into an $(n + 1)$-group, we need to pick a distinguished object, or 'basepoint'. So, an $(n + 1)$-group is essentially the same as a connected pointed $(n + 1)$-groupoid. If we translate this into the language of topology, we see that an $(n + 1)$-group amounts to a connected pointed homotopy $(n + 1)$-type. This is usually called $B|G|$, the **classifying space** of the topological group $|G|$.

These two viewpoints are closely related. The homotopy groups of $B|G|$ are the same as those of $|G|$, just shifted:

$$\pi_{j+1}(B|G|) = \pi_j(|G|).$$

You may think this is unduly complicated. Why bother thinking about an $(n + 1)$-group in *two different ways* using topology? In fact, both are important. Given your n-groupoid F, it's good to use both tricks just described to study the $(n + 1)$-group $\mathrm{AUT}(F)$.

The first trick gives a topological group $|\mathrm{AUT}(F)|$ that happens to be a homotopy n-type. This group is often called the group of **homotopy self-equivalences** of $|F|$, the homotopy n-type corresponding to F. The reason is that its elements are homotopy equivalences $f \colon |F| \xrightarrow{\sim} |F|$.

The second trick gives a connected pointed homotopy $(n + 1)$-type $B|\mathrm{AUT}(F)|$. We can use this to classify fibrations with F as fiber.

Aaron Lauda: *How does that work?*

JB: Well, we've seen that fibrations whose base B and fiber F are n-groupoids should be classified by $(n + 1)$-functors

$$k \colon B \to \mathrm{AUT}(F)$$

where $\mathrm{AUT}(F)$ is a $(n + 1)$-groupoid with one object. But normal people think about this using topology. So, they turn B into a homotopy n-type, say $|B|$. They turn $\mathrm{AUT}(F)$ into a connected pointed homotopy $(n + 1)$-type: the classifying space $B|\mathrm{AUT}(F)|$. And, they turn k into a map, say

$$|k| \colon |B| \to B|\mathrm{AUT}(F)|.$$

So, instead of thinking of the Postnikov invariant as an equivalence class of $(n + 1)$-functors

$$[k] \in H(B, \mathrm{AUT}(F))$$

they think of it as a homotopy class of maps:

$$[|k|] \in [|B|, B|\mathrm{AUT}(F)|]$$

where now the square brackets mean 'homotopy classes of maps' — that's what equivalence classes of j-functors become in the world of topology. And, they show fibrations with base $|B|$ and fiber $|F|$ are classified by this sort of Postnikov invariant.

Since I'm encouraging you to freely hop back and forth between the language of n-groupoids and the language of topology, from now on I won't write $|\cdot|$ to describe the passage from n-groupoids to spaces, or n-functors to maps. I just wanted to sketch how it worked, here.

Aaron Lauda: *So, all this is part of some highbrow approach to cohomology... but how does this relate to plain old cohomology, like the kind you first learn about in school?*

JB: Right. Suppose we're playing the Postnikov tower game. We have a homotopy n-type E, and somehow we know its homotopy groups π_j. So, we get this tower of fibrations

$$K(\pi_j, j) = F_j \to E_j \xrightarrow{p_j} E_{j-1}$$

where E_n is the space with started with and E_{-1} is just a point. To classify our space E just need to classify all these fibrations. That's what the Postnikov invariants do:

$$[k_j] \in H(E_{j-1}, \mathrm{AUT}(K(\pi_j, j))).$$

Now I'm using the language of topology, where AUT stands for the group of homotopy self-equivalences. But in the language of topology, the Postnikov invariants are homotopy classes of maps

$$k_j \colon E_{j-1} \to B\,\mathrm{AUT}(K(\pi_j, j)).$$

So, in general, our cohomology involves the space $B\,\mathrm{AUT}(K(\pi_j, j))$, which sounds pretty complicated. But we happen to know some very nice automorphisms of $K(\pi_j, j)$. It's an abelian topological group, at least for $j > 1$, so it can act on itself by left translations. Thus, sitting inside $B\,\mathrm{AUT}(K(\pi_j, j))$ we actually have $BK(\pi_j, j)$, which is actually the same as $K(\pi_j, j+1)$, since applying B shifts things up one level.

If the map k_j happens to land in this smaller space, at least up to homotopy, we call our space **simple**. Then we can write the Postnikov invariant as

$$[k_j] \in [E_{j-1}, K(\pi_j, j+1)]$$

and the thing on the right is just what people call the **ordinary cohomology** of our space E_{j-1} with coefficients in the group π_j, at least if $j > 1$. So, they write

$$[k_j] \in H^{j+1}(E_{j-1}, \pi_j).$$

Note that by now the indices are running all the way from $j-1$ to $j+1$, since we've played so many sneaky level-shifting tricks.

MS: *Actually, Postnikov towers have a nice interpretation in terms of cohomology even for spaces that aren't 'simple'. The trick is to use 'cohomology with local coefficients'. Given a space X and an abelian group A together with an action ρ of $\pi_1(X)$ on A, you can define cohomology groups $H_\rho^n(X,A)$ where the coefficients are 'twisted' by ρ. It then turns out that*

$$H(X, \mathrm{AUT}(K(\pi,j))) = [X, B\,\mathrm{AUT}(K(\pi,j))]$$
$$= \coprod_\rho H_\rho^{j+1}(X,\pi).$$

So for a space that isn't necessarily simple, a topologist would consider its Postnikov invariants to live in some cohomology with local coefficients. In the simple case, the action ρ is trivial, so we don't need local coefficients.

4. A low-dimensional example.

4.1. Review of Postnikov towers. Last time we discussed a big idea; this time let's look at an example. Let's start with a single fibration:

$$F \to E \to B.$$

This means that we have some point $* \in B$ and $F = p^{-1}(*)$ is the homotopy fiber, or 'essential preimage' over $*$. This won't depend on the choice of $*$ if B is connected. Let's restrict ourselves to that case: this is no great loss, since any base is a disjoint union of connected components.

We can then classify these fibrations via their 'classifying maps'

$$k\colon B \to \mathrm{AUT}(F)$$

where $\mathrm{AUT}(F)$ is an $(n+1)$-group if F is an n-groupoid. A lowbrow way to state this classification is that there's a notion of equivalence for both these guys, and the equivalence classes of each are in one-to-one correspondence. We could also try to state a highbrow version, which asserts that $\hom(B, \mathrm{AUT}(F))$ is equivalent to some $(n+1)$-category of fibrations with B as base and F as fiber. But let's be lowbrow today.

In both cases, hopefully the notion of equivalence is sort of obvious. 'Equivalence of fibrations' looks a lot like equivalence for extensions of groups — which are, in fact, a special case. In other words, fibrations are equivalent when there exists a vertical map making this diagram commute (weakly):

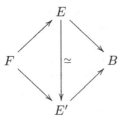

On the other side, the notion of equivalence for classifying maps

$$k, k' \colon B \to \mathrm{AUT}(F)$$

is equivalence of $(n + 1)$-functors, if we think of E and F as n-groupoids, or homotopy of maps if we think of E and F spaces.

This sounds great, but of course we're using all sorts of concepts from n-category theory that haven't been made precise yet. So, today we'll do an example, where we cut things down to a low enough level that we can handle it.

But first — why are we so interested in this? I hope you remember why it's so important. We have a grand goal: we want to classify n-groupoids. This is a Sisyphean task. We'll never actually complete it — but nonetheless, we can learn a lot by trying.

For example, consider the case $n = 1$: the classification of groupoids. Every groupoid is a disjoint union of groups, so we just need to classify groups. Let's say we start by trying to classify finite groups. Well, it's not so easy — after 10,000 pages of work people have only managed to classify the finite *simple* groups. Every finite group can be built up out of those by repeated extensions:

$$1 \to F \to E \to B \to 1.$$

These extensions are just a special case of the fibrations we've been talking about. So we can classify them using cohomology, at least in principle. But it won't be easy, because each time we do an extension we get a new group whose cohomology we need to understand. So, we'll probably never succeed in giving a useful classification of all finite groups. Luckily, even what we've learned so far can help us solve a lot of interesting problems.

Now suppose we want to classify n-groupoids for $n > 1$. We do it via their Postnikov towers, which are iterated fibrations. Given an n-groupoid E, we successively squash it down, dimension by dimension, until we get a single point:

At each step, we're **decategorifying**: to get the $(j-1)$-groupoid E_{j-1} from the j-groupoid E_j, we promote all the j-isomorphisms to *equations*. That's really what's going on, although I didn't emphasize it last time. Last time I emphasized that the map

$$p_j \colon E_j \to E_{j-1}$$

'forgets purely j-stuff'. What that means here is that decategorification throws out the top level, the j-morphisms, while doing as little damage as possible to the lower levels.

So, we get fibrations

$$F_j \to E_j \to E_{j-1}$$

where each homotopy fiber F_j, which records the stuff that's been thrown out, is a j-groupoid with only nontrivial j-morphisms: it has at most one i-morphism for any $i < j$. If you look at the periodic table, you'll see this means F_j is secretly a *group* for $j = 1$, and an *abelian group* for $j \geq 2$. Another name for this group is $\pi_j(E)$, which is more intuitive if you think of E as a space. If you think of F_j as a space, then it's the Eilenberg–Mac Lane space $K(\pi_j(E), j)$.

What do we learn from this business? That's where the basic principle of Galois theory comes in handy. We take the fibrations

$$F_j \to E_j \xrightarrow{\;p_j\;} E_{j-1}$$

and describe them via their classifying maps

$$k_j \colon E_{j-1} \to \mathrm{AUT}(F_j).$$

These give cohomology classes

$$[k_j] \in H(E_{j-1}, \mathrm{AUT}(F_j))$$

called **Postnikov invariants**.

So, we ultimately classify n-groupoids by a list of groups, namely π_1, ..., π_n, and all these cohomology classes $[k_j]$. What I want to do is show you how this works in detail, in a very low-dimensional case.

4.2. Example: The classification of 2-groupoids. Let's illustrate this for $n = 2$ and classify connected 2-groupoids. Since we're assuming things are connected, we might as well, for the purposes of classification, consider our connected 2-groupoids to be 2-groups. These have one object, a bunch of 1-morphisms from it to itself which are weakly invertible, and a bunch of 2-morphisms from these to themselves which are strictly invertible.

We can classify these using cohomology. Here's how. Given a 2-group, take a skeletal version of it, say E, and form these four things:

1. The group $G = \pi_1(E) =$ the group of '1-loops', i.e. 1-morphisms that start and end at the unique object. Composition of these would, a priori, only be associative up to isomorphism, but we said we picked a *skeletal* version, so these isomorphic objects have to be, in fact, equal.

2. The group $A = \pi_2(E) =$ the group of '2-loops', i.e. 2-morphisms which start and end at the identity 1-morphism 1_*. They form a group more obviously, and the Eckmann–Hilton argument shows this group is abelian.

3. An action ρ of G on A, where $\rho(g)(a)$ is defined by 'conjugation' or 'whiskering':

You can think of the loops as starting and ending at anything, if you like, by doing more whiskering. Then you have to spend a year figuring out whether you want to use left whiskering or right whiskering. This is supposed to be familiar from topology: there π_1 always acts on π_2.

4. The associator

$$\alpha_{g_1 g_2 g_3} \colon (g_1 g_2) g_3 \to g_1 (g_2 g_3)$$

gives a map

$$\alpha \colon G^3 \to A$$

as follows. Take three group elements and get an interesting automorphism of $g_1 g_2 g_3$. Automorphisms of anything can be identified with automorphisms of the identity, by whiskering. Explicitly, we cook up an element of A as follows:

(Don't ask why I put the whisker on the left instead of the right; you can do it either way and it doesn't really matter, though various formulas work out slightly differently.)

That's the stuff and structure, but there's also a property: the associator satisfies the pentagon identity, which means that α satisfies some equation. You all know the pentagon identity. It turns out this equation on α is something that people had been talking about for ages, before Mac Lane invented the pentagon identity. In fact, one of the people who'd been talking about it for ages was Mac Lane himself, because he'd also helped invent cohomology of groups. It's called the **3-cocycle equation** in group cohomology:

$$\rho(g_0)\alpha(g_1, g_2, g_3) - \alpha(g_0 g_1, g_2, g_3) + \alpha(g_0, g_1 g_2, g_3)$$
$$- \alpha(g_0, g_1, g_2 g_3) + \alpha(g_0, g_1, g_2) = 0.$$

Here we write the group operation in A additively, since it's abelian.

This equation is secretly just the pentagon identity satisfied by the associator; that's why it has 5 terms. But, people in group cohomology often write it simply as $d\alpha = 0$, because they know a standard trick for getting function of $(n + 1)$ elements of G from a function of n elements, and this trick is called the 'differential' d in group cohomology. If you have trouble remembering this trick, just think of a bunch of kids riding a school bus, but today there's one more kid than seats on the bus. What can we do? Either the first kid can jump out and sit on the hood, or the first two kids can squash into the first seat, and so on... or you can throw the last kid out the back window. That's a good way to remember the formula I just wrote.

Note that our *skeletal* 2-group is not necessarily *strict*! Making isomorphic objects equal doesn't mean making isomorphisms into identities. The associator isomorphism is still nontrivial, but it just happens in this case to be an *automorphism* from one object to the same object.

THEOREM 7 (Sinh). *Equivalence classes of 2-groups are in one-to-one correspondence with equivalence classes of 4-tuples*

$$(G, A, \rho, \alpha)$$

consisting of a group, an abelian group, an action, and a 3-cocycle.

The equivalence relation on the 4-tuples is via isomorphisms of G and A which get along with ρ, and get along with α up to a coboundary. Since cohomology is precisely cocycles modulo coboundaries, we really get the traditional notion of group cohomology showing up.

Just as the 3-cocycle equation comes from the pentagon identity for monoidal categories, the coboundary business comes from the notion of a monoidal *natural transformation*: monoidally equivalent monoidal categories won't have the same α, but their α's will differ by a coboundary.

This particular kind of group cohomology is called the 'third' cohomology since α is a function of three variables. We say that

$$[\alpha] \in H_\rho^3(G, A),$$

the third group cohomology of G with coefficients in a G-module A (where the action is defined by ρ). Sometimes this is called 'twisted' cohomology.

So, in short, once we fix G, A, and ρ, the equivalence classes of 2-groups we can build are in one-to-one correspondence with $H_\rho^3(G, A)$.

4.3. Relation to the general case. Now I'm going to show why this stuff is a special case of the general notion of cohomology we introduced last time. Why is $H_\rho^3(G, A)$ a special case of what we were calling $H(B, \text{AUT}(F))$?

Consider a little Postnikov tower where we start with a 2-group E and decategorify it getting a group B. We get a fibration $F \to E \to B$. To relate this to what we were just talking about, think of B as the group G. The 2-group F is the 2-category with one object, one morphism and some abelian group A as 2-morphisms. So, seeing how the 2-group E is built out of the base B and the fiber F should be the same as seeing how its built out of G and A. We want to see how the classifying map

$$k \colon B \to \mathrm{AUT}(F)$$

is the same as an element of $H^3_\rho(G, A)$.

It's good to think about this using a little topology. As a *space*, B is called $K(G, 1)$. It is made by taking one point:

•

one edge for element $g_1 \in G$:

a triangle for each pair of elements:

a tetrahedron for each triple:

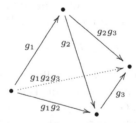

and so on. The fiber F is A regarded as a 2-group with only an identity 1-cell and 0-cell. So, as a space, F is called $K(A, 2)$, built with one point, one edge, one triangle for every element of A, one tetrahedron whenever $a_1 + a_2 = a_3 + a_4$, and so on.

Now, $\mathrm{AUT}(F)$ is a *3-group* which looks roughly like this. It has one object (which we can think of as 'being' F), one 1-morphism for every automorphism $f\colon F \to F$, one 2-morphism

$$F \underset{f'}{\overset{f}{\Longrightarrow}}_{\gamma} F$$

for each pseudonatural isomorphism, and one 3-morphism for each modification.

That seems sort of scary; what you have to do is figure out what that actually amounts to in the case when F is as above. Let me just tell you. It turns out that in fact, in our case $\mathrm{AUT}(F)$ has

- one object;
- its morphisms are just the group $\mathrm{Aut}(A)$;
- only identity 2-morphisms (as you can check);
- A as the endo-3-morphisms of any 2-morphism.

As a space, this is called '$B(\mathrm{AUT}(F))$', and it is made from one point, an edge for each automorphism f, a triangle for each equation $f_1 f_2 = f_3$, and tetrahedrons whose boundaries commute which are labeled by arbitrary elements of A.

Now, we can think about our classifying map as a weak 3-functor

$$k\colon B \to \mathrm{AUT}(F)$$

but we can also think of it as a map of spaces

$$k\colon K(G,1) \to B(\mathrm{AUT}(F)).$$

Let's just do it using spaces — or actually, simplicial sets. We have to map each type of simplex to a corresponding type. Here's how it goes:

- This map is boring on 0-cells, since there's only one choice.
- We get a map $\rho\colon G \to \mathrm{Aut}(A)$ for the 1-cells, which is good because that's what we want.
- The map on 2-cells says that this a group homomorphism, since it sends equations $g_1 g_2 = g_3$ to equations $\rho(g_1)\rho(g_2) = \rho(g_3)$.
- The map on 3-cells sends tetrahedra in $K(G,1)$, which are determined by triples of elements of G, to elements of A. This gives the may $\alpha\colon G^3 \to A$.
- The map on 4-cells is what forces α to be a 3-cocycle.

Our map $B \to \mathrm{AUT}(F)$ is *weak*, which is all-important. Even though our 2-morphisms are trivial, which makes the action on 1-morphisms actually a strict homomorphism, and our domain has no interesting 3-morphisms, we also get the higher data which gives the 3-cocycle $G^3 \to A$:

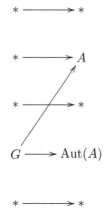

In fact, all sorts of categorified algebraic gadgets should be classified as 'layer cakes' built using Postnikov invariant taking values in the cohomology theory for that sort of gadget. We get group cohomology when we classify n-groupoids. Similarly, to classify categorified Lie algebras, which are called L_∞ algebras, we need Lie algebra cohomology — Alissa Crans has checked this in the simple case of an L_∞ algebra with only two nonzero chain groups. The next case is A_∞ algebras, which are categorified associative algebras. I bet that classifying these involves Hochschild cohomology — but I haven't ever sat down and checked it. And, it should keep on going. There should be a general theorem about this. That's what I mean by the 'layer cake philosophy' of cohomology.

4.4. Questions and comments.

4.4.1. Other values of n.

Aaron Lauda: *what about other H^n?*

JB: Imagine an alternate history of the world in which people knew about n-categories and had to learn about group cohomology from them.

We can figure out H^3 using 2-groups.

$$\vdots$$
$$*$$
$$\pi_2$$
$$\pi_1$$
$$*$$

To get the classical notion of H^4, we would have to think about classifying 3-groups that only have interesting 3-morphisms and 1-morphisms.

$$\vdots$$
$$*$$
$$\pi_3$$
$$*$$
$$\pi_1$$
$$*$$

We still get the whiskering action of 1-morphisms on 3-morphisms, and the pentagonator gives a 4-cocycle. Group cohomology, as customarily taught, is about classifying these 'fairly wimpy' Postnikov towers in which there are just two nontrivial groups: $H^{n+2}(\pi_1, \pi_n)$. This is clearly just a special case of something, and that something is a lot more complicated.

MS: *What about H^2?*

JB: Ah, that's interesting! In general, $H^n(G, A)$ classifies ways of building an $(n - 1)$-group with G as its bottom layer (what I was just calling π_1) and A as its top layer (namely π_{n-1}). So, it's all about layer-cakes with two nontrivial layers: the first and $(n-1)$st layers. For simplicity I'm assuming the action of G on A is trivial here.

But the case $n = 2$ is sort of degenerate: now our layer cake has only *one* nontrivial layer, the first layer, built by squashing A right into G. More precisely, $H^2(G, A)$ classifies ways of building a 1-group — an ordinary group — by taking a *central extension* of G by A. We've seen that 3-cocycles come from the associator, so it shouldn't be surprising that 2-cocycles come from something more basic: multiplication, where *two* elements of G give you an element of A.

Ironically, this weird degenerate low-dimensional case is the highest-dimensional case of group cohomology that ordinary textbooks bother to give any clean conceptual interpretation to. They say that H^2 classifies certain ways build a group out of two groups, but they don't say that H^n classifies certain ways to build an $(n - 1)$-group out of two groups. They don't say that extensions are just degenerate layer-cakes. And, it gets even more confusing when people start using H^2 to classify 'deformations' of algebraic structures, because they don't always admit that a deformation is just a special kind of extension.

4.4.2. The unit isomorphisms.

MS: *What happened to the unit isomorphisms?*

JB: When you make a 2-group skeletal, you can also make its unit isomorphisms equal to the identity. However, you can't make the associator be the identity — that's why it gives some interesting data in our classification of 2-groups, namely the 3-cocycle. This 3-cocycle is the only obstruction to a 2-group being both skeletal and strict.

5. Appendix: Posets, fibers, and n-topoi. This appendix is a hodgepodge of (proposed) answers to questions that arose during the lectures, some musings about higher topos theory, and some philosophy about the distinction between pointedness and connectedness.

5.1. Enrichment and posets. We observed in §2.5.3 that while (-1)-categories are truth values, having only a 0-category (that is, a set) of them is a bit unsatisfactory, since it doesn't allow us to talk about *implication* between truth values. The notetaker believes the best resolution to this problem is to extend the notion of '0-category' to include not just sets but *posets*. Then we can say that truth values form a poset with two elements, true and false, and a single nonidentity implication false $\Rightarrow$ true. A set can then be regarded as a discrete poset, or equivalently a poset in which every morphism is invertible; that is, a '0-groupoid'.

One way to approach a general theory including posets is to start from very low-dimensional categories and build up higher-dimensional ones using *enrichment*. We said in §2 that one of the general principles of n-category theory is that big n-categories have lots of little n-categories inside them. Another way of expressing this intuition is to say that *An n-category is a category enriched over $(n-1)$-categories.*

What does 'enriched' mean? Roughly speaking, a **category enriched over** V consists of

- A collection of objects $x, y, z, \ldots$
- For each pair of objects x, y, an object in V called $\hom(x, y)$.
- For each triple of objects x, y, z, a morphism in V called

$$\circ\colon \hom(x, y) \times \hom(y, z) \longrightarrow \hom(x, z).$$

- Units, associativity, etc.

Remember than in the world of n-categories, it doesn't make sense to talk about anything being strictly equal except at the top level. So when V is, for instance, the $(n+1)$-category of n-categories, the composition in a V-enriched category should only be associative and unital up to coherent equivalence. For example:

- 1-categories are categories enriched over sets;
- weak 2-categories are categories weakly enriched over categories;
- weak 3-categories are categories weakly enriched over weak 2-categories.

Making this precise for $n > 2$ is tricky, but it's a good intuition.

We may also say that a **V-enriched functor** $p\colon E \to B$ between such categories consists of:

- A function sending objects of E to objects of B;
- Morphisms of V-objects $\hom(x, y) \to \hom(px, py)$;
- various other data (again, as weak as appropriate).

And that's as far up as we need to go. We'll say that a V-**enriched groupoid** is a V-enriched category such that 'every morphism is invertible' in a suitably weak sense.

Let's investigate this notion in our very-low-dimensional world, starting with (-2)-categories, which we take to all be trivial by definition. What is a category enriched over (-2)-categories? Well, it has a collection of objects, together with, for every two objects, the unique (-2)-category as $\mathrm{hom}(x, y)$, and composition maps which are likewise unique. Thus a (-2)-category-enriched category is either:

- empty (has no objects), or
- has some number of objects, each of which is uniquely isomorphic to every other.

Thus it is either empty (false) or contractible (true), agreeing with the notion of (-1)-category that we got from topology. In particular, every (-1)-category is a groupoid.

Our general notion of functor now says that there should be a (-1)-functor from false to true, which we can call 'implication'. This is in line with topology: there is also a continuous map from the empty space to a contractible one.

Continuing on, a category enriched over (-1)-categories has a collection of objects together with, for every pair of objects x, y, a truth value $\mathrm{hom}(x, y)$, and for every triple x, y, z, a morphism

$$\mathrm{hom}(x, y) \times \mathrm{hom}(y, z) \longrightarrow \mathrm{hom}(x, z).$$

Now, the product of two (-1)-categories is empty (that is, false) if and only if one of the factors is. Thus, when we interpret (-1)-categories as truth values, the product $\times$ becomes the logical operation 'and', so if we interpret the truth of $\mathrm{hom}(x, y)$ as meaning $x \leq y$, we see that a category enriched over (-1)-categories is precisely a **poset**. (A non-category theorist would call this a *preordered set* since we don't have antisymmetry, but from a category theorist's perspective that's asking for equality of objects instead of isomorphism, which is perverse.)

Thus, from the enrichment point of view, perhaps '0-category' should mean a poset, rather than a set. As remarked above, we can view a set as a discrete poset (that is, one in which $x \leq y$ only when $x = y$). Categorically, a poset is *equivalent* to a discrete one whenever $x \leq y$ implies $y \leq x$, which is essentially the condition for it to be a *0-groupoid*: since composites are uniquely determined in a poset, a morphism $x \leq y$ is an isomorphism precisely when $y \leq x$ as well.

In a way, it's not surprising that our intuition may have been a little off in this regard, since a lot of it was coming from topology where *everything* is a groupoid.

What happens at the next level? Well, if we enrich over sets (0-groupoids), we get what are usually called categories, or 1-categories.

If we enrich instead over posets, we get what could variously be called **poset-enriched categories, locally posetal 2-categories**, or perhaps **2-posets**.

At the 0-level, we had one extra notion arising: instead of just sets, we got posets as well. At the 1-level, in addition to categories and poset-enriched categories, we also have a third notion: groupoids. These different levels correspond to the different levels of invertibility one can impose. If we start with a 2-poset and make its 2-morphisms all invertible, we get just a category. Then if we go ahead and make its 1-morphisms also invertible, we end up with a groupoid.

We can then go ahead and consider categories enriched over each of these three things, obtaining respectively 3-posets, 2-categories, and locally groupoidal 2-categories. And again there is an extra level that comes in: if we also make the 1-morphisms invertible, we get 2-groupoids.

All these various levels of invertibility can be fit together into the 'enrichment table' below. A dotted arrow $X \dashrightarrow Y$ means that a Y is a category enriched over Xs. The horizontal arrows denote inclusions; as we move to the left along any given line, we make more and more levels of morphisms invertible, coming from the top down. In general, the nth level will have $n + 2$ different levels of invertibility stretching off to the left.

THE ENRICHMENT TABLE

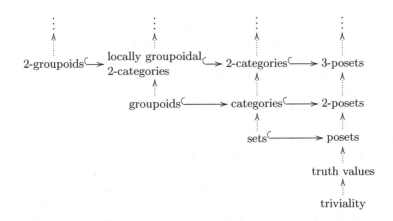

What does the enrichment table have to do with the periodic table? Recall that the n in 'n-categories' labels the *columns* of the periodic table, while the rows are labeled with the amount of monoidal structure. Thus we could, if we wanted to, combine the two into a three-dimensional table, replacing the line across the top of the periodic table with the whole table of enrichment.

The enrichment table only includes n-categories for finite n, but we can obtain various different '∞' notions by thinking about passing to some sort of 'limit' in various directions. Of course, these aren't actually limits in any formal sense. For example, it makes intuitive sense to say that the 'vertical limit' along the column

$$\text{sets} \dashrightarrow \text{1-categories} \dashrightarrow \text{2-categories} \dashrightarrow \cdots$$

should be the ∞-categories. Moreover, this should also be the limit along any other column. This is because in (say) the mth column, all the cells of the top m dimensions are invertible, but in the limit all these invertible cells get pushed off to infinity and we end up with noninvertible cells of all dimensions.

We can also consider 'diagonal limits'. It makes intuitive sense to say that the limit along the far-left diagonal, consisting of n-groupoids for increasing n, is the ∞-groupoids, aka homotopy types (à la Grothendieck). The limit along the next diagonal will be the ∞-categories with all morphisms above level 1 invertible. These are often called $(\infty, 1)$-categories (but sometimes also $(1, \infty)$-categories); see Bergner's survey article for an introduction to them.

By the way, the term '$(\infty, 1)$-categories' may sound strange, but it is just the most frequently used case of a general terminology. An **(n,m)-category** is an n-category all of whose j-morphisms for $j > m$ are invertible. Thus a n-category may also be called an (n, n)-category, an n-groupoid may be called an $(n, 0)$-category, and a locally groupoidal 2-category may be called a $(2, 1)$-category.

To stretch this terminology to its logical limit, we can call a poset-enriched category a $(1, 2)$-category, a poset a $(0, 1)$-category, and so on for the right-hand column of the enrichment table. If, instead of regarding an n-category as enriched over $(n - 1)$-categories, we return to regarding it as an ∞-category in which all cells of dimension $> n$ are identities, we can give the following characterization of (n, m)-categories which includes the case of posets as well.

DEFINITION 8. *An **(n,m)-category** is an ∞-category such that*
- *All j-morphisms for $j > n + 1$ exist and are unique wherever possible. In particular, this implies that all parallel $(n+1)$-morphisms are 'equal'.*
- *All j-morphisms for $j > m$ are invertible.*

In the next section we'll consider at length one reason that including the n-posets in the periodic table is important. Here's a different, simpler reason. Let E be a category, and consider its Postnikov tower:

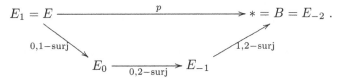

As we said in §3.1, E_0 is what we get by making parallel morphisms in E equal if they become equal in B; but here $B = *$, so this just means we identify *all* parallel morphisms. This precisely makes E into a *poset*—not necessarily a set. Thus in order for E_j to be a j-category in the factorization of an n-category which isn't a groupoid, we have to consider posets as a sort of 0-category, poset-enriched categories as a sort of 1-category, and generally j-posets as a sort of j-category.

5.2. Fibers and fibrations. Consider the fibers (or, rather, homotopy fibers) of a functor $p: E \to B$; we saw in §2.4 that their 'dimension' should reflect how much the functor p forgets. We'd like a generalization of Fact 3 there that applies to categories in addition to groupoids, but it turns out that for this we'll need to include the n-posets again. Consider first the following examples.

EXAMPLE 9. *We know that the functor* $p:$ **AbGp** $\to$ **Gp** *forgets only properties. What is the (essential) preimage* $p^{-1}(G)$ *for some group* G? *It is the category of all abelian groups equipped with isomorphisms to* G, *and morphisms which preserve the given isomorphisms. This category is contractible if* G *is abelian, and empty otherwise; in other words, it is essentially a* (-1)-*category.*

EXAMPLE 10. *Even more simply, consider an equivalence of categories* $p: E \to B$, *which forgets nothing. The the preimage* $p^{-1}(b)$ *is nonempty (since p is essentially surjective), and contractible (since p is full and faithful); thus it is essentially a* (-2)-*category.*

These examples, along with the groupoid case we considered in Fact 3, lead us to guess that a functor will forget 'at most n-stuff' precisely when its essential preimages are all n-categories. We consider properties to be (-1)-stuff, structure to be 0-stuff, ordinary stuff to be 1-stuff, eka-stuff to be 2-stuff, and so on.

However, this guess is not quite right, as we can see by considering some examples that forget structure.

EXAMPLE 11. *Consider the usual forgetful functor* $p:$ **Gp** $\to$ **Set**, *which we know forgets at most structure. Given a set, such as the 4-element set, its essential preimage* $p^{-1}(4)$ *is the category of 4-element labeled groups (since their underlying sets are equipped with isomorphisms to the given set 4), and homomorphisms that preserve the labeling.*

What does this look like? Well, given two labeled 4-element groups, there's exactly one function between them that preserves the labeling and either it's a group homomorphism or it isn't. Since the function preserving labeling is necessarily a bijection, if it is a homomorphism, then it is in fact a group isomorphism; thus this category is (equivalent to) a set.

In this example, we got what we expected, but we had to use a special property of groups: that a bijective homomorphism is an isomorphism. For many other types of structure, this won't be the case.

EXAMPLE 12. *Consider the forgetful functor $p\colon \mathbf{Top} \to \mathbf{Set}$ sending a topological space to its underlying set of points, which also forgets at most structure (in fact, purely structure). In this case, the essential preimage of the 4-element set is the collection of labeled 4-point topological spaces and continuous maps that preserve the labeling. Again, between any two there is exactly one function preserving the labeling, and either it is continuous or it isn't, so this category is a* poset. *In general, however, it won't be a set, since a continuous bijection is not necessarily a homeomorphism.*

Thus, in order to get a good characterization of levels of forgetfulness by using essential preimages, we really need to include the n-posets as n-categories.

Let's look at a couple of examples involving higher dimensions.

EXAMPLE 13. *We have a forgetful 2-functor*

$$[monoidal\ categories] \longrightarrow [categories]$$

which forgets at most stuff (since it is locally faithful, i.e. 3-surjective). Here the fiber over a category C is the category of ways to add a monoidal structure to C. There are lots of different ways to do this, and in between them we have monoidal functors that are the identity on objects (up to a specified equivalence, if we use the essential preimage), and in between those we have monoidal transformations whose components are identities (or specified isomorphisms). Now, there's at most one natural transformation from one functor to another whose components are identities, and either it's monoidal or it isn't. This shows that this collection is in fact a locally posetal 2-category, or a '2-poset', but in fact these monoidal natural transformations are automatically invertible when they exist, so it is in fact it is a 1-category.

EXAMPLE 14. *Let V be a nice category to enrich over, and consider the 'underlying ordinary category' functor*

$$(-)_0\colon V\text{-}\mathbf{Cat} \longrightarrow \mathbf{Cat}.$$

The category C_0 has the same objects as C, and $C_0(X,Y) = V(I, C(X,Y))$. What this functor forgets depends a lot on V:

- *In many cases, such as topological spaces, simplicial sets, categories, it is 0-surjective (any ordinary category can be enriched), but in others, such as abelian groups, it is not.*
- *In general it is not 1-surjective: not every ordinary functor can be enriched.*
- *In general, it is not 2-surjective: not every natural transformation is V-natural. It is 2-surjective, however, whenever the functor $V(I, -)$ is faithful, as for topological spaces and abelian groups. But when V is, say, simplicial sets, the functor $V(I, -)$ is not faithful, since a simplicial map is not determined by its action on vertices.*
- *It is always 3-surjective: a V-natural transformation is determined uniquely by its underlying ordinary natural transformation.*

Thus in general, $(-)_0$ *forgets at most stuff, but when* $V(I,-)$ *is faithful, it forgets at most structure.*

Now, what is the fiber over an ordinary category C? *Its objects are enrichments of* C, *its morphisms are* V-functors whose underlying ordinary functors are the identity, and its 2-cells are V-natural transformations whose components are identities. Such a 2-cell is merely the assertion that two V-functors are equal, so in general this is a 1-category. This is what we expect, since $(-)_0$ forgets at most stuff. However, when $V(I,-)$ is faithful, a V-functor is determined by its underlying functor, so the fiber is in fact a poset, as we expect it to be since in this case the functor forgets at most structure.*

It would be nice to have a good example of a 2-functor which forgets at most stuff and whose fibers are 2-posets that are not 1-categories, but I haven't thought of one.

EXAMPLE 15. *In order to do an example that forgets 2-stuff, consider the forgetful 2-functor*

$$[pairs\ of\ categories] \longrightarrow [categories].$$

This functor is not j-surjective for any $j \leq 3$, so it forgets at most 2-stuff. And here the (essential) fiber over a category is a genuine 2-category: we can have arbitrary functors and natural transformations living on that extra category we forgot about.*

Can we make this formal and use it as an alternate characterization of how much a functor forgets? The answer is: 'sometimes.' Here's what's true always in dimension one:

- If a functor is an equivalence, then all its essential fibers are contractible ((-2)-categories);
- If it is full and faithful, then all its essential fibers are empty or contractible ((-1)-categories);
- If it is faithful, then all its essential fibers are posets;
- and of course, if it is arbitrary, then its essential fibers can be arbitrary categories.

However, in general none of the implications above can be reversed. This is because a statement about the essential fibers really tells us only about the arrows which live over isomorphisms, while full and faithful tell us something about *all* the arrows.

There are, however, two cases in which the above implications *are* reversible:

1. When all categories involved are *groupoids*. This is because in this case, all arrows live over isomorphisms, since they all *are* isomorphisms.
2. If the functor is a *fibration* in the categorical sense.

Being a fibration in the categorical sense is like being a fibration in the topological sense, except that (1) we allow ourselves to lift arrows that have direction, since our categories have such arrows, and (2) we don't allow ourselves to take just any old lift, but require that the lift satisfy a nice universal property. I won't give the formal definition here, since you can find it in many places; instead I want to try to explain what it means.

The notion essentially means that the extra properties, structure, or stuff that lives upstairs in E can be 'transported' along arrows downstairs in B in a *universal* way. When we're transporting along arrows downstairs that are invertible, like paths in topology or arrows in an n-groupoid, this condition is unnecessary since the invertibility guarantees that we aren't making any irreversible changes. My favorite example is the following.

EXAMPLE 16. *Let B be the category of rings and ring homomorphisms. Let E be the category whose objects are pairs (R, M) where R is a ring and M is an R-module, and whose morphisms are pairs $(f, \varphi) \colon (R, M) \to (S, N)$ where $f \colon R \to S$ is a ring homomorphism and $\varphi \colon M \to N$ is an 'f-equivariant map', i.e. $\varphi(rm) = f(r)\varphi(m)$. Then if $f \colon R \to S$ is a ring homomorphism and N is an S-module, there is a canonical associated R-module f^*N—namely, M with R acting through f—and a canonical f-equivariant map $f^*N \to N$—namely the identity map. This map is 'universal' in a suitable sense, and is clearly what we should mean by 'transporting' N backwards along f.*

The formal definition of fibration simply makes this notion precise.

The introduction of directionality here also means that we get different things by transporting objects along arrows backwards and forwards. In the above example, the dual construction would be to take an R-module M and construct an S-module $f_! M = S \otimes_R M$ by 'extending scalars' to S. Again this comes with a canonical f-equivariant map $M \to f_! M$. Thus there are actually two notions of categorical fibration; for historical reasons, the 'backwards' one is usually called a **fibration** and the 'forwards' one an **opfibration** (or a 'cofibration', but we eschew that term because it carries the wrong topological intuition). Either one works equally well for the characterization of forgetfulness by fibers.

Another nice thing about the notion of categorical fibration is that while the principle of Galois theory does not apply, in general, to functors between arbitrary categories, it does apply to fibrations. Recall that in the groupoid case, fibrations over a base space (n-groupoid) B with fiber F are equivalent to functors $B \to \mathrm{AUT}(F)$. One can show that for a base category B, fibrations over B are equivalent to (weak) functors $B^{op} \to \mathbf{Cat}$. The way to think of this is that since our arrows are no longer necessarily invertible, the induced morphisms of fibers are no longer necessarily automorphisms, nor are all the fibers necessarily the same. Thus instead of the automorphism n-group of 'the' fiber, we have to use the whole *category* of possible fibers: in this case, $\mathbf{Cat}$, since the fibers are categories.

Fibrations also have the nice property that the essential preimage is equivalent to the literal or 'strict' preimage. Since many forgetful functors, like those above, are fibrations, in such cases we can use the strict preimage instead of the essential one. In fact, a much weaker property than being a fibration is enough for this; it suffices that objects upstairs can be transported along 'equivalences' downstairs (which coincides with the notion of fibrations in the n-groupoid case, when all morphisms are equivalences). This is true in many examples which are not full-fledged fibrations.

This advantage also implies, however, that there is a sense in which the notion of categorical fibration is 'not fully weak'. Ross Street has defined a weaker notion of fibration which does not have this property, and which makes sense in any (weak) 2-category. It is easy to check that this weaker notion of fibration also suffices for the characterization of forgetfulness via fibers. Of course, unlike for traditional fibrations, in this case it is essential that we use essential preimages, rather than strict ones, since the two are no longer equivalent.

There ought to be a notion of categorical fibration for n-categories. Some people have studied particular cases of this. Claudio Hermida has studied 2-fibrations between 2-categories. André Joyal, Jacob Lurie, and others have studied various notions of fibration between quasi-categories, which are one model for $(\infty, 1)$-categories; see Joyal's introduction to quasi-categories in this volume for more details.

Let's end this section by formulating a hypothesis about the behavior of fibers for n-categories. Generalizing an idea from the first lecture, let's say that a functor **forgets at most k-stuff** if it is j-surjective for $j > k+1$.

HYPOTHESIS 17. *If a functor between n-categories forgets at most k-stuff, then its fibers are k-categories (which we take to include poset-enriched k-categories). The converse is true for n-groupoids and for n-categorical fibrations.*

We've checked this hypothesis above for $n = 1$ and for n-groupoids (modulo Grothendieck). Actually, we only checked it for $(1,1)$-categories, while to be really consistent, we should check it for $(1,2)$-categories too, but I'll leave that to you. For $n = 0$ it says that an isomorphism of posets has contractible fibers (obvious) and that an inclusion of a sub-poset has fibers which are empty or contractible (also obvious). Surely someone can learn about 2-fibrations and check this hypothesis for $n = 2$ as well.

As one last note, recall that in the topological case, when we studied Postnikov towers in §3.2, we were able, by the magic of homotopy theory, to convert all the maps in our factorization into fibrations. It would be nice if a similiar result were true for categorical fibrations. It isn't true as long as we stick to plain old categories, but there's a sense in which it becomes true once we generalize to things called 'sites' and their corresponding 'topoi'. I won't say any more about this, but it leads us into the next topic.

5.3. n-topoi. Knowing about the existence of n-posets and how they fit into the enrichment table also clarifies the notion of topos, and in particular of n-topos.

Topos theory (by which is usually meant what we would call 1-topos theory and 0-topos theory; I'll explain later) is a vastly beautiful and interconnected edifice of mathematics, which can be quite intimidating for the newcomer, not least due to the lack of a unique entry point. In fact, the title of Peter Johnstone's epic compendium of topos theory, *Sketches of an Elephant*, compares the many different approaches to topos theory to the old story of six blind men and an elephant. (The six blind men had never met an elephant before, so when one was brought to them, they each felt part of it to determine what it was like. One felt the legs and said "an elephant is like a tree," one felt the ears and said "an elephant is like a banana leaf," one felt the trunk and said "an elephant is like a snake," and so on. But of course, an elephant is all of these things and none of them.)

So what is a topos anyway? For now, I want you to think of a (1-)topos as *a 1-category that can be viewed as a generalized universe of sets.* What this turns out to mean is the following:

- A topos has limits and colimits;
- A topos is cartesian closed; and
- A topos has a 'subobject classifier'.

It turns out that this much does, in fact, suffice to allow us to more or less replace the category of sets with any topos, and build all of mathematics using objects from that topos instead of our usual notion of sets. (There's one main caveat I will bring up below.)

Now, how can we generalize this to n-topoi for other values of n? I'm going to instead ask the more general question of how we can generalize it to (n, m)-topoi for other values of n and m. I claim that a sensible generalization should allow us to assert that:

> An (n, m)-topos is an (n, m)-category that can be viewed
> as a generalized universe of $(n - 1, m - 1)$-categories.

One thing this tells us is that we shouldn't expect to have much of a notion of topos for n-groupoids: we don't want to let ourselves drop off the enrichment table. Inspecting the definition of 1-topos confirms this: groupoids generally don't have limits or colimits, let alone anything fancier like exponentials or subobject classifiers. The only groupoid that is a topos is the trivial one. This is a bit unfortunate, since it means we can't test our hypotheses using homotopy theory in any obvious way, but we'll press on anyway.

Let's consider our lower-dimensional world. What should we mean by a '$(0, 1)$-topos'? (I'm going to abuse terminology and call this a '0-topos', since as we saw above, we expect the only $(0, 0)$-topos to be trivial.) Well, our general philosophy tells us that it should be a poset that can be viewed as a generalized universe of truth values. At this point you may think you know what it's going to turn out to be—and you may be right, or you may not be.

What do the characterizing properties of a 1-topos say when interpreted for posets? Limits in a poset are meets (greatest lower bounds), and colimits are joins (least upper bounds), so our 0-topoi will be complete lattices. Being cartesian closed for a poset means that for any elements b, c there exists an object $b \Rightarrow c$ such that

$$a \leq (b \Rightarrow c) \quad \text{if and only if} \quad (a \wedge b) \leq c.$$

As we expected, this structure makes our poset look like a generalized collection of truth values: we have a conjunction operation $\wedge$, a disjunction operation $\vee$, and an implication operation $\Rightarrow$. We can define a negation operator by $\neg a = (a \Rightarrow \bot)$, which turns out to behave just as we expect, except that in general $\neg\neg a \neq a$. Thus the logic we get is not *classical* logic, but *constructive* logic, in which the principle of double negation is denied (as are equivalent statements such as the 'law of excluded middle', $a \vee \neg a$). Boolean algebras, which model classical logic, are a special case of these cartesian closed posets, which are called **Heyting algebras**. Thus, a 0-topos is essentially just a complete Heyting algebra.

Now, one of the most exciting things in topos theory is that Heyting algebras turn up in topology! Namely, the lattice of open sets $\mathscr{O}(X)$ of any topological space X is a complete Heyting algebra, and any continuous map $f \colon X \to Y$ gives rise to a map of posets $f^* \colon \mathscr{O}(Y) \to \mathscr{O}(X)$ which preserves finite meets and arbitrary joins (but not $\Rightarrow$). Thus, we can view complete Heyting algebras as a sort of 'generalized topological space'. When we do this, we call them **locales**. So a more correct thing to say is that 0-topoi are the same as locales.

Now, I didn't mention the subobject classifier. In fact, it doesn't turn out to mean anything interesting for posets. This makes us wonder what its appearance for 1-topoi means. One answer is that *it allows us to apply the principle of Galois theory inside a 1-topos.*

What should that mean? Well, what does the principle of Galois theory (suitably generalized to nonidentical fibers) say for sets? It says, first of all, that functions $p \colon E \to B$, for sets E and B, are equivalent to functors $B \to \mathbf{Set}$. This is straightforward: we take each $b \in B$ to the fiber over it.

But what if we reduce the dimension of the fibers? A function $p \colon E \to B$ whose fibers are (-1)-categories, i.e. truth values, is just a subset of B, and the principle of Galois theory says that these should be equivalent to functors from B to the category of truth values, which is the poset $\{\text{false} \leq \text{true}\}$, often written $\mathbf{2}$. This is just the correspondence between subsets and their *characteristic functions*.

Now, a subobject classifier is a categorical way of saying that you have on object Ω which acts like $\mathbf{2}$: it is a target for characteristic functions of subobjects (monomorphisms). Thus, this condition in the definition of 1-topos essentially tells us that we can apply the principle of Galois theory inside the topos. (It turns out that the unrestricted version for arbitrary functions $p \colon E \to B$ is also true in a topos, once you figure out how to interpret it correctly.)

Now, since a 1-topos is a generalized universe of sets and contains an object Ω which acts as a generalization of the poset **2** of truth values, we naturally expect Ω to be a generalized universe of truth values, i.e. a 0-topos. This is in fact the case, although there are couple of different ways to make this precise.

One such way is to consider the **subterminal objects** of the topos, which are the objects U such that for any other object E there is *at most* one map $E \to U$. They are called 'subterminal' because they are the subobjects of the terminal object 1, which are by definition the same as the maps $1 \to \Omega$, or the 'points' of Ω. They can also be described as the objects which are 'representably (-1)-categories', since each hom-set $C(E, U)$ has either 0 or 1 element, so it is precisely a truth value. Thus it makes sense that the collection of subterminal objects turns out to be a 0-topos, whose elements are the 'internal truth values' in our given 1-topos. Since in general the logic of a 0-topos is constructive, not classical, the internal logic of a 1-topos is also in general constructive; this is the one caveat I mentioned earlier for our ability to redo all of mathematics in an arbitrary topos.

Thus every 1-topos, or universe of sets, contains inside it a 0-topos, or universe of truth values. We can also go in the other direction: given a locale X (a 0-topos), we can construct its category $\mathrm{Sh}(X)$ of **sheaves**, by an obvious generalization of the notion of sheaves on a topological space, and this turns out to be a 1-topos, which we regard as 'the category of sets in the universe parametrized by X'. As we expect, the subobject classifier in $\mathrm{Sh}(X)$ turns out to be $\mathscr{O}(X)$. In fact, this embeds the category of locales in the (2-)category of topoi, which leads us to consider any 1-topos as a vastly generalized kind of topological space.

As a side note, recall that in §5.1 we observed that a set is a groupoid enriched over truth values. Thus you might expect that the objects of $\mathrm{Sh}(X)$, which intuitively are 'sets in the universe where the truth values are $\mathscr{O}(X)$', could be defined as 'groupoids enriched over $\mathscr{O}(X)$'. This is almost right; the problem is that all the objects of such a groupoid turn out to have 'global extent', while an arbitrary sheaf can have objects which are only 'partially defined'. We can, however, make it work if we consider instead groupoids enriched over a suitable *bicategory* constructed from $\mathscr{O}(X)$.

Anyway, these relationships between 0-topoi and 1-topoi lead us to hope that in higher dimensions, each (n, m)-topos will contain within it topoi of lower dimensions, and in turn will embed in topoi of higher dimensions via a suitable categorification of sheaves (usually called 'n-stacks' or simply 'stacks'). Notions of (n, m)-topos have already been studied for a few other values of n and m. For instance, there has also been a good deal of interest lately in something that people call '∞-topoi', although from our point of view a better name would be $(\infty, 1)$-topoi. These are special $(\infty, 1)$-categories that can be considered as a generalized universe of homotopy types (i.e. ∞-groupoids). And for a long time algebraic geometers have been studying 'stacks of groupoids', which are pretty close to what we would call a '$(2,1)$-topos'.

Where does the interest in higher topoi come from? In topology, the principle of Galois theory already works very nicely, and people were working with fibrations, homotopy groups, Postnikov towers, and cohomology long before Grothendieck came along to tell them they were really working with ∞-groupoids. A fancy way to say this is that the category of spaces is already an $(\infty, 1)$-topos.

But in algebraic geometry, the Galois theory fails, because the category under consideration is 'too rigid'. The n-groups $\mathrm{AUT}(F)$ just don't exist. So what the algebraic geometers do is to take their category and *embed* it in a larger category in which the desired objects *do* exist; we would say that they embed it in a $(2, 1)$-topos (if they're only interested in one level of automorphisms) or an $(\infty, 1)$-topos (if they're interested in the full glory of homotopy theory). The way they do this is with a suitable generalization of the sheaf construction to arbitrary categories.

In the case $m > 1$, it is not clear whether there exists a single notion of (n, m)-topos that shares most of the good properties of 1-topoi. Several people have studied this question, however, with some partial encouraging results; the 'fibrational cosmoi' of Ross Street can be viewed as generalized universes of 1-categories, and more recently Mark Weber has studied certain special cosmoi under the name '2-topos'. One of the defining properties of a cosmos is the existence of 'presheaf objects' which allow the application of the principle of Galois theory to internal fibrations in the 2-category (suitably defined). Some people speak of this as "considering sets to be generalized truth values".

5.4. Geometric morphisms, classifying topoi, and n-stuff. In this section we'll see that morphisms between topoi admit similar 'Postnikov' factorizations, which in turn tell us interesting things about the logical theories they 'classify'. This section will probably be most interesting to readers with some prior acquaintance with topos theory, but I've tried to make it as accessible as possible.

Recall that a continuous map $f\colon X \to Y$ of topological spaces gives rise to a function $f^*\colon \mathcal{O}(Y) \to \mathcal{O}(X)$ which preserves finite meets and arbitrary joins. Let X and Y be locales and $\mathcal{O}(X)$ and $\mathcal{O}(Y)$ the corresponding complete Heyting algebras; we *define* a map of locales $f\colon X \to Y$ to be a function $f^*\colon \mathcal{O}(Y) \to \mathcal{O}(X)$ preserving finite meets and arbitrary joins. We distinguish notationally between the locale X and its poset of 'open sets' $\mathcal{O}(X)$ because the maps go in the opposite direction, even though the locale X technically consists of nothing but $\mathcal{O}(X)$.

Similarly, let X and Y be topoi, and $\mathscr{S}(X)$ and $\mathscr{S}(Y)$ their corresponding 1-categories. We define a map of 1-topoi $f\colon X \to Y$ to be a functor $f^*\colon \mathscr{S}(Y) \to \mathscr{S}(X)$ which preserves finite limits and arbitrary colimits; these maps are called **geometric morphisms** for historical reasons.

Now, it turns out that for any (small) category C, the category $\mathbf{Set}^C$ of functors from C to $\mathbf{Set}$ is a topos, and functors $f\colon C \to D$ give rise to geometric morphisms $\widehat{f}\colon \mathbf{Set}^C \to \mathbf{Set}^D$. We can thus ask how properties of the functor f are reflected in properties of the geometric morphism $\widehat{f}$. It turns out that we have the following dictionary (at least, 'modulo splitting idempotents', which is something I don't want to get into—just remember that this is all morally true, but there are some details).

f is full and faithful	$\sim$	$\widehat{f}$ is an 'inclusion'
f is essentially surjective	$\sim$	$\widehat{f}$ is a 'surjection'
f is faithful	$\sim$	$\widehat{f}$ is 'localic'
f is full and essentially surjective	$\sim$	$\widehat{f}$ is 'hyperconnected'.

What do all those strange terms on the right mean? I'm certainly not going to define them! But I'll try to give you some idea of how to think about them. The notions of 'inclusion' and 'surjection' are suitable generalizations of the correspondingly named notions for topological spaces. Moreover, just as is true for spaces, any geometric morphism factors uniquely as a surjection followed by an inclusion; this also parallels one of our familiar factorizations for functors. This part of the correspondence should make some intuitive sense.

To explain the term 'localic', consider a geometric morphism $p\colon E \to S$. It turns out that we can think of this either as a map between two topoi in the universe of sets, or we can use it to think of E as an *internal* topos in the generalized universe supplied by the topos S. We say that the morphism p is 'localic' if this internal topos is equivalent to the sheaves on some internal locale in S. It turns out that there is another sort of morphism called 'hyperconnected' such that every geometric morphism factors uniquely as a hyperconnected one followed by a localic one, and this too corresponds to a factorization we know and love for functors.

Moreover, every inclusion is localic, and every hyperconnected morphism is a surjection, and it follows that every geometric morphism factors as a hyperconnected morphism, followed by a surjective localic map, followed by an inclusion. This should also look familiar in the world of functors.

Now I want to explain why these classes of geometric morphisms in fact have an *intrinsic* connection to the notions of properties, structure, and stuff, but to do that I have to talk about 'classifying topoi'.

The basic idea of classifying topoi is that we can apply the principle of Galois theory once again, only this time we apply it in the 2-category *of topoi*, and we apply it to classify *models of logical theories*. Let $\mathbb{T}$ be a **typed logical theory**; thus it has some collection of 'types', some 'function and relation symbols' connecting these types, and some 'axioms' imposed on the behavior of these symbols. An example is the theory of categories, which has two types O ('objects') and A ('arrows'), three function symbols $s, t\colon A \to O$, $i\colon O \to A$, a relation symbol c of type $A \times A \times A$

(here $c(f, g, h)$ is intended to express the assertion that $h = g \circ f$), and various axioms, such as

$$(t(f) = s(g)) \Rightarrow \exists! h \, c(f, g, h)$$

(which says that any two composable arrows have a unique composite). A model of such a theory assigns a set to each type and a function or relation to each symbol, such that the axioms are satisfied; thus a model of the theory of categories is just a small category.

The fact that a topos is a generalized universe of sets implies that we can consider models of such a theory in *any* topos, not just the usual topos of sets. It turns out that for suitably nice theories $\mathbb{T}$ (called 'geometric' theories), there exists a topos $[\mathbb{T}]$ such that for any other topos B, the category of models of $\mathbb{T}$ in B is equivalent to $\mathrm{hom}(B, [\mathbb{T}])$, the category of geometric morphisms from B to $[\mathbb{T}]$ (remember that 1-topoi form a 2-category). Thus, once again, some structure 'in' or 'over' B can be classified by functors from B to a 'classifying object'.

Now suppose that we have two theories $\mathbb{T}$ and $\mathbb{T}'$ such that $\mathbb{T}'$ is $\mathbb{T}$ with some extra types, symbols, and/or axioms added. Since this means that any model of $\mathbb{T}'$ gives, by neglect of structure, a model of $\mathbb{T}$, by the Yoneda lemma we have a geometric morphism $p \colon [\mathbb{T}'] \to [\mathbb{T}]$. It turns out that

- p is an *inclusion* when $\mathbb{T}'$ adds only extra axioms to $\mathbb{T}$
- p is *localic* when $\mathbb{T}'$ adds extra functions, relations, and axioms to $\mathbb{T}$, but no new types
- p is a *surjection* when $\mathbb{T}'$ adds extra types, symbols, and axioms to $\mathbb{T}$, but no new properties of the existing types and symbols in $\mathbb{T}$ are implied by this new structure.
- p is *hyperconnected* when $\mathbb{T}'$ adds extra types to $\mathbb{T}$, along with symbols and axioms relating to these new types, but no new functions, relations, or axioms on the existing types in $\mathbb{T}$ are implied by this new structure.

(There are various ways to make these notions precise, which I'm not going to get into.) Thus, these classes of geometric morphisms actually directly encode the notions of forgetting properties (axioms), structure (function and relation symbols), and/or stuff (types).

Notice that localic morphisms are those that add no new types; this is consistent with the fact that locales are 0-topoi, and 0-categories know only about properties $((-1)$-stuff) and structure (0-stuff), not stuff (1-stuff). In particular, a classifying topos $[\mathbb{T}]$ is equivalent to a topos of sheaves on a locale precisely when the theory $\mathbb{T}$ has no types. Such a theory, which consists only of propositions and axioms, is called a **propositional theory**; from our point of view, we might also call it a '0-theory', with the more general typed theories considered above being '1-theories'. As far as I know, there has been very little work on notions of n-theories for higher values of n.

Now, given the correspondence between theories and classifying topoi, any factorization for geometric morphisms leads to a factorization for geometric theories. These factorizations are mostly what we would expect, but can be slightly different due to the requirement that all theories in sight be geometric.

EXAMPLE 18.   *Consider the forgetful map from monoids to semigroups. (A semigroup is a set with an associative binary operation.) Considered as a functor* **Mon** $\to$ **SGp**, *it is faithful, but not essentially surjective (since not every semigroup has an identity) or full (since not every semigroup homomorphism between monoids preserves the identity). If we factor it into a full-and-essentially-surjective functor followed by a full-and-faithful one, the intermediate category we obtain is the category of 'semigroups with identity', i.e. the category whose objects are monoids but whose morphisms do not necessarily preserve the identity.*

Now, the theories $\mathbb{M}$ *of monoids and* $\mathbb{S}$ *of semigroups are both geometric, so they have classifying topoi* $[\mathbb{M}]$ *and* $[\mathbb{S}]$, *and as we expect there is a geometric morphism* $[\mathbb{M}] \to [\mathbb{S}]$ *which is localic. If we factor it into a surjection followed by an inclusion, however, the intermediate topos we obtain is not the classifying topos for semigroups with identity, because that theory is not geometric. Instead, the intermediate topos we get is the classifying topos for semigroups such that for any finite set of elements, there is an element which behaves as an identity for them. In general, however, the 'identities' for different finite sets could be different.*

This theory is, in a sense, the 'closest geometric approximation' to the theory of semigroups with identity. This notion is in accord with the general principle (which we have not mentioned) that geometric logic is the 'logic of finite observation'. In this case, it is evident that if we can only 'observe' finitely many elements of the semigroup, we can't tell the difference between such a model of our weird intermediate geometric theory and a semigroup that has an actual identity.

These considerations may lead us to speculate that morphisms of higher topoi, once defined, will have similar 'Postnikov factorizations'. However, in the absence of confidence that good notions of (n, m)-topos exist for $m > 1$, this must remain a speculation.

5.5. Monomorphisms and epimorphisms.

A question was asked at one point (in §3.3.1) about whether notions like essential surjectivity can be defined purely 2-categorically, and thereby interpreted in any 2-category, the way that epimorphisms and monomorphisms make sense in any 1-category. This section is an attempt to partially answer that question.

The definitions of monomorphism and epimorphisms in 1-categories are 'representable' in the following sense:

- $m: A \to B$ is a monomorphism if for all X, the function

$$C(X, m)\colon C(X, A) \to C(X, B)$$

is injective.

- $e\colon E \to B$ is an epimorphism if for all X, the function

$$C(e, X)\colon C(B, X) \to C(A, X)$$

is injective.

Note that both notions invoke *injectivity* of functions of sets. Thus, the natural notions to consider first are functors which are 'representably' faithful or full-and-faithful. It is easy to check that this works in the covariant direction:

- A functor $p\colon A \to B$ is faithful if and only if it is representably faithful, i.e. all functors

$$\mathbf{Cat}(X, p)\colon \mathbf{Cat}(X, A) \to \mathbf{Cat}(X, B)$$

are faithful; and

- A functor is full and faithful if and only if it representably full and faithful.
- A functor is an equivalence if and only if it is representably an equivalence.

Thus, it makes sense to define a 1-morphism in a 2-category to be **faithful** or **full and faithful** when it is representably so.

We may generalize this (hypothetically) by saying that a functor between n-categories is j-**monic** if it is k-surjective for all $k > j$ (note that this is equivalent to saying that it 'forgets at most $(j-1)$-stuff'), and that a 1-morphism $p\colon A \to B$ in an $(n+1)$-category C is j-**monic** if all functors $C(X, p)$ are j-monic. By analogy with the above observation, we expect that these definitions will be equivalent for the $(n+1)$-category of n-categories.

Thus, every functor is 2-monic, the 1-monic functors are the faithful ones, the 0-monic functors are the full and faithful ones, and the (-1)-monic functors are the equivalences. More degenerately, in a 1-category, every map is 1-monic, the 0-monic morphisms are the usual monomorphisms, and the (-1)-monic morphisms are the isomorphisms.

We may define, dually, a 1-morphism $p\colon E \to B$ in an $(n+1)$-category C to be j-**epic** if all the functors

$$C(p, X)\colon C(B, X) \to C(E, X)$$

are $(n-1-j)$-monic. For example, in a 1-category, every morphism is (-1)-epic, the 0-epic morphisms are the usual epimorphisms, and the 1-epic morphisms are the isomorphisms.

The 'inversion' of numbering here may look a little strange if we remember that every n-category is secretly an ∞-category; when did it suddenly start to matter which n we are using? But it turns out that the transformation above is actually exactly what is required to make the notion independent of n. For example, if $p\colon E \to B$ is a surjective function in **Set**, then $\mathbf{Set}(p, X)$ is injective, hence 0-monic, for any set X; thus p is 0-epic in the 1-category **Set**. But now consider p as a functor between discrete categories. When X is a nondiscrete category, $\mathbf{Cat}(p, X)$ is faithful, but not full; hence it is only 1-monic, but by our definition this is just what is required so that p is again 0-epic.

It is easy to check that in a 2-category, every morphism is (-1)-epic, and the 2-epic morphisms are the equivalences. However, even in the 2-category **Cat**, the 0-epic and 1-epic morphisms are not that well-behaved. Here is what is true (proofs are left to the reader):

- If a functor $p\colon E \to B$ is essentially surjective, then it is 0-epic.
- Similarly, if it is full and essentially surjective, then it is 1-epic.

However, neither implication is reversible. For example, the inclusion of the category 2, which has two objects and one nonidentity morphism between them, into the category $\mathscr{I}$, which has two uniquely isomorphic objects, is 1-epic, but not full. And if $p\colon E \to B$ has the property that every object of B is a retract of an object in the image of p, then p is 0-epic, but it need not be essentially surjective.

We thus seek for other characterizations of surjective functions in **Set** which will generalize better to **Cat**. It turns out that the best-behaved notion is the following:

DEFINITION 19. *An epimorphism $p\colon E \to B$ in a 1-category is a* **strong epimorphism** *if it is 'left orthogonal' to monomorphisms, i.e. for any monomorphism $m\colon X \to Y$, every commutative square*

$$
\begin{array}{ccc}
E & \longrightarrow & X \\
{\scriptstyle e}\downarrow & \nearrow & \downarrow {\scriptstyle m} \\
B & \longrightarrow & Y
\end{array}
$$

has a unique diagonal filler.

In a category with equalizers, the orthogonality property implies that p is already an epimorphism. In **Set**, every epimorphism is strong, but in general this is not true.

Notice that saying $p\colon E \to B$ is left orthogonal to $m\colon X \to Y$ in the category C is equivalent to saying that the following square is a pullback:

$$
\begin{array}{ccc}
C(B, Y) & \longrightarrow & C(E, Y) \\
\downarrow & & \downarrow \\
C(B, X) & \longrightarrow & C(E, X)
\end{array}
$$

Therefore, we generalize this to 2-categories as follows.

DEFINITION 20. *A 1-morphism* $p\colon E \to B$ *in a 2-category* C *is* **left orthogonal** *to another* $m\colon X \to Y$ *if the square*

$$\begin{array}{ccc} C(B,Y) & \longrightarrow & C(E,Y) \\ \downarrow & & \downarrow \\ C(B,X) & \longrightarrow & C(E,X) \end{array}$$

is a pullback (in a suitable 2-categorical sense).

We can now check that

- A functor $p\colon E \to B$ is essentially surjective if and only if it is left orthogonal to all full and faithful functors, and
- It is full and essentially surjective if and only if it is left orthogonal to all faithful functors.

The forward directions are exercises in category theory. The idea is that we must progressively 'lift' objects, morphisms, and equations (to show functoriality and naturality) from 'downstairs' to 'upstairs'. In both cases, for each j, one of the two functors is j-surjective, so we can use that functor to lift the j-morphisms. The reverse directions are easy using the Postnikov factorization.

We are thus motivated to define, hypothetically, a j-epimorphism in an $(n+1)$-category to be a **strong** j-**epimorphism** if it is left orthogonal (in a suitably weak sense) to all j-monic morphisms. We have just shown that in **Cat**, the strong 1-epics are precisely the full and essentially surjective functors, while the strong 0-epics are the essentially surjective functors. Clearly all functors are strong (-1)-epic, while only equivalences are strong 2-epic.

We can also prove that in 2-categories with finite limits, any morphism which is left orthogonal to j-monomorphims is automatically a j-epimorphism; we use 2-categorical limits such as 'inserters' and 'equifiers' to take the place of equalizers in the 1-dimensional version. As we have seen, even in **Cat**, not every j-epimorphism is strong.

This leads us to formulate the following hypothesis.

HYPOTHESIS 21. *The strong j-epics in nCat (that is, functors which are left orthogonal to all j-monic functors) are precisely the functors which are k-surjective for $k \le j$. Not every j-epic is strong, even in n**Cat**.*

Since j-monic functors are those that forget 'at most $(j-1)$-stuff', we might say that the strong j-epics are the functors which 'forget no less than j-stuff'. For example, the strong 0-epics in **Cat** are the essentially surjective functors, which do not forget properties $((-1)$-stuff), although they may forget structure (0-stuff) and 1-stuff. Similarly, the strong 1-epics, being essentially surjective and full, do not forget properties or structure, although they may forget 1-stuff.

What does this look like for n-groupoids? For a functor between n-groupoids, being k-surjective is equivalent to inducing a surjection on π_k

and an *injection* on π_{k-1}. Why? Well, remember that π_k of an n-groupoid consists of the automorphisms of the identity $(k-1)$-morphism, modulo the $(k+1)$-morphisms. Thus being surjective on k-morphisms implies being surjective on π_k (although there might be new $(k+1)$-morphisms appearing preventing it from being an isomorphism), but also being injective on π_{k-1}, since everything we quotient by downstairs has to already be quotiented by upstairs (although here there might be entirely new $(k-1)$-morphisms appearing downstairs). This is also equivalent to saying that π_k of the homotopy fiber is trivial.

Thus, a functor between n-groupoids is j-monic if it induces isomorphisms on π_k for $k > j$ and an injection on π_j. Our above conjecture then translates to say that the strong j-epics should be the maps $A \to W$ inducing isomorphisms on π_k for $k < j$ and a surjection on π_j. This is precisely what topologists call a j-**equivalence** or a j-**connected map**, since it corresponds to the vanishing of the 'relative homotopy groups' $\pi_k(W, A)$ for $k \leq j$.

Using this identification, we can then prove our conjecture for n-groupoids (modulo Grothendieck). Suppose we have a square

of maps between n-groupoids (i.e. topological spaces), in which p is a j-equivalence and m is j-monic.

By magic of homotopy theory, similar to the way we can transform any map into a fibration, we can transform the map p into a 'relative cell complex'. This means that W is obtained from A by attaching 'cells' D^k along their boundaries S^{k-1}. Since p is a j-equivalence, we can assume that we are only attaching cells of dimension $k > j$. Thus our problem is reduced to defining a lift on each individual D^k. But our assumption on m guarantees that since (inductively) we have a lift of the boundary S^{k-1}, the whole cell D^k must also lift, up to homotopy.

This only shows that single maps lift, but we can also show that the appropriate square is a homotopy pullback by considering various modified squares. (The fanciest way to make this precise is to construct a 'Quillen model structure'.) Thus j-connected maps are in fact left 'homotopy' orthogonal to j-monic maps.

Just as before, we can prove the converse using our knowledge of factorizations. Suppose we have a map $p\colon A \to W$ which is left orthogonal to all j-monic maps. By picking out a particular part of the Postnikov factorization of p, we get a factorization $A \xrightarrow{f} E \xrightarrow{g} W$ in which g is j-monic and f is j-connected. Then in the square

$$
\begin{array}{ccc}
A & \xrightarrow{\ f\ } & E \\
{\scriptstyle p}\downarrow & {\scriptstyle h}\nearrow & \downarrow{\scriptstyle g} \\
W & =\!=\!= & W
\end{array}
$$

there exists a diagonal lift h. Consider any $k \leq j$; then $\pi_k(f)$ is either an isomorphism (if $k < j$) or surjective (if $k = j$), by assumption. This implies that $\pi_k(h)$ must also be surjective. But $\pi_k(g)\pi_k(h)$ is the identity, so in fact $\pi_k(h)$ must be an isomorphism. Therefore, since $hp \simeq f$, $\pi_k(p)$ must be an isomorphism if $k < j$ and surjective if $k = j$, since $\pi_k(f)$ is so. Thus p was already j-connected.

So, modulo Grothendieck's dream, we have proved Hypothesis 21 in the case of n-groupoids. Joyal's paper on quasi-categories, in this volume, includes a theory of factorization systems in $(\infty, 1)$-categories, generalizing the above arguments for ∞-groupoids to objects of any $(\infty, 1)$-category.

5.6. Pointedness versus connectedness. This final section will be even more philosophical, and perhaps controversial, than the others. The central point I wish to make is that the operations of looping and delooping should only be applied to *pointed* n-categories, just as they are only applied in homotopy theory to pointed topological spaces. When we do this, various problems with the periodic table resolve themselves.

What sort of problems? It's well-known that the hypothesis "a k-monoidal n-category is a k-degenerate $(n + k)$-category" is false, even in low dimensions, if you interpret 'is' as referring to a fully categorical sort of equivalence. The simplest example is that while a monoid 'is' a one-object category in a certain sense, the category of monoids is not equivalent to the (2-)category of categories-with-one-object. Similarly, the 2-category of monoidal categories is not equivalent to the (3-)category of one-object bicategories, and so on. In general, the objects and morphisms turn out mostly correct, but the higher-level transformations and so on are wrong. Eugenia Cheng and Nick Gurski have investigated in detail what happens and how you can often carefully chop things off at a particular level in the middle to get an equivalence, but here I want to consider a different point of view.

Let's consider the case of groupoids. The topological version of a one-object groupoid is a $K(G, 1)$, so the periodic table leads us to expect that the homotopy theory of $K(G, 1)$s should be equivalent to the category of groups. This is true, but only if we interpret the $K(G, 1)$s as *pointed* spaces and the corresponding homotopy theory likewise. Otherwise, we get the theory of groups and group homomorphisms modulo conjugation.

A related issue is that the homotopy groups π_n, and in particular π_1, are really only defined on *pointed* spaces. While it's true that different choices of basepoint give rise to isomorphic groups (at least for a connected space), the isomorphism is not canonical. In particular, this means that π_n is not *functorial* on the category of unpointed spaces.

Thus, by analogy with topology, we are motivated to consider 'pointed categories'. A **pointed n-category** is an n-category A equipped with a functor $1 \to A$ from the terminal n-category (which has exactly one j-morphism for every j). Note that this is essentially the same as choosing an object in A. A **pointed functor** between two pointed n-categories is a functor $A \to B$ such that

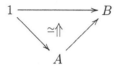

commutes up to a specified natural equivalence. A **pointed transformation** is a transformation α such that

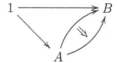

commutes with the specified equivalences up to an invertible modification. And so on for higher data.

What does this look like in low dimensions? A pointed set is just a set with a chosen element, and a pointed function between such sets is a function preserving the basepoints. More interestingly, a pointed category is a category A with a chosen base object $* \in A$, a pointed functor is a functor $f \colon A \to B$ equipped with an isomorphism $f* \cong *$, and a pointed natural transformation is a natural transformation $\alpha \colon f \to g$ such that

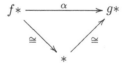

commutes.

We have only required our basepoints to be preserved up to coherent equivalence, in line with general n-categorical philosophy, but we now observe that we can always 'strictify' a pointed functor to preserve the basepoints on the nose. Define f' to be f on all objects except $f'* = *$, with the action on arrows defined by conjugating with the given isomorphism $f* \cong *$. Then $f'* = *$, and f' is isomorphic to f via a pointed natural isomorphism. Thus the 2-category of pointed categories, pointed functors, and pointed transformations is biequivalent to the 2-category of pointed categories, *strictly* pointed functors, and pointed transformations. We expect this to be true in higher dimensions as well. Observe that if f and g are strictly pointed functors, then a pointed natural transformation

$\alpha\colon f \to g$ is just a natural transformation $f \to g$ such that the component $\alpha_* = 1_*$.

Now we can define a functor Ω from the 2-category of pointed categories to the category of monoids. We take our pointed functors to be strict for convenience, since as we just saw there is no loss in doing so; otherwise we would just have to conjugate by the isomorphisms $f* \cong *$. We define $\Omega A = A(*, *)$ on objects, and for a strictly pointed functor $f\colon A \to B$, we get a monoid homomorphism $A(*, *) \to B(f*, f*) = B(*, *)$. Finally, since as we observed above a pointed transformation between strictly pointed functors is the identity on $*$, these transformations induce simply identities, which is good since those are the only 2-cells we've got in our codomain!

In the other direction, we construct a functor B from the category of monoids to the 2-category of pointed categories, sending a monoid M to the category BM with one object $*$ and $BM(*, *) = M$, and a monoid homomorphism to the obvious (strictly) pointed functor. We now observe that B is left adjoint to Ω (in a suitable sense), and that moreover the adjunction restricts to an adjoint biequivalence between the category of monoids (regarded as a locally discrete 2-category) and the 2-category of pointed categories with exactly one isomorphism class of objects (which we may call 'pointed and connected').

Similarly, one can construct 'adjoint' functors B and Ω between monoidal categories and pointed bicategories, and show that they restrict to inverse 'triequivalences' between monoidal categories and pointed bicategories with one equivalence class of objects ('pointed connected bicategories'). We can also do this for commutative monoids and 'pointed monoidal categories', but in fact here the word 'pointed' becomes redundant: every monoidal category has an essentially unique basepoint, namely the unit object. (Similarly, any monoid has a unique basepoint, namely its identity. This is because the terminal monoid and the terminal monoidal category are also 'initial' in a suitable sense.) We then obtain an adjoint biequivalence between commutative monoids and connected monoidal categories.

Composing these two adjunctions, we obtain an adjoint pair $B^2 \dashv \Omega^2$ between commutative monoids and pointed bicategories. This restricts to a biequivalence between commutative monoids and pointed bicategories with one equivalence class of objects and one isomorphism class of 1-morphisms ('pointed 1-connected bicategories').

All of these equivalences carry over in an obvious way to the groupoid cases, so that groups are equivalent to pointed groupoids with one isomorphism class of objects, groupal groupoids (2-groups) are equivalent to pointed 2-groupoids with one equivalence class of objects, and abelian groups are equivalent to 2-groups with one isomorphism class of objects, and also to 2-groupoids with one equivalence class of objects and one isomorphism class of morphisms. These are well-known topological results.

This suggests the following pointed version of the correspondence described in the periodic table. Say that an n-category is i-**connected** if it has exactly one equivalence class of j-morphisms for $0 \leq j \leq i$.

HYPOTHESIS 22 (Delooping Hypothesis). *There is an adjoint pair* $B^i \dashv \Omega^i$ *between k-monoidal n-categories and (pointed) $(k-i)$-monoidal $(n+i)$-categories, which restricts to an equivalence between k-monoidal n-categories and (pointed) $(i-1)$-connected $(k-i)$-monoidal $(n+i)$-categories.* We have placed "pointed" in parentheses because it is expected to be redundant for $k > i$.

We have called these functors Ω and B by analogy with the corresponding topological constructions of loop space and delooping (or 'classifying space'). Note that topologists usually say that the left adjoint of Ω is the 'suspension' functor Σ, rather than the 'delooping' functor B. This is because they often consider the functor Ω to take its values just in spaces, rather than monoidal spaces (say, A_∞-spaces). We would get a corresponding adjoint pair in our situation by composing the two adjunctions

$$n\text{-categories} \underset{U}{\overset{F}{\rightleftarrows}} \perp \begin{array}{c} \text{monoidal} \\ n\text{-categories} \end{array} \underset{\Omega}{\overset{B}{\rightleftarrows}} \perp \begin{array}{c} \text{pointed} \\ (n+1)\text{-categories} \end{array}$$

where $F \dashv U$ is the free-forgetful adjunction. Then $\Sigma A = BF(A)$, the delooping of the free monoidal n-category on A, is what deserves to be called the 'suspension' of A.

Note that there is also a forgetful functor from pointed $(n+1)$-categories to unpointed $(n+1)$-categories, which has a *left* adjoint $(-)_+$ called 'adding a disjoint basepoint'.

Let us investigate further the question of 'connectedness'. Recall from §2.3 that for a space X we say that $\pi_j(X)$ **vanishes for all basepoints** if given any $f\colon S^j \to X$, there exists $g\colon D^{j+1} \to X$ extending f. When X is nonempty, this is equivalent to requiring that the actual groups $\pi_j(X)$ vanish for all base points. Topologists define a nonempty space X to be k-**connected** if $\pi_i(X)$ is trivial for $j \leq k$ and all basepoints. (We'll deal with the empty set later.)

We can generalize this to n-categories in a straightforward way, but we use a different terminology because unlike n-groupoids, n-categories are not characterized by a list of homotopy groups. We say that an n-category **has no j-homotopy** when any two parallel j-morphisms are equivalent. Another way to say this, which is closer to the topology, is to define S^j to be the n-category consisting of two parallel j-morphisms, and D^{j+1} to consist of a $(j+1)$-equivalence between two parallel j-morphisms; then X has no j-homotopy just when all maps $S^j \to X$ extend to D^{j+1}. We can then define an n-category to be k-**connected** if it has no j-homotopy for $j \leq k$.

Note that since we don't know what a (-1)-morphism is, the category S^{-1} can only be empty. And since in general D^j is generated by a single j-equivalence, D^0 should just consist of a single object. Thus an n-category has no (-1)-homotopy just when it is nonempty. Similarly, S^{-2} and D^{-1} should both be empty, so every n-category has no (-2)-homotopy. Therefore, just as for groupoids, an n-category is always (-2)-connected, is (-1)-connected when it is nonempty, and for $k \geq 0$ it is k-connected when it has precisely one isomorphism class of j-cells for $0 \leq j \leq k$. Thus, this definition of connectedness agrees with the one we gave just before Hypothesis 22. Moreover, the new definition allows that hypothesis to make sense even for $i = 0$, in which case it says that all k-monoidal n-categories are nonempty and come equipped with an essentially unique basepoint (the unit object).

Now, what about that pesky empty set? By classical topological definitions, the empty set is unquestionably both connected (it is not the disjoint union of two nonempty open sets) and path-connected (any two points in it are connected by a path). But by our definitions, although it has no 0-homotopy, it is not 0-connected, because it does not have no (-1)-homotopy. (Of course, the empty set is the only space with no 0-homotopy which is not 0-connected.)

This disagreement is perhaps a slight wart on our definitions. However, it is worth pointing out that π_0 of the empty set must also be empty. In particular, it is not equal to $0 = \{0\}$. So the only way the empty set can be 0-connected, if we use the topological definition that X is k-connected $\pi_j(X) = *$ for all $0 \leq j \leq k$, is if we maintain that $\pi_0(X)$, just like the other π_j, requires a base point to be defined (in which case it is a pointed set). In this case, since the empty set has no basepoints, it is still vacuously true that $\pi_0(\emptyset) = 0$ for all basepoints.

We would like to emphasize the crucial distinction between *connected* (having precisely one equivalence class of objects) and being *pointed* (being *equipped* with a chosen object). Clearly, every connected n-category can be pointed in a way which is unique up to equivalence, *but not up to unique equivalence*. Similarly, functors between connected n-categories can be made pointed, but not in a unique way, while transformations and higher data can *not* in general be made pointed at all. Thus the $(n + 1)$-categories of connected n-categories and of pointed connected n-categories are not equivalent; the latter is equivalent to the n-category of monoidal $(n-1)$-categories, but the former is not.

This distinction explains an observation due to David Corfield that the periodic table seems to be missing a row. If in the periodic table we replace 'k-monoidal n-categories' by '$(k-1)$-connected $(n+k)$-categories', then the first row is seen to be the (-2)-connected things (that is, no connectivity imposed) while the second row is the 0-connected things. Thus there appears to be a row missing, consisting of the (-1)-connected, or nonempty, things, and morover the top row should be shifted over one to keep the diagonals moving correctly. So we should be looking at a table like this:

THE CONNECTIVITY PERIODIC TABLE

	$n = -1$	$n = 0$	$n = 1$
$k = -1$	truth values	sets	categories
$k = 0$	nonempty sets	nonempty categories	nonempty 2-categories
$k = 1$	connected categories	connected 2-categories	connected 3-categories
$k = 2$	1-connected 2-categories	1-connected 3-categories	1-connected 4-categories

In this table, the objects in the spot labeled k and n have nontrivial j-homotopy for only $n + 2$ consecutive values of j, starting at $j = k$. The column $n = -2$, which is not shown, consists entirely of trivialities, since if you have nontrivial j-homotopy for zero consecutive values of j, it doesn't matter at what value of j you start counting.

So we have two different periodic tables, and it isn't that one is right and one is wrong, but rather that one is talking about monoidal structures (or equivalently, by the delooping hypothesis, pointed *and* connected things) and the other is talking about connectivity. Note that unlike the monoidal periodic table, the connectivity periodic table does not stabilize.

Finally, here's another reason to make the distinction between 'connected' and 'pointed'. We observed above that for ordinary n-categories in the universe of sets, every connected n-category can be made pointed in a way unique up to (non-unique) equivalence. However, this can become false if we pass to n-categories in some other universe (topos), such as 'sheaves' of n-categories over some space.

Consider, for instance, the relationship between groups and connected groupoids. A 'sheaf of connected groupoids' is something called a 'gerbe' (a "locally connected locally nonempty stack in groupoids"), while a 'sheaf of groups' is a well-known thing, but very different. Every sheaf of groups gives rise to a gerbe, by delooping (to get a prestack of groupoids) and then 'stackifying', but it's reasonably fair to say that the whole interest of gerbes comes from the fact that most of them *don't* come from a sheaf of groups. The ones that do are called 'trivial', and a gerbe is trivial precisely when it has a basepoint (a global section). So the equivalence of groups with *pointed* connected groupoids is true even in the world of sheaves, but in this case not every 'connected' groupoid can be given a basepoint.

If we move down one level, this corresponds to the statement that not every well-supported sheaf has a global section. Thus in the world of sheaves, not every 'nonempty' set can be given a basepoint. So one cause of the confusion between connectedness and pointedness is what we might call 'Set-centric-ness': the two notions are quite similar in the topos of sets, but in other topoi they are much more distinct.

6. Annotated bibliography. The following bibliography should help the reader find more detailed information about some topics mentioned in the talks and Appendix. It makes no pretense to completeness, and we apologize in advance to all the authors whose work we fail to cite.

1.1 Galois Theory. For a treatment of Galois theory that emphasizes the analogy to covering spaces, try:

Adrien Douady and Régine Douady, *Algèbre et Théories Galoisiennes*, Cassini, Paris, 2005.

To see where the analogy between commutative algebras and spaces went after the work of Dedekind and Kummer, try this:

Dino Lorenzini, *An Invitation to Arithmetic Geometry*, American Mathematical Society, Providence, Rhode Island, 1996.

Finally, for a very general treatment of Galois theory, try this:

Francis Borceux and George Janelidze, *Galois Theories*, Cambridge Studies in Advanced Mathematics **72**, Cambridge U. Press, Cambridge, 2001.

1.2 The fundamental group. The fundamental group is covered in almost every basic textbook on algebraic topology. This introduction conveys some of the spirit of modern homotopy theory:

Marcelo Aguilar, Samuel Gitler, and Carlos Prieto, *Algebraic Topology from a Homotopical Viewpoint*, Springer, Berlin, 2002.

It also provides a good concrete introduction to classifying spaces, via covering spaces and then vector bundles. However, the treatment of some topics (such as homology) may strike more traditional algebraic topologists as perverse.

1.3 The fundamental groupoid. It is possible that a good modern introduction to algebraic topology should start with the fundamental groupoid rather than the fundamental group. Ronnie Brown has written a text that takes this approach:

Ronald Brown, *Topology and Groupoids*, Booksurge Publishing, North Charleston, South Carolina, 2006.

1.4 Eilenberg–Mac Lane spaces. There are a lot of interesting ideas packed in Eilenberg and Mac Lane's original series of papers on the cohomology of groups, starting around 1942 and going on until about 1955:

Samuel Eilenberg and Saunders Mac Lane, *Eilenberg–Mac Lane: Collected Works*, Academic Press, Orlando, Florida, 1986.

These papers are a bit tough to read, but they repay the effort even today. The spaces $K(G, n)$ appear implicitly in their 1945 paper 'Relations between homology and the homotopy groups of spaces', though much more emphasis is given on the corresponding chain complexes. The concept of k-invariant, so important for Postnikov towers, shows up in the 1950 paper 'Relations between homology and the homotopy groups of spaces, II'. The three papers entitled 'On the groups $H(\Pi, n)$, I, II, III' describe the bar construction and how to compute, in principle, the cohomology groups of any space $K(G, n)$ (where of course G is abelian for $n > 1$).

1.5 Grothendieck's dream. The classification of general extensions of groups goes back to Schreier:

O. Schreier, Über die Erweiterung von Gruppen I, *Monatschefte für Mathematik and Physik* **34** (1926), 165–180. Über die Erweiterung von Gruppen II, *Abh. Math. Sem. Hamburg* **4** (1926), 321–346.

But, the theory was worked out more thoroughly by Dedecker:

P. Dedecker, Les foncteuers Ext_Π, H^2_Π and H^2_Π non abeliens, *C. R. Acad. Sci. Paris* **258** (1964), 4891–4895.

To really understand our discussion of Schreier theory, one needs to know a bit about 2-categories. These are good introductions:

G. Maxwell Kelly and Ross Street, Review of the elements of 2-categories, Springer Lecture Notes in Mathematics **420**, Springer, Berlin, 1974, pp. 75-103.

Ross Street, Categorical structures, in *Handbook of Algebra*, vol. 1, ed. M. Hazewinkel, Elsevier, Amsterdam, 1996, pp. 529–577.

What we are calling 'weak 2-functors' and 'weak natural transformations', they call 'pseudofunctors' and 'pseudonatural transformations'.

Our treatment of Schreier theory used a set-theoretic section $s\colon B \to E$ in order to get an element of $H(B, \text{AUT}(F))$ from an exact sequence $1 \to F \to E \to B \to 1$. The arbitrary choice of section is annoying, and in categories other than **Set** it may not exist. Luckily, Jardine has given a construction that avoids the need for this splitting:

J. F. Jardine, Cocycle categories, sec. 4: Group extensions and 2-groupoids, available at ⟨http://www.math.uiuc.edu/K-theory/0782/⟩.

The generalization of Schreier theory to higher dimensions has a long and tangled history. Larry Breen generalized it 'upwards' from groups to 2-groups:

Lawrence Breen, Theorie de Schreier superieure, *Ann. Sci. Ecole Norm. Sup.* **25** (1992), 465-514. Also available at ⟨http://www.numdam.org/numdam-bin/feuilleter?id=ASENS19924255⟩.

It has also been generalized 'sideways' from groups to groupoids:

V. Blanco, M. Bullejos and E. Faro, Categorical non abelian cohomology, and the Schreier theory of groupoids, available as math.CT/0410202.

However, the latter generalization is already implicit in the work of Grothendieck: he classified all groupoids fibered over a groupoid B in terms of weak 2-functors from B to **Gpd**, the 2-groupoid of groupoids. The point is that **Gpd** contains $\text{AUT}(F)$ for any fixed groupoid F:

Alexander Grothendieck, *Revêtements Étales et Groupe Fondamental (SGA1)*, chapter VI: Catégories fibrées et descente, Lecture Notes in Mathematics 224, Springer, Berlin, 1971. Also available as math.AG/0206203.

A categorified version of Grothendieck's result can be found here:

Claudio Hermida, Descent on 2-fibrations and strongly 2-regular 2-categories, *Applied Categorical Structures*, **12** (2004), 427–459. Also available at ⟨http://maggie.cs.queensu.ca/chermida/papers/2-descent.pdf⟩.

While Grothendieck was working on fibrations and 'descent', Giraud was studying a closely related topic: nonabelian cohomology with coefficients in a gerbe:

Jean Giraud, *Cohomologie Non Abélienne*, Die Grundlehren der mathematischen Wissenschaften **179**, Springer, Berlin, 1971.

Nonabelian cohomology and n-categories came together in Grothendieck's letter to Quillen. This is now available online along with many other works by Grothendieck, thanks to the 'Grothendieck Circle':

Alexander Grothendieck, *Pursuing Stacks*, 1983. Available at ⟨http://www.grothendieckcircle.org/⟩.

Unfortunately we have not explained how these ideas are related to 'n-stacks' (roughly weak sheaves of n-categories) and 'n-gerbes' (roughly weak sheaves of n-groupoids that are locally connected and nonempty). For more modern work on n-stacks, nonabelian cohomology and their relation to Galois theory, try these and the many references therein:

André Hirschowitz and Carlos Simpson, Descente pour les n-champs, available as math.AG/9807049.

Bertrand Toen, Toward a Galoisian interpretation of homotopy theory, available as math.AT/0007157.

Bertrand Toen, Homotopical and higher categorical structures in algebraic geometry, Habilitation thesis, Université de Nice, 2003, available as math.AG/0312262.

Also see the material on topoi and higher topoi in the bibliography for Section §5.

2.   The Power of Negative Thinking.   The theory of weak n-categories (and ∞-categories) is in a state of rapid and unruly development, with many alternate approaches being proposed. For a quick sketch of the basic ideas, try:

John C. Baez, An introduction to n-categories, in *7th Conference on Category Theory and Computer Science,* eds. E. Moggi and G. Rosolini, Lecture Notes in Computer Science **1290**, Springer, Berlin, 1997.

John C. Baez and James Dolan, Categorification, in *Higher Category Theory,* eds. E. Getzler and M. Kapranov, Contemp. Math. **230**, American Mathematical Society, Providence, Rhode Island, 1998, pp. 1–36.

For a tour of ten proposed definitions, try:

Tom Leinster, A survey of definitions of n-category, available as math.CT/0107188.

For more intuition on these definitions work, see this book:

Eugenia Cheng and Aaron Lauda, *Higher-Dimensional Categories: an Illustrated Guide Book,* available at ⟨http://www.dpmms.cam.ac.uk/~elgc2/guidebook/⟩.

Another useful book on this nascent subject is:

Tom Leinster, *Higher Operads, Higher Categories,* London Math. Soc. Lecture Note Series **298**, Cambridge U. Press, Cambridge, 2004. Also available as math.CT/0305049.

In the present lectures we implicitly make use of the 'globular' weak ∞-categories developed by Batanin:

Michael A. Batanin, Monoidal globular categories as natural environment for the theory of weak n-categories, *Adv. Math.* **136** (1998), 39–103.

For recent progress on the homotopy hypothesis in this approach, see:

Denis-Charles Cisinski, Batanin higher groupoids and homotopy types, available as math.AT/0604442.

However, the most interesting questions about weak n-categories, including the stabilization hypothesis, homotopy hypothesis and other hypotheses mentioned in these lectures, should ultimately be successfully addressed by every 'good' approach to the subject. At the risk of circularity, one might even argue that this constitutes part of the criterion for which approaches count as 'good'.

The stabilization hypothesis is implicit in Larry Breen's work on higher gerbes:

Lawrence Breen, On the classification of 2-gerbes and 2-stacks, *Astérisque* **225**, Société Mathématique de France, 1994.

But a blunt statement of this hypothesis, together with the Periodic Table, appears here:

John C. Baez and James Dolan, Higher-dimensional algebra and topological quantum field theory, *Jour. Math. Phys.* **36** (1995), 6073–6105. Also available as q-alg/9503002.

There has been a lot of progress recently toward precisely formulating and proving the stabilization hypothesis and understanding the structure of k-tuply monoidal n-categories and their relation to k-fold loop spaces:

Carlos Simpson, On the Breen–Baez–Dolan stabilization hypothesis for Tamsamani's weak n-categories, available as math.CT/9810058.

Michael A. Batanin, The Eckmann–Hilton argument and higher operads, available as math.CT/0207281.

Michael A. Batanin, The combinatorics of iterated loop spaces, available as math.CT/0301221.

Eugenia Cheng and Nick Gurski, The periodic table of n-categories for low dimensions I: degenerate categories and degenerate bicategories, in *Categories in Algebra, Geometry and Mathematical Physics*, eds. M. Batanin *et al*, Contemp. Math. **431**, American Mathematical Society, Providence, Rhode Island, 2007. Also available as arXiv:0708.1178

Eugenia Cheng and Nick Gurski, The periodic table of n-categories for low dimensions II: degenerate tricategories, available as arXiv:0706.2307.

The mathematical notion of 'stuff' was introduced here:

John C. Baez and James Dolan, From finite sets to Feynman diagrams, in *Mathematics Unlimited - 2001 and Beyond*, vol. 1, eds. Bjørn Engquist and Wilfried Schmid, Springer, Berlin, 2001, pp. 29–50. Also available as math.QA/0004133

where 'stuff types' (groupoids over the groupoid of finite sets and bijections) were used to explain the combinatorial underpinnings of the theory of Feynman diagrams. A more detailed study of this subject can be found here:

John C. Baez and Derek Wise, *Quantization and Categorification*, Quantum Gravity Seminar, U. C. Riverside, Spring 2004 lecture notes, available at ⟨http://math.ucr.edu/home/baez/qg-spring2004/⟩.

On this page you will find links to a pedagogical introduction to properties, structure and stuff by Toby Bartels, and also to a long online conversation in which (-1)-categories and (-2)-categories were discovered. See also:

Simon Byrne, *On Groupoids and Stuff*, honors thesis, Macquarie University, 2005, available at
⟨http://www.maths.mq.edu.au/~street/ByrneHons.pdf⟩
and
⟨http://math.ucr.edu/home/baez/qg-spring2004/ByrneHons.pdf⟩.

Jeffrey Morton, Categorified algebra and quantum mechanics, available as math.QA/0601458.

3. Cohomology: The Layer-Cake Philosophy. In topology, it's most common to generalize the basic principle of Galois theory from covering spaces to fiber bundles along these lines:

> **Principal G-bundles over a base space B are**
> **classified by maps from B to the classifying space BG.**

For example, if B is a CW complex and G is a topological group, then isomorphism classes of principal G-bundles over M are in one-to-one correspondence with homotopy classes of maps from B to BG. But another approach, more in line with higher category theory, goes roughly as follows:

> **Fibrations over a pointed connected base space B with fiber**
> **F are classified by homomorphisms sending based loops in**
> **B to automorphisms of F.**

Stasheff proved a version of this which is reviewed here:

James Stasheff, H-spaces and classifying spaces, I-IV, *AMS Proc. Symp. Pure Math.* **22** (1971), 247–272.

He treats the space ΩB of based loops in B as an A_∞ space, i.e. a space with a product that is associative up to a homotopy that satisfies the pentagon identity up to a homotopy that satisfies a further identity up to a homotopy... ad infinitum. He classifies fibrations over B with fiber F in terms of A_∞-morphisms from ΩB into the topological monoid $\mathrm{Aut}(F)$ consisting of homotopy equivalences of F.

Another version was proved here:

J. Peter May, *Classifying Spaces and Fibrations*, AMS Memoirs **155**, American Mathematical Society, Providence, 1975.

Moore loops in B form a topological monoid $\Omega_\mathrm{M} B$. May defines a **transport** to be a homomorphism of topological monoids from $\Omega_\mathrm{M} B$ to $\mathrm{Aut}(F)$. After replacing F by a suitable homotopy-equivalent space, he defines an equivalence relation on transports such that the equivalence classes are in natural one-to-one correspondence with the equivalence classes of fibrations over B with fiber F.

By iterating the usual classification of principal G-bundles over B in terms of maps $B \to BG$, we obtain the theory of Postnikov towers. Unfortunately most expository accounts limit themselves to 'simple' spaces, namely those which π_1 acts trivially on the higher homotopy groups. For the general case see:

C. Alan Robinson, Moore–Postnikov systems for non-simple fibrations, *Ill. Jour. Math.* **16** (1972), 234–242.

For a treatment of Postnikov towers based on simplicial sets rather than topological spaces, try:

J. Peter May, *Simplicial Objects in Algebraic Topology*, Van Nostrand, Princeton, 1968.

Here is the paper by Street on cohomology with coefficients in an ∞-category:

Ross Street, Categorical and combinatorial aspects of descent theory, available at math.CT/0303175.

The idea of cohomology with coefficients in an ∞-category seems to have originated here:

John E. Roberts, Mathematical aspects of local cohomology, in *Algèbres d'Opérateurs et Leurs Applications en Physique Mathématique*, CNRS, Paris, 1979, pp. 321–332.

4.   A Low-Dimensional Example. This section will make more sense if one is comfortable with the cohomology of groups. To get started, try:

Joseph J. Rotman, *An Introduction to Homological Algebra*, Academic Press, New York, 1979.

or this more advanced book with the same title:

Charles A. Weibel, *An Introduction to Homological Algebra*, Cambridge U. Press, Cambridge, 1995.

For more detail, we recommend:

Kenneth S. Brown, *Cohomology of Groups*, Graduate Texts in Mathematics **182**, Springer, Berlin, 1982.

The classification of 2-groups up to equivalence using group cohomology was worked out by a student of Grothendieck:

Hoang X. Sinh, *Gr-categories*, Université Paris VII doctoral thesis, 1975.

She called them **gr-categories** instead of 2-groups, and this terminology remains common in the French literature. Her thesis, while very influential, was never published. Later, Joyal and Street described the whole 2-category of 2-groups using group cohomology here:

André Joyal and Ross Street, Braided monoidal categories, Macquarie Mathematics Report No. 860081, November 1986. Also available at
⟨http://rutherglen.ics.mq.edu.au/~street/JS86.pdf⟩.

Joyal and Street call them **categorical groups** instead of 2-groups. Like Sinh's thesis, this paper was never published — the published paper with a similar title leaves out the classification of 2-groups and moves directly to the classification of braided 2-groups. These are also nice examples of the general 'layer-cake philosophy' we are discussing here. Since braided 2-groups are morally the same as connected pointed homotopy types with only π_2 and π_3 nontrivial, their classification involves $H^4(K(G,2),A)$ (for G abelian) instead of the cohomology group we are considering here, $H^3(G,A) = H^3(K(G,1),A)$. For details, see:

André Joyal and Ross Street, Braided tensor categories, *Adv. Math.* **102** (1993), 20–78.

Finally, since it was hard to find a clear treatment of the classification of 2-groups in the published literature, an account was included here:

John C. Baez and Aaron Lauda, Higher-dimensional algebra V: 2-Groups, *Th. Appl. Cat.* **12** (2004), 423–491. Also available as math.QA/0307200.

along with some more history of the subject. The analogous classification of Lie 2-algebras using Lie algebra cohomology appears here:

John C. Baez and Alissa S. Crans, Higher-dimensional algebra VI: Lie 2-Algebras, *Th. Appl. Cat.* **12** (2004), 492–528. Also available as math.QA/0307263.

This paper also shows that an element of $H^{n+1}_\rho(\mathfrak{g},\mathfrak{a})$ gives a Lie n-algebra with $\mathfrak{g}$ as the Lie algebra of objects and $\mathfrak{a}$ as the abelian Lie algebra of $(n-1)$-morphisms.

5. Appendix: Posets, Fibers, and n-Topoi. As mentioned in the notes to Section §1.5, fibrations as functors between categories were introduced by Grothendieck in SGA1. They were then extensively developed by Jean Bénabou, and his handwritten notes of a 1980 course *Des Categories Fibrées* have been very influential. Unfortunately these remain hard to get, so we suggest:

Thomas Streicher, Fibred categories à la Jean Bénabou, April 2005, available as
⟨http://www.mathematik.tu-darmstadt.de/~streicher/FIBR/FibLec.pdf.gz⟩.

Street's fully weakened definition of fibrations in general weak 2-categories can be found here:

Ross Street, Fibrations in bicategories, *Cah. Top. Geom. Diff. Cat.* **21** (1980), 111–160. Errata, *Cah. Top. Geom. Diff. Cat.* **28** (1987), 53–56.

There are several good introductions to topos theory; here are a couple:

Saunders Mac Lane and Ieke Moerdijk, *Sheaves in Geometry and Logic: a First Introduction to Topos Theory*, Springer, New York, 1992.

Colin McLarty, *Elementary Categories, Elementary Toposes*, Oxford U. Press, Oxford, 1992.

The serious student will eventually want to spend time with the *Elephant*:

Peter Johnstone, *Sketches of an Elephant: a Topos Theory Compendium*, Oxford U. Press, Oxford. Volume 1, comprising Part A: Toposes as categories, and Part B: 2-Categorical aspects of topos theory, 2002. Volume 2, comprising Part C: Toposes as spaces, and Part D: Toposes as theories, 2002. Volume 3, comprising Part E: Homotopy and cohomology, and Part F: Toposes as mathematical universes, in preparation.

The beginning of part C is a good introduction to locales and Heyting algebras. We eagerly await part E for more illumination on one of the main themes of this paper, namely the foundations of cohomology theory.

The idea of defining sheaves as categories enriched over a certain bicategory is due to Walters:

Robert F. C. Walters, Sheaves on sites as Cauchy-complete categories, *J. Pure Appl. Algebra* **24** (1982), 95–102

Street has some papers on cosmoi:

Ross Street and Robert F. C. Walters, Yoneda structures on 2-categories, *J. Algebra* **50** (1978), 350–379.

Ross Street, Elementary cosmoi, I, *Category Seminar*, Lecture Notes in Math. **420**, Springer, Berlin, 1974, pp. 134–180.

Ross Street, Cosmoi of internal categories, *Trans. Amer. Math. Soc.* **258** (1980), 271–318.

And Weber's 2-topos paper can now be found here:

Mark Weber, Yoneda structures from 2-toposes, *Appl. Cat. Struct.* **15** (2007), 259–323. Slightly different version available as math.CT/0606393.

The theory of ∞-topoi — or what we prefer to call (∞, 1)-topoi, to emphasize the room left for further expansion — is new and still developing. A book just came out on this subject:

Jacob Lurie, Higher topos theory, available as math.CT/0608040.

For a good introduction to ∞-topoi which takes a slightly nontraditional approach to topoi, see:

Charles Rezk, Toposes and homotopy toposes, available at ⟨http://www.math.uiuc.edu/~rezk/homotopy-topos-sketch.dvi⟩.

This paper is an overview of ∞-topoi using model categories and Segal categories:

Bertrand Toen, Higher and derived stacks: a global overview, available as math.AG/0604504.

The correspondence between properties of small functors and properties of geometric morphisms is, to our knowledge, not written down all together anywhere. Johnstone summarized it in his talk at the Mac Lane memorial conference in Chicago in 2006. One can extract this information from *Sketches of an Elephant* if one looks at the examples in the sections on various types of geometric morphism in part C, and always think 'modulo splitting idempotents'.

Classifying topoi are explained somewhat in Mac Lane and Moerdijk, and more in part D of the *Elephant*. We also recommend having a look at the version of classifying topoi in part B of sketches, which uses 2-categorical limits to construct them. This makes the connection with n-stuff a little clearer.

Another good introduction to classifying topoi, and their relationship to topology, is:

Steven Vickers, Locales and toposes as spaces, available at ⟨http://www.cs.bham.ac.uk/~sjv/LocTopSpaces.pdf⟩.

A SURVEY OF $(\infty, 1)$-CATEGORIES

JULIA E. BERGNER*

Abstract. In this paper we give a summary of the comparisons between different definitions of so-called $(\infty, 1)$-categories, which are considered to be models for
∞-categories whose n-morphisms are all invertible for $n > 1$. They are also, from the
viewpoint of homotopy theory, models for the homotopy theory of homotopy theories.
The four different structures, all of which are equivalent, are simplicial categories, Segal
categories, complete Segal spaces, and quasi-categories.

AMS(MOS) 2000 Subject Classifications. 18-02, 55-02.

1. Introduction. The intent of this paper is to summarize some of
the progress that has been made since the IMA workshop on n-categories on
the topic of $(\infty, 1)$-categories. Heuristically, a $(\infty, 1)$-category is a weak ∞-
category in which the n-morphisms are all invertible for $n > 1$. Practically
speaking, there are several ways in which one could encode this information. In fact, at the moment, there are four models for $(\infty, 1)$-categories:
simplicial categories, Segal categories, complete Segal spaces, and quasi-
categories. They have arisen out of different motivations in both category
theory and homotopy theory, but work by the author and Joyal-Tierney
has shown that they are all equivalent to one another, in that they can be
connected by chains of Quillen equivalences of model categories.

From the viewpoint of higher category theory, these comparisons provide a kind of baby version of the comparisons which are being attempted
between various definitions of weak n-category. In [37], Toën actually axiomatizes a theory of $(\infty, 1)$-categories and proves that any such theory
is equivalent to the theory of complete Segal spaces. In [36], he sketches
arguments for proving the equivalences between the four structures used
in this paper. Although some of the functors he suggests do not appear to
give the desired Quillen equivalences (or at any rate are not used in the
known proofs), he gives a good overview of the problem. Another good
introduction of the problem can be found in a preprint by Porter [24], and
a nice description of the idea behind $(\infty, 1)$-categories can be found in [34].

From another point of view, these comparisons are of interest in homotopy theory, as what we are here calling a $(\infty, 1)$-category can be considered
to be a model for the homotopy theory of homotopy theories, a concept
which will be made more precise in the section on simplicial categories. The
idea is that a simplicial category is in some way "naturally" a homotopy
theory but functors between such are not particularly easy to work with.
The goal of finding an equivalent but nicer model led to Rezk's complete

---

*Department of Mathematics, University of California, Riverside, CA 92521
(bergnerj@member.ams.org).

J.C. Baez, J.P. May (eds.), *Towards Higher Categories*, The IMA Volumes
in Mathematics and its Applications 152, DOI 10.1007/978-1-4419-1524-5_2,
© Springer Science+Business Media, LLC 2010

Segal spaces [27] and the task of showing that they were essentially the same as simplicial categories.

It should be noted that there are other proposed models, including that of A_∞-categories. Joyal and Tierney briefly discuss some other approaches to this idea in their epilogue [21].

Furthermore, we should also mention that these structures are of interest in areas beyond homotopy theory and higher category theory. For instance, there are situations in algebraic geometry in which simplicial categories have been shown to provide information that the more commonly used derived category cannot. Given a scheme X, for example, its derived category $\mathcal{D}(X)$ does not seem to determine its K-theory spectrum, whereas its simplicial localization $\mathcal{L}(X)$ does [38]. Furthermore, the simplicial category $\mathcal{L}(X)$ forms a stack, which $\mathcal{D}(X)$ does not [16]. Similar work is also being done using dg categories, which are in many ways analogous to simplicial categories [35]. In particular, the category of dg categories has a model category structure which is defined using the same essential ideas as the model structure on the category of simplicial categories [32].

Also motivated by ideas in algebraic geometry, Lurie uses quasi-categories and their relationship with simplicial categories in his work on higher stacks [22]. The first chapter of his manuscript is also a good introduction to many of the ideas of $(\infty, 1)$-categories.

Another application of the model category of simplicial categories can be found in recent work of Douglas on twisted parametrized stable homotopy theory [6]. He uses diagrams of simplicial categories weakly equivalent to the simplicial localization of the category of spectra (i.e., equivalent as homotopy theories to the homotopy theory of spectra) in order to define an appropriate setting in which to do Floer homotopy theory.

In this paper, we will provide some background on each of the four structures and then describe the various Quillen equivalences.

2. Background on model categories and simplicial sets. Since this paper is meant to be an overview, we are not going to go deeply into the details here, and the reader familiar with model categories and simplicial sets may skip to the next section. For the non-expert, we will give the basic ideas behind both model category structures and simplicial sets, as well as other simplicial objects.

The motivation for a model category structure is a common occurrence in many areas of mathematics. Suppose that we have a category whose morphisms include some, called weak equivalences, which we would like to think of as isomorphisms but do not necessarily have inverses. Two classical examples are the category of topological spaces and (weak) homotopy equivalences between them, and the category of chain complexes of modules over a ring R and the quasi-isomorphisms, or morphisms which induce isomorphisms on all homology groups. In order to make these maps actually isomorphisms, we could formally invert them. Namely, we could formally add in an inverse to every such map and then add all the necessary composites so that the result would actually be a category. The problem with this approach is that often this process results in a category such that the morphisms between any two given objects form a proper class rather than a set. While it is common to work in categories with a proper class of objects, it is generally assumed that there is only a set of morphisms between any two objects, even if there is a proper class of morphisms altogether.

Imposing the structure of a model category allows us formally to invert the weak equivalences while keeping the morphisms under control. A model category $\mathcal{M}$ is a category with three distinguished classes of morphisms, the weak equivalences as already described, plus fibrations and cofibrations, satisfying several axioms. We refer the reader to [13], [15], [17], or the original [25] for these axioms. The importance of these axioms is that they allow us to work with particularly nice objects in the category, called fibrant-cofibrant objects, between which we can define "homotopy classes of maps" even in a situation where the traditional notion of homotopy class (as in topological spaces) no longer makes sense. The axioms of a model category guarantee that every object of $\mathcal{M}$ has a fibrant-cofibrant replacement, and thus we can define the *homotopy category* of $\mathcal{M}$, denoted $\text{Ho}(\mathcal{M})$, to be the category whose objects are the same as those of $\mathcal{M}$ and whose morphisms are homotopy classes of maps between the respective fibrant-cofibrant replacements. In fact, this construction is independent of the choice of such replacements, and the homotopy category, up to equivalence, is independent of the choice of fibrations and cofibrations. There are many examples of categories with two (or more) different model category structures, each leading to the same homotopy category because the weak equivalences are the same even if the fibrations and cofibrations are defined differently.

As an example of a model category, consider the category of topological spaces and the subcategory of weak homotopy equivalences, or maps which induce isomorphisms on all homotopy groups. There is a natural choice of fibrations and cofibrations such that this category has a model category structure. (There is also a model category structure on this category where the weak equivalences are the homotopy equivalences, but the former is considered to be the standard model structure.)

One can then define what it means to have a map between model categories, namely, a functor which preserves essential properties of the model structures. It is convenient to use adjoint pairs of functors to work with model structures, where the left adjoint preserves cofibrations and the right adjoint preserves fibrations. Such an adjoint pair is called a Quillen pair. There is also the notion of Quillen equivalence between model categories, where the adjoint functors preserve the essential homotopical information. In particular, a Quillen equivalence induces an equivalence of homotopy categories, but it is in fact much stronger, in that it preserves higher-order information, as we will discuss further in the next section.

Given this kind of structure and interaction between structures, one can ask the following question, one that can, in fact, be considered a motivation for the research described in this paper. Given a model category $\mathcal{M}$, is there a model category $\mathcal{N}$ which is Quillen equivalent to $\mathcal{M}$ but which more easily provides information that is difficult to obtain from $\mathcal{M}$ itself? The properties one might look for in $\mathcal{N}$ depend very much on the question being asked. Thus, it is hoped that the four model structures given in this paper will be able to provide information about one another.

An important illustration is that of topological spaces and simplicial sets. Heuristically, simplicial sets provide a combinatorial model for topological spaces, and the fact that they are a "model" here means that there is a model category structure on the category of simplicial sets which is Quillen equivalent to the standard model category structure on the category of topological spaces. Simplicial sets are frequently (but not always) easier to work with because they are just combinatorial objects.

To give a formal definition of a simplicial set, consider the category Δ of finite ordered sets $[n] = \{0 \to 1 \to 2 \to \cdots \to n\}$ for each $n \geq 0$, and order-preserving maps between them. (As the notation suggests, one can also consider n to be a small category.) Let Δ^{op} denote the opposite category, where we reverse the direction of all the morphisms. Then a simplicial set is a functor $X : \Delta^{op} \to \mathcal{S}ets$. There is a geometric realization functor between the category $\mathcal{S}\mathcal{S}ets$ of simplicial sets and the category of topological spaces. Specifically, an element of X_0 is assigned to a point, an element of X_1 is assigned to a geometric 1-simplex, and so forth, where identifications are given by the face maps of the simplicial set. Thus, a simplicial set can be regarded as a generalization of a simplicial complex, where the simplices are not required to form "triangles" and a given simplex of degree n is regarded as a degenerate k-simplex for each $k > n$ [14, I.2].

In fact, we can perform this kind of construction in categories other than sets. A simplicial object in a category $\mathcal{C}$ is just a functor $X : \Delta^{op} \to \mathcal{C}$. The primary example we will consider in this paper is that of simplicial spaces, or functors $\Delta^{op} \to \mathcal{S}\mathcal{S}ets$. To emphasize the fact that they are simplicial objects in the category of simplicial sets, they are often also called bisimplicial sets.

Given these main ideas, we can now turn to the four different models for $(\infty, 1)$-categories, or homotopy theories.

3. Simplicial categories. The first of the four categories we consider is that of small simplicial categories. By a simplicial category, we mean what is often called a simplicially enriched category, or a category with a simplicial set of morphisms between any two objects. Given two objects a and b in a simplicial category $\mathcal{C}$, this simplicial set is denoted $\mathrm{Map}_{\mathcal{C}}(a, b)$. This terminology is potentially confusing because the term "simplicial category" can also be used to describe a simplicial object in the category of all small categories. We recover our sense of the term if the face and degeneracy maps are all required to be the identity map on objects.

Simplicial categories have been studied for a variety of reasons, but here we will focus on their importance in homotopy theory, and, in particular, on how a simplicial category can be considered to be a homotopy theory.

We should note here that although for set-theoretic reasons we restrict ourselves to small simplicial categories, or those with only a set of objects, in practice many of the simplicial categories one cares about are large. The standard approach to this problem is to assume that one is working in a larger universe for set theory in which the given category is indeed "small."

Given a model category $\mathcal{M}$, we can consider its homotopy category Ho($\mathcal{M}$). For many applications, it is sufficient to work in the homotopy category, but it is important to remember that in passing from the original model category to the homotopy category we have lost a good deal of information. Of course, part of the goal was formally to invert the weak equivalences, but in addition the model category possessed higher-homotopical information that the homotopy category has lost. For example, the model category contains the tools needed to take homotopy limits and homotopy colimits.

In a series of papers, Dwyer and Kan develop the theory of simplicial localizations, in which, given a model category $\mathcal{M}$, one can obtain a simplicial category which still holds this higher homotopical information. In fact, they construct two different such simplicial categories from $\mathcal{M}$, the standard simplicial localization LM [11] and the hammock localization $L^H M$ [10], but the two are equivalent to one another [8, 2.2]. Furthermore, taking the component category $\pi_0 LM$, which has the objects of LM and the morphisms the components of the simplicial hom-sets of LM, is equivalent to the homotopy category Ho($\mathcal{M}$) [11].

Furthermore, there is a natural notion of "equivalence" of simplicial categories, which is often called a *Dwyer-Kan equivalence* or simply *DK-equivalence*. It is a generalization of the definition of equivalence of categories to the simplicial setting. In particular, a DK-equivalence is a simplicial functor $f : \mathcal{C} \to \mathcal{D}$ between two simplicial categories satisfying the following two conditions:

1. For any objects a, b of $\mathcal{C}$, the map of simplicial sets

$$\mathrm{Map}_{\mathcal{C}}(a, b) \to \mathrm{Map}_{\mathcal{D}}(fa, fb)$$

   is a weak equivalence.

2. The induced functor on component categories $\pi_0 f : \pi_0 \mathcal{C} \to \pi_0 \mathcal{D}$ is an equivalence of categories.

A Quillen equivalence between model categories then induces a DK-equivalence between their simplicial localizations.

More generally, if one is not concerned with set-theoretic issues, we can take the simplicial localization of any category with weak equivalences. Since, in a fairly natural sense, a "homotopy theory" is really some category with "weak equivalences" that we would like to invert, a homotopy theory gives rise to a simplicial category. In fact, the converse is also true: given any simplicial category, it is, up to DK-equivalence, the simplicial localization of some category with weak equivalences [9, 2.5]. Thus, the study of simplicial categories is really the study of homotopy theories.

A first approach to applying the techniques of homotopy theory to a category of simplicial categories itself was first given by Dwyer and Kan [11]. In this paper, they define a model category structure on the category of simplicial categories with a fixed set $\mathcal{O}$ of objects. The idea was then proposed that the homotopy theory of (all) simplicial categories was essentially the "homotopy theory of homotopy theories." Dwyer and Spalinski mention this concept at the end of their survey paper [13], and the idea was further explored by Rezk [27], whose ideas we will return to in the next section. The author then showed in [2] that the category of all small simplicial categories with the DK-equivalences has a model category structure, thus formalizing the idea.

In order to define the fibrations in this model structure, we need the following notion. If $\mathcal{C}$ is a simplicial category and x and y are objects of $\mathcal{C}$, a morphism $e \in \mathrm{Map}_{\mathcal{C}}(x, y)_0$ is a *homotopy equivalence* if the image of e in $\pi_0 \mathcal{C}$ is an isomorphism.

THEOREM 3.1. *[2, 1.1] There is a model category structure on the category $\mathcal{SC}$ of small simplicial categories defined by the following three classes of morphisms:*

1. *The weak equivalences are the DK-equivalences.*
2. *The fibrations are the maps $f \colon \mathcal{C} \to \mathcal{D}$ satisfying the following two conditions:*
   - *For any objects x and y in $\mathcal{C}$, the map*

$$\mathrm{Map}_{\mathcal{C}}(x, y) \to \mathrm{Map}_{\mathcal{D}}(fx, fy)$$

   is a fibration of simplicial sets.
   - *For any object x_1 in $\mathcal{C}$, y in $\mathcal{D}$, and homotopy equivalence $e \colon fx_1 \to y$ in $\mathcal{D}$, there is an object x_2 in $\mathcal{C}$ and homotopy equivalence $d \colon x_1 \to x_2$ in $\mathcal{C}$ such that $fd = e$.*
3. *The cofibrations are the maps which have the left lifting property with respect to the maps which are fibrations and weak equivalences.*

The advantage of this model category is that its objects are fairly straightforward. As mentioned above, there is a reasonable argument for saying that simplicial categories really are homotopy theories. The disadvantage here lies in the weak equivalences, in that they are difficult to identify. Thus, it was natural to look for a model with nicer weak equivalences.

4. Complete Segal spaces. Complete Segal spaces are probably the most complicated objects to define of the four models described in this paper, but from the point of view of homotopy theory, they might be the easiest to use because the corresponding model structure gives what Dugger calls a presentation for the homotopy theory [7]. They are defined by Rezk [27] whose purpose was explicitly to find a nice model for the homotopy theory of homotopy theories.

A complete Segal space is first a simplicial space. It should be noted that we require that certain of our objects be fibrant in the Reedy model structure on the category of simplicial spaces [26]. This structure is defined by levelwise weak equivalences and cofibrations, but its importance here is that several of our constructions will be homotopy invariant because the objects involved satisfy this condition.

We begin by defining Segal spaces, for which we need the Segal map assigned to a simplicial space. As one might guess from its name, the Segal map is first defined by Segal in his work with Γ-spaces [29]. Let $\alpha^i : [1] \to [k]$ be the map in $\mathbf{\Delta}$ such that $\alpha^i(0) = i$ and $\alpha^i(1) = i+1$, defined for each $0 \leq i \leq k - 1$. We can then define the dual maps $\alpha_i : [k] \to [1]$ in $\mathbf{\Delta}^{op}$. For $k \geq 2$, the Segal map is defined to be the map

$$\varphi_k \colon X_k \to \underbrace{X_1 \times_{X_0} \cdots \times_{X_0} X_1}_{k}$$

induced by the maps

$$X(\alpha_i) \colon X_k \to X_1.$$

DEFINITION 4.1. *[27, 4.1] A Reedy fibrant simplicial space W is a Segal space if for each $k \geq 2$ the map φ_k is a weak equivalence of simplicial sets. In other words, the Segal maps*

$$\varphi_k \colon W_k \to \underbrace{W_1 \times_{W_0} \cdots \times_{W_0} W_1}_{k}$$

are weak equivalences for all $k \geq 2$.

A nice property of Segal spaces is the fact that they can be regarded as analogous to simplicial categories, in that we can define their "objects" and "morphisms" in a meaningful way. Given a Segal space W, its set of objects, denoted $\mathrm{ob}(W)$, is the set of 0-simplices of the space W_0, namely, the set $W_{0,0}$. Given any two objects x, y in $\mathrm{ob}(W)$, the mapping space $\mathrm{map}_W(x, y)$ is the fiber of the map $(d_1, d_0) \colon W_1 \to W_0 \times W_0$ over (x, y). Given a 0-simplex x of W_0, we denote by id_x the image of the degeneracy map $s_0 \colon W_0 \to W_1$. We say that two 0-simplices of $\mathrm{map}_W(x, y)$, say f and g, are homotopic, denoted $f \sim g$, if they lie in the same component of the simplicial set $\mathrm{map}_W(x, y)$.

Given $f \in \mathrm{map}_W(x, y)_0$ and $g \in \mathrm{map}_W(y, z)_0$, there is a composite $g \circ f \in \mathrm{map}_W(x, z)_0$, and this notion of composition is associative up to homotopy. The homotopy category $\mathrm{Ho}(W)$ of W, then, has as objects the set $\mathrm{ob}(W)$ and as morphisms between any two objects x and y, the set $\mathrm{map}_{\mathrm{Ho}(W)}(x, y) = \pi_0 \mathrm{map}_W(x, y)$.

Finally, a map g in $\mathrm{map}_W(x, y)_0$ is a homotopy equivalence if there exist maps $f, h \in \mathrm{map}_W(y, x)_0$ such that $g \circ f \sim \mathrm{id}_y$ and $h \circ g \sim \mathrm{id}_x$. Any map in the same component as a homotopy equivalence is itself a homotopy equivalence [27, 5.8]. Therefore we can define the space W_{hoequiv} to be the subspace of W_1 given by the components whose zero-simplices are homotopy equivalences.

We then note that the degeneracy map $s_0 : W_0 \to W_1$ factors through W_{hoequiv} since for any object x the map $s_0(x) = \mathrm{id}_x$ is a homotopy equivalence. Therefore, we have the following definition:

DEFINITION 4.2. *[27, §6] A complete Segal space is a Segal space W for which the map $s_0 : W_0 \to W_{hoequiv}$ is a weak equivalence of simplicial sets.*

We can now consider some particular kinds of maps between Segal spaces. Note that, as the name suggests, these maps are very similar in spirit to the weak equivalences in $\mathcal{SC}$.

DEFINITION 4.3. *A map $f : U \to V$ of Segal spaces is a DK-equivalence if*

1. *for any pair of objects $x, y \in U_0$, the induced map*

$$\mathrm{map}_U(x, y) \to \mathrm{map}_V(fx, fy)$$

   is a weak equivalence of simplicial sets, and
2. *the induced map $\mathrm{Ho}(f) : \mathrm{Ho}(U) \to \mathrm{Ho}(V)$ is an equivalence of categories.*

We are now able to describe the important features of the complete Segal space model category structure.

THEOREM 4.1. *[27, 7.2, 7.7] There is a model structure $\mathcal{CSS}$ on the category of simplicial spaces such that*

1. *The weak equivalences between Segal spaces are the DK-equivalences.*
2. *The cofibrations are the monomorphisms.*
3. *The fibrant objects are the complete Segal spaces.*

What makes the model category $\mathcal{CSS}$ so nice to work with is the fact that the weak equivalences between the fibrant objects, the complete Segal spaces, are easy to identify.

PROPOSITION 4.1. *[27, 7.6] A map $f : U \to V$ between complete Segal spaces is a DK-equivalence if and only if it is a levelwise weak equivalence.*

To avoid further technical detail, we have not defined what a general weak equivalence is in $\mathcal{CSS}$, but the interested reader can find it in Rezk's paper [27, §7]. The important point is that, when working with the complete Segal spaces, the weak equivalences are especially convenient.

5. Segal categories. We now turn to our third model, that of Segal categories. These are natural generalizations of simplicial categories, in that they can be regarded as simplicial categories with composition only given up to homotopy. They first appear in the literature in a paper of Dwyer, Kan, and Smith [12], where they are called special Δ^{op}-diagrams of simplicial sets. In particular, Segal categories are again a kind of simplicial space.

We begin with the definition of a Segal precategory.

DEFINITION 5.1. *A* Segal precategory *is a simplicial space* X *such that* X_0 *is a discrete simplicial set.*

As with the Segal spaces in the previous section, we can use the Segal maps to define Segal categories.

DEFINITION 5.2. *A* Segal category X *is a Segal precategory* $X :$ $\Delta^{op} \to SSets$ *such that for each* $k \geq 2$ *the Segal map*

$$\varphi_k \colon X_k \to \underbrace{X_1 \times_{X_0} \cdots \times_{X_0} X_1}_{k}$$

is a weak equivalence of simplicial sets.

A model category structure $SeCat_c$ for Segal categories is given by Hirschowitz and Simpson [16]. In fact, they generalize the definition to that of a Segal n-category and give a model structure for Segal n-categories for any $n \geq 1$. The idea behind this generalization is used for both the Simpson and Tamsamani definitions of weak n-category [30], [33].

The author gives a new proof of this model structure, just for the case of Segal categories, from which it is easier to characterize the fibrant objects [3]. It should be noted that, as in the case of CSS, this model structure is actually defined on the larger category of Segal precategories. However, the fibrant-cofibrant objects are Segal categories [1].

To define the weak equivalences in $SeCat_c$, we first note that there is a functor L_c assigning to every Segal precategory a Segal category [3, §5]. Then, if we are working with a Segal category X, we can define its "objects," "mapping spaces," and "homotopy category" just as we did for a Segal space.

THEOREM 5.1. *There is a cofibrantly generated model category structure* $SeCat_c$ *on the category of Segal precategories with the following weak equivalences, fibrations, and cofibrations.*

- *Weak equivalences are the maps* $f \colon X \to Y$ *such that the induced map* $map_{L_c X}(x, y) \to map_{L_c Y}(fx, fy)$ *is a weak equivalence of simplicial sets for any* $x, y \in X_0$ *and the map* $Ho(L_c X) \to Ho(L_c Y)$ *is an equivalence of categories.*
- *Cofibrations are the monomorphisms. (In particular, every Segal precategory is cofibrant.)*
- *Fibrations are the maps with the right lifting property with respect to the maps which are both cofibrations and weak equivalences.*

It should not be too surprising that the weak equivalences in this case are again called DK-equivalences, since the same idea underlies the definition in each of the three categories we have considered.

Furthermore, there is also second model structure $Se\mathcal{C}at_f$ on this same category, with the same weak equivalences but different fibrations and cofibrations. Thus each leads to the same homotopy theory, and in fact they are Quillen equivalent, but the slight difference between the two is key in comparing Segal categories with the other models. We will not define the fibrations and cofibrations in $Se\mathcal{C}at_f$ here, as they are technical and unenlightening in themselves, but we refer the interested reader to [3, §7] for the details. As the subscript suggests, the initial motivation was to find a model structure with the same weak equivalences but in which the fibrations, rather than the cofibrations, were given by levelwise fibrations of simplicial sets. As it turns out, such a description does not work, but one is not far off thinking of the fibrations as being levelwise.

6. Quasi-categories. Quasi-categories are perhaps the most mysterious, as far as why they are equivalent to the others. While each of the other models consists of simplicial spaces, or objects easily related to simplicial spaces, quasi-categories are simplicial sets, and thus simpler than any of the others. Thus, they may also provide a good model to use when we actually want to compute something for a given homotopy theory. They were first defined by Boardman and Vogt [4] and are sometimes called weak Kan complexes.

Recall that in the category of simplicial sets we have several particularly important objects. For each $n \geq 0$, there is the n-simplex $\Delta[n]$ and its boundary $\dot{\Delta}[n]$. If we remove the kth face of $\dot{\Delta}[n]$, we get the simplicial set denoted $V[n,k]$. Given any simplicial set X, a horn in X is a map $V[n,k] \to X$. A Kan complex is a simplicial set such that every horn factors through the inclusion map $V[n,k] \to \Delta[n]$. A quasi-category is then a simplicial set X such that every horn $V[n,k] \to X$ factors through $\Delta[n]$ for each $0 < k < n$. Such horns are called the inner horns. Further details on quasi-categories can be found in Joyal's papers [18] and [20].

Like the cases for Segal categories and complete Segal spaces, the model structure $\mathcal{Q}\mathcal{C}at$ for quasi-categories is defined on a larger category, in this case the category of simplicial sets. To define the weak equivalences, we need some definitions.

First, consider the nerve functor $N : \mathcal{C}at \to \mathcal{SS}ets$, where $\mathcal{C}at$ denotes the category of small categories. This functor has a left adjoint $\tau_1 : \mathcal{SS}ets \to \mathcal{C}at$. Given a simplicial set X, the category $\tau_1(X)$ is called its fundamental category. We can then define a functor $\tau_0 : \mathcal{SS}ets \to \mathcal{S}ets$, where $\tau_0(X)$ is the set of isomorphism classes of objects of the category $\tau_1(X)$. Now, if X^Y denotes the simplicial set of maps $Y \to X$, for any pair (X, Y) of simplicial sets we define $\tau_0(Y, X) = \tau_0(X^Y)$. A weak categorical equivalence is a map

$A \to B$ of simplicial sets such that the induced map $\tau_0(B, X) \to \tau_0(A, X)$ is an isomorphism of sets for any quasi-category X.

THEOREM 6.1. *[20] There is a model category structure* QCat *on the category of simplicial sets in which the weak equivalences are the weak categorical equivalences and the cofibrations are the monomorphisms. The fibrant objects of* QCat *are the quasi-categories.*

7. Quillen equivalences. The origins of the comparisons between these various structures seem to be in various places. The question of simplicial categories and Segal categories is a fairly natural one, since a Segal category is essentially a simplicial category up to homotopy. It is addressed partially by Dwyer, Kan, and Smith [12], but they do not give a Quillen equivalence, partly because their work predates both model structures by several years. Schwänzl and Vogt also address this question, using topological rather than simplicial categories [28].

Rezk defines complete Segal spaces with the comparison with simplicial categories in mind [27]. While his functor from simplicial categories to complete Segal spaces naturally factors through Segal categories, he does not mention this fact as such. Initially, there did not seem to be a need to bring in the Segal categories from this point of view, but further investigation led to skepticism that his functor had the necessary adjoint to give a Quillen equivalence.

Toën mentions all four models and conjectures the relationships between them in [36]. As mentioned in the introduction, some but not all of these functors are the ones used in the known proofs. Toën further axiomatizes the notion of "theory of $(\infty, 1)$-categories" in [37]. He gives six axioms for a model structure to satisfy in order to be such a theory, and he shows that any such model category structure is Quillen equivalent to Rezk's complete Segal space model structure. As far as the author knows, these axioms have not been verified for the other three models given in this paper, but it seems likely that they should hold.

Some of the comparisons are also mentioned in work by Simpson, who sketches an argument for comparing the Segal categories and complete Segal spaces [31].

The author showed in [3] that simplicial categories are equivalent to Segal categories, which are in turn equivalent to the complete Segal spaces. However, the adjoint pairs go in opposite directions and therefore cannot be composed into a single Quillen equivalence. Further work by Joyal and Tierney has shown that there are Quillen equivalences between quasi-categories and each of the other three models.

We now look in more detail at these Quillen equivalences. Let us begin with the Segal categories and complete Segal spaces. Since the underlying category of CSS is the category of simplicial spaces and the underlying category of $SeCat_c$ is the category of simplicial spaces with a discrete space in degree zero, there is an inclusion functor $I \colon SeCat_c \to CSS$. This functor has a right adjoint $R \colon \mathcal{CSS} \to Se\mathcal{C}at_c$ which acts as a discretization functor. In particular, if it is applied to a complete Segal space W, the result is a Segal category which is DK-equivalent to it.

THEOREM 7.1. *[3, 6.3] The adjoint pair*

$$I \colon Se\mathcal{C}at_c \rightleftarrows \mathcal{CSS} : R$$

is a Quillen equivalence.

Then, as we mentioned in the section on Segal categories, the two model structures $Se\mathcal{C}at_c$ and $Se\mathcal{C}at_f$ are Quillen equivalent.

THEOREM 7.2.   *[3, 7.5] The identity functor induces a Quillen equivalence*

$$id \colon Se\mathcal{C}at_f \rightleftarrows Se\mathcal{C}at_c : id.$$

Turning to simplicial categories, the nerve functor $N \colon \mathcal{SC} \to Se\mathcal{C}at_f$ has a left adjoint F, which can be considered a rigidification functor.

THEOREM 7.3. *[3, 8.6] The adjoint pair*

$$F \colon Se\mathcal{C}at_f \rightleftarrows \mathcal{SC} : N$$

is a Quillen equivalence.

Turning to the quasi-categories, Joyal and Tierney have shown that there are in fact two different Quillen equivalences between $\mathcal{QCat}$ and $\mathcal{CSS}$. For the first of these equivalences, the map i_1^*, which associates to a complete Segal space W the simplicial set W_{*0}, has a left adjoint p_1^*.

THEOREM 7.4. *[21] The adjoint pair of functors*

$$p_1^* \colon \mathcal{QCat} \rightleftarrows \mathcal{CSS} : i_1^*$$

is a Quillen equivalence.

The second Quillen equivalence between these two model categories is given by a total simplicial set functor $t_! \colon \mathcal{CSS} \to \mathcal{QCat}$ and its right adjoint $t^!$.

THEOREM 7.5. *[21] The adjoint pair*

$$t_! \colon \mathcal{CSS} \rightleftarrows \mathcal{QCat} : t^!$$

is a Quillen equivalence.

Even one of these Quillen equivalences would be sufficient to show that all four of our model categories are equivalent to one another, but, interestingly, Joyal and Tierney go on to prove that there are also two different Quillen equivalences directly between $\mathcal{QCat}$ and $Se\mathcal{C}at_c$. The first of these functors is analogous to the pair given in Theorem 7.4; the right adjoint functor $j^* \colon Se\mathcal{C}at_c \to \mathcal{QCat}$ assigns to a Segal precategory X the simplicial set X_{*0}. Its left adjoint is denoted q^*.

THEOREM 7.6. *[21] The adjoint pair*

$$q^* : \mathcal{QCat} \rightleftarrows \mathcal{SeCat}_c : j^*$$

is a Quillen equivalence.

The second Quillen equivalence between these two model categories is given by the map $d^* : \mathcal{SeCat}_c \to \mathcal{QCat}$, which assigns to a Segal precategory its diagonal, and its right adjoint d_*.

THEOREM 7.7. *[21] The adjoint pair*

$$d^* : \mathcal{SeCat}_c \rightleftarrows \mathcal{QCat} : d_*$$

is a Quillen equivalence.

Finally, Joyal has also related the quasi-categories to the simplicial categories directly. There is the coherent nerve functor $\widetilde{N} : \mathcal{SC} \to \mathcal{QCat}$, first defined by Cordier and Porter [5]. Given a simplicial category X and the simplicial resolution $C_*[n]$ of the category $[n] = (0 \to \cdots \to n)$, the coherent nerve $\widetilde{N}(X)$ is defined by

$$\widetilde{N}(X)_n = \mathrm{Hom}_{\mathcal{SC}}(C_*[n], X).$$

This functor has a left adjoint $J : \mathcal{QCat} \to \mathcal{SC}$.

THEOREM 7.8. *[19] The adjoint pair*

$$J : \mathcal{QCat} \rightleftarrows \mathcal{SC} : \widetilde{N}$$

is a Quillen equivalence.

Thus, we have the following diagram of Quillen equivalences of model categories:

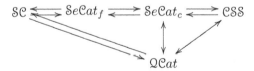

The single double-headed arrows indicate that in these cases either direction can be chosen to be a left (or right) adjoint, depending on which Quillen equivalence is used.

Acknowledgements. I would like to thank Bill Dwyer, Chris Douglas, André Joyal, Jacob Lurie, Peter May, and Bertrand Toën for reading early drafts of this paper, making suggestions, and sharing their work in this area.

REFERENCES

[1] J.E. BERGNER, A characterization of fibrant Segal categories, *Proc. Amer. Math. Soc.* (2007), **135**:4031–4037.

[2] J.E. BERGNER, A model category structure on the category of simplicial categories, *Trans. Amer. Math. Soc.* (2007), **359**:2043–2058.

[3] J.E. BERGNER, Three models for the homotopy theory of homotopy theories, *Topology* (2007), **46**:397–436.

[4] J.M. BOARDMAN AND R.M. VOGT, *Homotopy invariant algebraic structures on topological spaces.* Lecture Notes in Mathematics, **347**, Springer-Verlag, 1973.

[5] J.M. CORDIER AND T. PORTER, Vogt's theorem on categories of homotopy coherent diagrams, *Math. Proc. Camb. Phil. Soc.* (1986), **100**:65–90.

[6] C.L. DOUGLAS, Twisted parametrized stable homotopy theory, preprint available at math.AT/0508070.

[7] D. DUGGER, Universal homotopy theories, *Adv. Math.* (2001), **164**(1):144–176.

[8] W.G. DWYER AND D.M. KAN, Calculating simplicial localizations, *J. Pure Appl. Algebra* (1980), **18**:17–35.

[9] W.G. DWYER AND D.M. KAN, Equivalences between homotopy theories of diagrams, *Algebraic topology and algebraic K-theory* (Princeton, N.J.), 1983, 180–205, *Ann. of Math. Stud.*, **113**, Princeton Univ. Press, Princeton, NJ, 1987.

[10] W.G. DWYER AND D.M. KAN, Function complexes in homotopical algebra, *Topology* (1980), **19**:427-440.

[11] W.G. DWYER AND D.M. KAN, Simplicial localizations of categories, *J. Pure Appl. Algebra* (1980), **17**(3):267–284.

[12] W.G. DWYER, D.M. KAN, AND J.H. SMITH, Homotopy commutative diagrams and their realizations. *J. Pure Appl. Algebra* (1989), **57**:5–24.

[13] W.G. DWYER AND J. SPALINSKI, Homotopy theories and model categories, in *Handbook of Algebraic Topology*, Elsevier, 1995.

[14] P.G. GOERSS AND J.F. JARDINE, *Simplicial Homotopy Theory, Progress in Math*, Vol. **174**, Birkhauser, 1999.

[15] P.S. HIRSCHHORN, *Model Categories and Their Localizations, Mathematical Surveys, and Monographs*, **99**, AMS, 2003.

[16] A. HIRSCHOWITZ AND C. SIMPSON, Descente pour les *n*-champs, preprint available at math.AG/9807049.

[17] MARK HOVEY, *Model Categories, Mathematical Surveys and Monographs*, **63**. American Mathematical Society 1999.

[18] A. JOYAL, Quasi-categories and Kan complexes, *J. Pure Appl. Algebra* (2002), **17**(5):207–222.

[19] A. JOYAL, Simplicial categories vs quasi-categories, in preparation.

[20] ANDRÉ JOYAL, The theory of quasi-categories I, in preparation.

[21] ANDRÉ JOYAL AND MYLES TIERNEY, Quasi-categories vs Segal spaces, *Contemp. Math.* (2007), **431**:277–326.

[22] JACOB LURIE, Higher topos theory, preprint available at math.CT/0608040.

[23] SAUNDERS MAC LANE, *Categories for the Working Mathematician, Second Edition, Graduate Texts in Mathematics 5*, Springer-Verlag, 1997.

[24] TIMOTHY PORTER, S-categories, S-groupoids, Segal categories and quasicategories, preprint available at math.AT/0401274.

[25] DANIEL QUILLEN, *Homotopical Algebra*, Lecture Notes in Math **43**, Springer-Verlag, 1967.

[26] C.L. REEDY, Homotopy theory of model categories, unpublished manuscript, available at http://www-math.mit.edu/~psh.

[27] CHARLES REZK, A model for the homotopy theory of homotopy theory, *Trans. Amer. Math. Soc.*, **353**(3):973–1007.

[28] R. SCHWÄNZL AND R.M. VOGT, Homotopy homomorphisms and the hammock localization, *Bol. Soc. Mat. Mexicana* (2) (1992), **37**(1–2):431–448.

[29] GRAEME SEGAL, Categories and cohomology theories, *Topology* (1974), **13**: 293–312.

[30] CARLOS SIMPSON, A closed model structure for n-categories, internal Hom, n-stacks, and generalized Seifert-Van Kampen, preprint, available at math.AG/9704006.

[31] CARLOS SIMPSON, A Giraud-type characterization of the simplicial categories associated to closed model categories as $infty$-pretopoi, preprint available at math.AT/9903167.

[32] GONCALO TABUADA, Une structure de catégorie de modèles de Quillen sur la catégorie des dg-catégories, *C.R. Math. Acad. Sci. Paris* (2005), **340**(1):15–19.

[33] Z. TAMSAMANI, Sur les notions de n-categorie et n-groupoíde non-stricte via des ensembles multi-simpliciaux, preprint available at alg-geom/9512006.

[34] BERTRAND TOËN, Higher and derived stacks: a global overview, preprint available at math.AG/0604504.

[35] BERTRAND TOËN, The homotopy theory of dg-categories and derived Morita theory, *Invent. Math.* (2007), **167**(3):615–667.

[36] BERTRAND TOËN, Homotopical and Higher Categorical Structures in Algebraic Geometry (A View Towards Homotopical Algebraic Geometry), preprint available at math.AG/0312262.

[37] BERTRAND TOËN, Vers une axiomatisation de la théorie des catégories supérieures, *K-Theory* (2005), **34**(3):233–263.

[38] BERTRAND TOËN AND GABRIELE VEZZOSI, Remark on K-theory and S-categories, *Topology* (2004), **43**(4):765–791.

INTERNAL CATEGORICAL STRUCTURES IN HOMOTOPICAL ALGEBRA

SIMONA PAOLI*

Abstract. This is a survey on the use of some internal higher categorical structures in algebraic topology and homotopy theory. After providing a general view of the area and its applications, we concentrate on the algebraic modelling of connected $(n+1)$-types through cat^n-groups.

1. Introduction. In this paper we present a survey of an area of mathematics which arose historically within algebraic topology in an attempt to describe higher order versions of the fundamental group in homotopy theory.

We have tried to highlight the interplay between the categorical input and the topological and homotopical one. With this in mind, we provide an overview of the area in Section 2. Here we indicate the main categorical structures used, the main areas of applications and the form in which these categorical structures arise in the applications. In the subsequent sections we concentrate on one of the main achievements in the area, which is the construction of an algebraic model of connected n-types. Section 3 contains a description of the structures used in the modelling of connected $(n + 1)$-types, the cat^n-groups while Section 4 deals with some more technical aspects of cat^n-groups as an algebraic category. Section 5 provides a concise exposition of the proof that cat^n-groups model connected $(n+1)$-types; we have followed the approach to this proof given by Bullejos, Cegarra and Duskin [12]. We conclude in Section 5 with a broad outline of recent developments.

2. Overview. This section contains a brief survey of the area of mathematics covered by the title of the paper. The topic is located between algebraic topology and category theory and we could refer to it as "strict higher-dimensional algebra."

As a branch of algebraic topology, it arose from the investigation of higher-order versions of the fundamental group in homotopy theory, as envisaged in Grothendieck's "pursuing stacks" program [30]. The algebraic modelling of homotopy n-types was thus a central achievement in the area.

The categorical input comes essentially from two types of strict higher categorical structures: strict n-categories and n-fold categories. In concrete algebraic-topological situations, these appear internalized in a fixed based category such as groups and algebras. The resulting internalized structures admit equivalent descriptions: strict n-categories (resp. strict ω-categories)

*Department of Mathematics, Macquarie University, NSW 2109, AUSTRALIA (simonap@ics.mq.edu.au).

J.C. Baez, J.P. May (eds.), *Towards Higher Categories*, The IMA Volumes in Mathematics and its Applications 152, DOI 10.1007/978-1-4419-1524-5_3,
© Springer Science+Business Media, LLC 2010

give rise to crossed n-complexes (resp. crossed complexes) [8] ; n-fold categories give rise to cat^n-groups (and cat^n-algebras), or equivalently crossed
n-cubes (of groups, algebras, etc.) [24]. Another structure widely used in
the area is that of n-hypercrossed complexes [16]. Hypercrossed complexes
of groups are equivalent to simplicial groups via the Moore normalization
functor, hence they provide a non-abelian version of the Dold-Kan theorem. n-Hypercrossed complexes of groups (resp. groupoids) are models
of connected (resp. non-connected) $(n + 1)$-types [16, 26]. A categorical
description of hypercrossed complexes as algebras for a monad has been
recently given in [5].

The reason for the internalization is best understood when considering
cat^n-groups as homotopy models. The Kan loop-group functor G from
reduced simplicial sets to simplicial groups has a right adjoint $\overline{W}$, and this
adjoint pair induces an equivalence between the corresponding homotopy
categories [40]. On the other hand, any path-connected space X is weakly
homotopy equivalent to the geometric realization of its reduced singular
complex $\mathcal{S}_0 X$. Hence, the functor $G\mathcal{S}_0$ from based topological spaces to
simplicial groups translates the problem of modelling homotopy types into
the realm of simplicial groups. In the non-path-connected case, one needs
to use groupoids instead of groups, and the analogue of the Kan loop group
functor is the construction in [23]. The following table summarizes the main
structures in the area and their applications:

STRICT HIGHER-DIMENSIONAL ALGEBRA

1) Crossed n-complexes; d), e).
2) cat^n-groups, cat^n-algebras
     a), b), c), f).
3) n-Hypercrossed complexes; a), g).

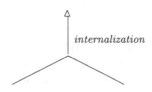

internalization

ALGEBRAIC TOPOLOGY
a) Modelling homotopy n-types.
b) Higher order Van Kampen theorem.
c) Non-abelian cohomology.
d) Homological algebra.
e) Combinatorial group theory.
f) Topological quantum field theory.
g) Homotopy 3-types and double loop spaces.

CATEGORY THEORY
1) n-cat, ω-cat.
2) Cat^n

The applications range throughout various areas of algebraic topology. Cat^n-groups and crossed n-complexes satisfy a higher-order Van Kampen theorem [7], and this was used to perform computations of homotopy groups [9]. Non-abelian cohomology theories naturally require cat^1-groups (equivalently crossed modules) as coefficient objects [6].

In homological algebra, n-crossed complexes were used in the interpretation of group cohomology [32, 31] and, more generally, of cohomology in a certain class of algebraic categories [52]. More recently, (co)homology theories of cat^n-groups have been developed [15, 13], [44] and cat^n-algebras have been described operadically in [34]. Combinatorial group theory uses crossed modules as a way to encode information about group presentations [10]. An application of algebraic models of n-types is found in topological quantum field theory [50]. Homotopy 3-types modelled by 2-crossed modules have been directly linked to double loop spaces in [3], whereas the analogous problem for $n > 3$ seems, so far, entirely open.

3. n-Fold categories in groups. The category of n-fold categories internalized in the category **Gp** of groups can be described in various equivalent ways. These arose in applications, as well as in independent proofs that they model homotopy $(n + 1)$-types in the path-connected case. This section is devoted to an exposition of some of these characterizations, starting with the case $n = 1$.

Given a category $\mathcal{D}$ with pullbacks, an *internal category in $\mathcal{D}$* consists of a diagram

$$\phi : D_1 \times_{D_0} D_1 \xrightarrow{\ c\ } D_1 \overset{\substack{\xrightarrow{d_0} \\ \xrightarrow{d_1}}}{\underset{\xleftarrow{\ i\ }}{}} D_0$$

where d_0, d_1 are "source" and "target" maps, i is the "identity" map, c is the "composition" map. We call D_1 the "arrow object," D_0 the "object object." These data satisfy the axioms of a category; that is, the following identities hold, where $\pi_0, \pi_1 : D_1 \times_{D_0} D_1 \to D_1$ are the two projections:

$$d_0 i = 1_{D_0} = d_1 i, \qquad d_1 \pi_1 = d_1 c, \qquad d_0 \pi_0 = d_0 c$$

$$c \begin{pmatrix} 1_{D_1} \\ i d_0 \end{pmatrix} = 1_{D_1} = \begin{pmatrix} i d_1 \\ 1_{D_1} \end{pmatrix}, \qquad c(1_{D_1} \times_{D_0} c) = c(c \times_{D_0} 1_{D_1}).$$

Given internal categories ϕ and ϕ', an *internal functor* $F : \phi \to \phi'$ consists of a pair of morphisms $F_0 : D_0 \to D_0'$, $F_1 : D_1 \to D_1'$ satisfying the axioms:

$$d_0 F_1 = F_0 d_0, \qquad\qquad d_1 F_1 = F_0 d_1$$
$$F_1 i = i F_0, \qquad\qquad F_1 c = c(F_1 \times_{F_0} F_1).$$

We denote by **Cat** $\mathcal{D}$ the category of internal categories and internal functors. If in addition $\mathcal{D}$ is semiabelian, then **Cat** $\mathcal{D}$ is equivalent to the category $\mathbf{CM}(\mathcal{D})$ of crossed modules in $\mathcal{D}$. At this level of generality, this fact was established fairly recently, in [33]. It was known earlier for a particular class of semiabelian categories, which includes most algebraic examples; these are the "categories of groups with operations" [43, 49]. We shall restrict to this context, as it is sufficient to handle the case of cat^n-groups.

Categories of groups with operations are a particular case of the general notion of variety of universal algebra. This includes familiar examples such as groups, rings, Lie algebras and many others. Recall [39, p. 124] that this is given by specifying a set S, a set of operators Ω acting on S and a set of identities E, in the following sense. There is a function $\Omega \rightarrow \mathbb{N}$ which assigns to each $\omega \in \Omega$ a natural number, called the arity of ω. An action of Ω on S ia a function assigning to each operator ω an n-ary operation $\omega_A : S^n = S \times \overset{n}{\cdots} \times S \rightarrow S$ (S^0 is understood to be the singleton). For example, in the case of the variety of groups we have three operators, product, inverse, identity, of arities 2,1,0 respectively: the corresponding operations take $(x, y) \rightarrow xy$, $x \rightarrow x^{-1}$, $* \rightarrow 1$.

Continuing with the general case, from the elements of Ω we can form the set of derived operators: given $\omega \in \Omega$ of arity n and n derived operators $\lambda_1 \cdots \lambda_n$ of arities $m_1, \ldots, m_n$, we can form a composite $\omega(\lambda_1 \cdots \lambda_n)$ of arity $m_1 + \cdots + m_n$; also given an operator λ of arity n and a function $f : \{1, \ldots, n\} \rightarrow \{1, \ldots, m\}$, we can use substitution to form an operator ϑ of arity m where $\vartheta(x_1 \ldots x_m) = \lambda(x_{f_1} \ldots x_{f_n})$. For instance, in the case of the variety of groups, taking $\omega = \lambda_1 = \lambda_2$ to be the product we obtain an operator $\omega(\lambda_1, \lambda_2)$ of order 4 and corresponding operations $G^4 \rightarrow G$, $(x_1, x_2, x_3, x_4) \rightarrow (x_1 x_2)(x_3 x_4)$; taking for λ the product and $f : \{1, 2\} \rightarrow \{1\}$ we obtain by substitution an operator of order 1, $\vartheta(x) = x^2$.

The set E of identities is a set of ordered pairs $< \lambda, \mu >$ of derived operators, where λ and μ have the same arity. An action A of Ω on S satisfies the identity $< \lambda, \mu >$ if $\lambda_A = \mu_A : S^n \rightarrow S$. For example, in the case of the variety of groups, E consists of the axiom for identity $1x = x = x1$, inverse $xx^{-1} = 1 = x^{-1}x$ and associative law $x(yz) = (xy)z$.

Summarizing, a *variety of universal algebras*, or $< \Omega, E >$-algebra, is a set S with an action A of Ω on S which satisfies all the identities of E. A morphism of $< \Omega, E >$-algebras is a function on the underlying sets which preserves all operators of Ω.

A *category $\mathcal{D}$ of groups with operations* is a variety of universal algebras which is, first of all, a variety of groups. Also, group identity is the only operator of arity 0. All other operators different from group multiplication, inverse and identity have order 1 and 2. We denote the set of these operators by Ω_1' and Ω_2' respectively. Finally, the following axioms hold: if $* \in \Omega_2'$, $*^0$ defined by $x *^0 y = y * x$ is in Ω_2'. Also, $a * (bc) = (a * b)(a * c)$ for all $* \in \Omega_2'$; $\omega(ab) = \omega(a)\omega(b)$, $\omega(a) * b = \omega(a * b)$ for all $\omega \in \Omega_1'$, $* \in \Omega_2'$.

In a category of groups with operations, there is a notion of action and of semidirect product. Given $A, B \in \mathbf{Ob}\mathcal{D}$, a B-structure A is a split extension of B by A in $\mathcal{D}$, that is the diagram in $\mathcal{D}$

$$A \overset{i}{\rightarrowtail} E \underset{s}{\overset{p}{\rightrightarrows}} B \qquad\qquad ps = \mathrm{id}_B. \tag{1}$$

If A is a B-structure, the semidirect product $A \widetilde{\times} B$ is the object of $\mathcal{D}$ which is $A \times B$ as a set, with operations

$$\omega(a, b) = (\omega(a), \omega(b))$$
$$(a', b')(a, b) = (a' s(b') \, a \, s(b')^{-1}, b'b)$$
$$(a', b') * (a, b) = (a' * a)(a' * s(b))(s(b') * a)$$

for all $a, a' \in A$, $b, b' \in B$, $\omega \in \Omega'_1$, $* \in \Omega'_2$. Given the split extension (1) there is an isomorphism $E \cong A \widetilde{\times} B$. For each $b \in B$, $a \in A$ we denote

$$^b a = s(b) \, a \, s(b)^{-1}$$
$$b * a = s(b) * a$$
$$a * b = a * s(b).$$

This notion of action allows us to formulate the notion of crossed module in $\mathcal{D}$.

DEFINITION 3.1. *Let $\mathcal{D}$ be a category of groups with operations. A crossed module in $\mathcal{D}$ consists of a triple (A, B, ϕ) where A is a B-structure, $\phi \in Mor(\mathcal{D})$ such that for all $a, a_1, a_2 \in A$, $b \in B$, $* \in \Omega'_2$,*

CM 1 $\phi(^b a) = b\phi(a)b^{-1}$,

CM 2 $^{\phi(a_1)} a_2 = a_1 a_2 a_1^{-1}$,

CM 3 $\phi(a_1) * a_2 = a_1 * a_2 = a_1 * \phi(a_2)$,

CM 4 $\begin{cases} \phi(b * a) = b * \phi(a) \\ \phi(a * b) = \phi(a) * b. \end{cases}$

A morphism from a crossed module (A, B, ϕ) to a crossed module (A', B', ϕ') consists of a pair of morphisms in $\mathcal{D}$, $f : A \to A'$ and $g : B \to B'$ such that $g\,\phi = f\,\phi'$ and, for all $a \in A$, $b \in B$,

$$f(^b a) = \, ^{g(b)} f(a), \qquad f(b * a) = g(b) * f(a).$$

Given a crossed module (A, B, Φ) in $\mathcal{D}$, consider the diagram

$$(A \widetilde{\times} B) \times_B (A \widetilde{\times} B) \underset{m}{\to} A \widetilde{\times} B \underset{s}{\overset{d_0}{\underset{d_1}{\rightrightarrows}}} B \tag{2}$$

where $s(b) = (1, b)$, $d_0(a, b) = b$, $d_1(a, b) = \phi(a)b$, $m((a, b), (c, \phi(a)b)) = (ca, b)$. Clearly d_0 and s are morphisms in $\mathcal{D}$. Axioms CM 1 and CM 4 imply that d_1 is a morphism in $\mathcal{D}$, while CM 2 and CM 3 imply that m is a morphism in $\mathcal{D}$. It is then straightforward to check that (2) is an internal category in $\mathcal{D}$.

Conversely, an internal category

$$C \times_B C \xrightarrow{m} C \overset{\overset{d_0}{\rightarrow}}{\underset{s}{\underset{\leftarrow}{\underset{d_1}{\rightarrow}}}} B \qquad\qquad (3)$$

in $\mathcal{D}$, determines a split extension of B by $\ker d_0$, so that $C \cong \ker d_0 \widetilde{\times} B$. After some straightforward identification, (3) has the same form as (2), hence $(\ker d_0, B, d_{1\,|\,\ker d_0})$ is a crossed module in $\mathcal{D}$. Clearly the two processes give an equivalence of categories. Notice also that, given the internal category (3) and $(a, b) \in \ker d_0 \widetilde{\times} B \cong C$, we have $m((a, b)(a^{-1}, \phi(a)b)) = (1, b) = s(b)$, $m((a^{-1}, \phi(a)b)(a, b)) = (1, \phi(a)b) = s(\phi(a)b)$. In other words, any arrow-object $c \in C$ is invertible in the sense that there is another arrow-object c' such that $m(c, c')$ and $m(c', c)$ are identities. Internal categories having this property are called *internal groupoids*. Summarizing:

THEOREM 3.1. [49] *Let $\mathcal{D}$ be a category of groups with operations. Then the categories $\mathbf{CM}(\mathcal{D})$ and $\mathbf{Cat}\,\mathcal{D}$ are equivalent. Also, every object of $\mathbf{Cat}\mathcal{D}$ is an internal groupoid.*

Particularizing Definition 3.1 to the case where $\mathcal{D}$ is the category of groups, we obtain that a crossed module in groups consists of a triple (T, G, μ) where $\mu : T \to G$ is a group homomorphism, G acts on T and, for all $t, t' \in G$, $g \in G$,

$$\mu(^g t) = g\mu(t)g^{-1}, \qquad\qquad ^{\mu(t)}t' = tt't^{-1}.$$

A morphism of crossed modules $(f_T, f_G) : (T, G, \mu) \to (T', G', \mu')$ consists of a pair of group homomorphisms $f_T : T \to T'$, $f_G : G \to G'$ such that $\mu' f_T = f_G \mu$, $f_T(^g t) = {}^{f_G(g)} f_T(t)$ for all $g \in G$, $t \in T$.

We list some common examples of crossed modules in groups;

(i) (Whitehead) [53] $(\pi_2(X, A, *), \pi_1(A, *), \partial)$,
    for $(X, A, *)$ a based pair of spaces, ∂ boundary map.
(ii) (N, G, i) $N \lhd G$, i inclusion, action of G on N by conjugation.
(iii) $(G, \mathrm{Aut}\,G, \mu)$, $\mu(g)(h) = ghg^{-1}$, $^\alpha g = \alpha(g)$, $\alpha \in \mathrm{Aut}\,G$,
    $g \in G$.
(iv) $(M, G, 0)$, M a G-module.

From Theorem 3.1, a crossed module in groups (T, G, μ) corresponds to an internal category in $\mathbf{Gp}$

$$(T \rtimes G) \times_G (T \rtimes G) \xrightarrow{m} T \rtimes G \overset{\overset{d_0}{\rightarrow}}{\underset{s}{\underset{\leftarrow}{\underset{d_1}{\rightarrow}}}} G \qquad\qquad (4)$$

where $d_0(t, g) = g$, $d_1(t, g) = \mu(t)g$, $s(g) = (1, g)$, $m((t, g), (t', \mu(t)g)) = (t't, g)$. An easy calculation shows that m is a group homomorphism if and only if the commutator $[\ker d_0, ker d_1]$ is trivial. Thus the internal category (4) is equivalent to the reflexive graph

$$T \rtimes G \underset{\underset{s}{\overset{d_1}{\leftarrow}}}{\overset{d_0}{\rightrightarrows}} G, \qquad d_0 s = \mathrm{id}_G, \qquad d_1 s = \mathrm{id}_G$$

with the extra condition that $[\ker d_0, ker d_1] = 1$. In turn, this corresponds to the triple $(T \rtimes G, d_0', d_1')$, where $d_0' = d_0 s$, $d_1' = d_1 s$ such that $d_0' d_1' = d_1'$, $d_1' d_0' = d_0'$, $[\ker d_0', \ker d_1'] = 0$. In the literature, this structure is called a cat^1-group. More precisely:

DEFINITION 3.2. *A cat^1-group consists of a triple (G, d, t) where $d, t : G \to G$ are group homomorphisms and*
  i) $dt = t$, $td = d$
  ii) $[\ker d, \ker t] = 1$.

A morphism of cat^1-groups $(G, d, t) \to (G', d', t')$ consists of a morphism $f : G \to G'$ such that $fd = d'f$, $ft = t'f$. Denoting by $\mathcal{C}^1\mathcal{G}$ the category of cat^1-groups, from the previous discussion we obtain

PROPOSITION 3.1. *The categories* **Cat(Gp)**, **CM(Gp)**, $\mathcal{C}^1\mathcal{G}$ *are equivalent.*

Other characterizations of internal categories in groups have been given; a simplicial one will be mentioned in next section, and one in terms of double groupoids with connections [11] will not be discussed here, but is very relevant in computations.

The notion of cat^1-group admits a natural generalization to higher dimensions.

DEFINITION 3.3. *The category $\mathcal{C}^n\mathcal{G}$ of* cat^n-groups *is defined as follows. A* cat^n-group *consists of a group G together with $2n$ endomorphisms $t_i, d_i : G \to G$, $1 \le i \le n$ such that for all $1 \le i, j \le n$*
  (i) $d_i t_i = t_i$, $t_i d_i = d_i$,
  (ii) $d_i t_j = t_j d_i$, $d_i d_j = d_j d_i$, $t_i t_j = t_j t_i$, $i \ne j$,
  (iii) $[\ker d_i, \ker t_i] = 1$.

A morphism of cat^n-groups $(G, d_i, t_i) \to (G', d_i', t_i')$ *is a group homomorphism $f : G \to G'$ such that $fd_i = d_i'f$, $ft_i = t_i'f$ $1 \le i \le n$.*

Notice that the identity (iii) in Definition 3.3 is equivalent to

  (iii)' $d_i(x)x^{-1}t_i(x)x^{-1} = t_i(x)x^{-1}d_i(x)x^{-1}$, $x \in G$.

Using this fact we can view $\mathcal{C}^n\mathcal{G}$ as a category of groups with operations as follows. Let $\Omega_0 = \{0\}$, $\Omega_1 = \{-\} \cup \{t_i, d_i\}$ $1 \le i \le n$, $\Omega_2 = \{+\}$; the set of identities comprises the group laws and the identities (i), (ii), (iii)' above. Hence we can speak of crossed modules in $\mathcal{C}^n\mathcal{G}$ or equivalently, by Theorem 3.1, of internal categories in $\mathcal{C}^n\mathcal{G}$. In the next theorem this notion is used to identify $\mathcal{C}^n\mathcal{G}$ with n-fold categories internal to **Gp**.

DEFINITION 3.4. *The category* **Cat^n(Gp)** *of n-fold categories in* **Gp** *is defined inductively by* **Cat^1(Gp)** $=$ **Cat(Gp)**, **Cat^n(Gp)** $=$ **Cat(Cat^{n-1}(Gp)**.

THEOREM 3.2. *For each $n \geq 1$ the categories $\mathcal{C}^n\mathcal{G}$, $\mathrm{CM}(\mathcal{C}^{n-1}\mathcal{G})$,* $\mathrm{Cat}^n(\mathrm{Gp})$ *are equivalent.* **Proof.** We first show that $\mathcal{C}^n\mathcal{G}$ and $\mathrm{CM}(\mathcal{C}^{n-1}\mathcal{G})$ are equivalent. Let $(T, \mathcal{G}, \mu)$ be an object of $\mathrm{CM}(\mathcal{C}^{n-1}\mathcal{G})$ with $T = (T, s_i, v_i)$, $\mathcal{G} = (G, d_i, t_i)$, $i = 1, \ldots, (n-1)$. Thus T is a $\mathcal{G}$-structure and, for all $g \in G$, $t, t' \in T$

$$\mu(\,{}^g t) = g\mu(t)g^{-1}, \qquad {}^{\mu(t)}t' = tt't^{-1}.$$

The object of $\mathcal{C}^n\mathcal{G}$ corresponding to $(T, \mathcal{G}, \mu)$ is $(T \rtimes G, r_i', w_i')$ where $r_i' = r_i$, $w_i' = w_i$ for $i = 1, \ldots, (n-1)$, $r_n'(t, g) = (1, g)$, $w_n'(t, g) = (1, \mu(t)g)$.

Conversely, given a cat^n-group $\mathcal{H} = (H, s_i, t_i)$, let $T = (\ker s_n, s_{i_|}, t_{i_|})$, $\mathcal{G} = (\mathrm{Im}\, s_n, s_{i_|}, t_{i_|})$ and $\mu = t_{n|\ker s_n}$, then T is a $\mathcal{G}$-structure and $(T, \mathcal{G}, \mu)$ is an object of $\mathrm{CM}(\mathcal{C}^{n-1}\mathcal{G})$ corresponding to $\mathcal{H}$. Hence

$$\mathcal{C}^n\mathcal{G} \simeq \mathrm{CM}(\mathcal{C}^{n-1}\mathcal{G}). \tag{5}$$

We prove by induction that $\mathcal{C}^n\mathcal{G} \simeq \mathrm{Cat}^n(\mathrm{Gp})$. By Proposition 3.1 we know this holds for $n = 1$. Suppose, inductively, that $\mathcal{C}^{n-1}\mathcal{G} \simeq \mathrm{Cat}^{n-1}(\mathrm{Gp})$. Then Theorem 3.1 and (5) imply $\mathcal{C}^n\mathcal{G} \simeq \mathrm{CM}(\mathcal{C}^{n-1}\mathcal{G}) \simeq \mathrm{Cat}(\mathcal{C}^{n-1}\mathcal{G}) \simeq \mathrm{Cat}(\mathrm{Cat}^{n-1}(\mathrm{Gp})) = \mathrm{Cat}^n(\mathrm{Gp})$, completing the inductive step.   □

There is also another characterization of n-fold categories in groups in terms of crossed n-cubes [24]. In the case $n = 2$ they are called "crossed squares" and were originally introduced in [36]. A crossed square is a "crossed module of crossed modules" and a crossed n-cube generalizes this notion to higher dimensions. The precise definition of crossed n-cube is quite lengthy and we refer the reader to [24] for further details.

4. Cat^n-groups as an algebraic category. In this section we discuss another aspect of cat^n-groups, which arises from the fact that they are internal structures in the algebraic category of groups. The material in this section is more technical than in the rest of this paper, but not necessary for the main line of development; it can therefore be omitted at first reading.

Recall that a category $\mathcal{C}$ is algebraic when there is a functor $U : \mathcal{C} \to$ **Set** which is monadic; this means that it has left adjoint $F : \mathbf{Set} \to \mathcal{C}$ and $\mathcal{C}$ is equivalent to the category $\mathbf{Set}^{\mathbb{T}}$ of $\mathbb{T}$-algebras for the monad $\mathbb{T} = (UF, \eta, F\varepsilon U)$ arising from the adjunction $F \dashv U$.

If $\mathcal{D}$ is a category of groups with operations, it follows from a general fact [39, Ch. VI §8, Theorem 1] that $\mathcal{D}$ is algebraic. From the above description of $\mathcal{C}^n\mathcal{G}$ as a category of groups with operations it follows that $\mathcal{C}^n\mathcal{G}$ is an algebraic category. This general theory, however, only provides us with existence of a left adjoint $\mathcal{F}_n : \mathbf{Set} \to \mathcal{C}^n\mathcal{G}$ to the underlying set functor. Recent applications have shown that explicit knowledge of $\mathcal{F}_n$ can be very useful; see for instance [15, 13, 44, 47, 46, 45].

For $n = 1$ the functor $\mathcal{F}_1 : \mathbf{Set} \to \mathrm{CM}(\mathrm{Gp})$ was first described explicitly in [15]. In [13, Lemma 9] there is a description of objects of $\mathcal{C}^n\mathcal{G}$

which are projective with respect to the class of regular epimorphisms, for each $n \geq 1$. From general theory [4, vol.II - Prop. 3.86], $\mathcal{F}_n(X)$ is projective with respect to regular epis, hence the result of [13] yields information about $\mathcal{F}_n(X)$. Here we give an explicit description of $\mathcal{F}_n$ for each $n \geq 0$ following [45]. A similar description was independently given in [14] using the language of crossed n-cubes.

Let $R_n : \mathcal{C}^n\mathcal{G} \to \mathbf{Gp}$ be the functor $R_n(G, d_i, t_i) = G$. The category $\mathcal{C}^n\mathcal{G}$ is (co)complete. In particular, given $\mathcal{T}, \mathcal{G} \in \mathcal{C}^n\mathcal{G}$, let $u_1 : \mathcal{T} \to \mathcal{T} \amalg \mathcal{G}$, $u_2 : \mathcal{G} \to \mathcal{T} \amalg \mathcal{G}$ be the two coproduct injections. Using the universality of coproducts, it is easily seen that every element of $R_n(\mathcal{T} \amalg \mathcal{G})$ has the form

$$u_1(t_1)u_2(g_1)u_1(t_2)u_2(g_2)\dots \quad \text{or} \quad u_2(g_1)u_1(t_1)u_2(g_2)u_1(t_2)\dots \quad (6)$$

where $t_i \in R_n\mathcal{T}$, $g_i \in R_n\mathcal{G}$ for all i.

Consider the functor $\mathcal{V}_n : \mathbf{CM}(\mathcal{C}^{n-1}\mathcal{G}) \to \mathcal{C}^{n-1}\mathcal{G}$ defined by

$$\mathcal{V}_n(\mathcal{T}, \mathcal{G}, \mu) = \mathcal{T} \times \mathcal{G}.$$

For any $\mathcal{H} \in \mathcal{C}^{n-1}\mathcal{G}$, let $p : \mathcal{H} \amalg \mathcal{H} \to \mathcal{H}$ be defined by $pu_1 = 0$, $pu_2 = \mathrm{id}$; let $\overline{\mathcal{H}} = \ker p$. Consider the functor $\mathcal{L}_n : \mathcal{C}^{n-1}\mathcal{G} \to \mathbf{CM}(\mathcal{C}^{n-1}\mathcal{G})$ defined by

$$\mathcal{L}_n(\mathcal{H}) = (\overline{\mathcal{H}}, \mathcal{H} \amalg \mathcal{H}, i)$$

where i is the inclusion. We are going to show that $\mathcal{L}_n$ is left adjoint to $\mathcal{V}_n$. Let $(\mathcal{T}, \mathcal{G}, \mu) \in \mathbf{CM}(\mathcal{C}^{n-1}\mathcal{G})$ and let $h : \mathcal{H} \to \mathcal{T}$, $g : \mathcal{H} \to \mathcal{G}$ be morphisms in $\mathcal{C}^{n-1}\mathcal{G}$ inducing a morphism $(h, g) : \mathcal{H} \to \mathcal{T} \times \mathcal{G} = \mathcal{V}_n(\mathcal{T}, \mathcal{G}, \mu)$. We are going to produce a unique morphism $(f_\mathcal{T}, f_\mathcal{G}) : \mathcal{L}_n(\mathcal{H}) \to (\mathcal{T}, \mathcal{G}, \mu)$ in $\mathbf{CM}(\mathcal{C}^{n-1}\mathcal{G})$ such that $(h, g) = (f_\mathcal{T} \times f_\mathcal{G})(u_1, u_2)$.

Consider the commutative diagram in $\mathcal{C}^{n-1}\mathcal{G}$:

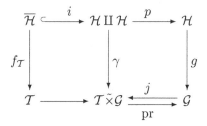

Here $\tilde{\times}$ is the semidirect product in the category of groups with operations $\mathcal{C}^{n-1}\mathcal{G}$; γ is uniquely determined by $\gamma u_1 = ih$, $\gamma u_2 = jg$; $f_\mathcal{T}$ is the restriction of γ. Let $f_\mathcal{G} : \mathcal{H} \amalg \mathcal{H} \to \mathcal{G}$ be determined by $f_\mathcal{G} u_1 = \mu h$, $f_\mathcal{G} u_2 = g$.

For any $x \in R_{n-1}\mathcal{H}$, $w \in R_{n-1}(\mathcal{H} \amalg \mathcal{H})$, $w' \in R_{n-1}\overline{\mathcal{H}}$, we have

$$f_\mathcal{T}(u_1(x)w'u_1(x)^{-1}) = h(x)f_\mathcal{T}(w')h(x)^{-1} = {}^{\mu h(x)}f_\mathcal{T}(w') = {}^{f_\mathcal{G} u_1(x)}f_\mathcal{T}(w')$$

$$f_\mathcal{T}(u_2(x)w'u_2(x)^{-1}) = \gamma(u_2(x)w'u_2(x)^{-1}) = jg(x)f_\mathcal{T}(w')jg(x)^{-1}$$
$$= {}^{g(x)}f_\mathcal{T}(w') = {}^{f_\mathcal{G} u_2(x)}f_\mathcal{T}(w').$$

Since every element of $R_{n-1}(\mathcal{H} \amalg \mathcal{H})$ has the form (6), we conclude that, for all $w \in R_{n-1}(\mathcal{H} \amalg \mathcal{H})$, $w' \in R_{n-1}\overline{\mathcal{H}}$,

$$f_{\mathcal{T}}(^w w') = f_{\mathcal{T}}(ww'w^{-1}) = {}^{f_{\mathcal{G}}(w)} f_{\mathcal{T}}(w') \tag{7}$$

and therefore also

$$\begin{aligned}
\mu f_{\mathcal{T}}(wu_1(x)w^{-1}) &= \mu({}^{f_{\mathcal{G}}(w)} h(x)) = f_{\mathcal{G}}(w)\mu h(x)f_{\mathcal{G}}(w^{-1}) \\
&= f_{\mathcal{G}}(w)f_{\mathcal{G}}u_1(x)f_{\mathcal{G}}(w)^{-1} = f_{\mathcal{G}}(wu_1(x)w^{-1}).
\end{aligned} \tag{8}$$

On the other hand, it is not difficult to prove that $R_{n-1}\overline{\mathcal{H}}$ is the subgroup of $R_n(\mathcal{H} \amalg \mathcal{H})$ generated by elements of the form $wu_1(x)w^{-1}$, $x \in R_{n-1}\mathcal{H}$, $w \in R_{n-1}(\mathcal{H} \amalg \mathcal{H})$. Hence from (8) we deduce that

$$\mu f_{\mathcal{T}} = f_{\mathcal{G}}i. \tag{9}$$

Thus (7) and (9) prove that $(f_{\mathcal{T}}, f_{\mathcal{G}})$ is a morphism in $\mathbf{CM}(\mathcal{C}^n\mathcal{G})$. It is clear that $(f_{\mathcal{T}}, f_{\mathcal{G}})(u_1, u_2) = (h, g)$ and that $(f_{\mathcal{T}}, f_{\mathcal{G}})$ is unique. This concludes the proof that $\mathcal{L}_n \dashv \mathcal{V}_n$.

Let $U : \mathbf{Gp} \to \mathbf{Set}$ be the forgetful functor and F its left adjoint. Let $\alpha_n : \mathbf{CM}(\mathcal{C}^{n-1}\mathcal{G}) \to \mathcal{C}^n\mathcal{G}$ be the functor realizing the equivalence of categories as in the proof of Theorem 3.2, and let β_n be its pseudo-inverse. We can always choose $\alpha_n \dashv \beta_n$. Let $\mathcal{U}_1 : \mathcal{C}^1\mathcal{G} \to \mathbf{Set}$, $\mathcal{U}_1 = U\mathcal{V}_1\beta_1$ and, inductively, $\mathcal{U}_n = \mathcal{U}_{n-1}\mathcal{V}_n\beta_n$. Then $\mathcal{U}_1$ has left adjoint $\mathcal{F}_1 = \alpha_1\mathcal{L}_1 F$; inductively, if $\mathcal{U}_{n-1}$ has left adjoint $\mathcal{F}_{n-1}$, then $\mathcal{U}_n$ has left adjoint $\mathcal{F}_n = \alpha_n\mathcal{L}_n\mathcal{F}_{n-1}$. Using Linton's criteria for monadicity [38], one can show that $\mathcal{U}_n$ is monadic. Summarizing:

THEOREM 4.1. *Let $U : \mathbf{Gp} \to \mathbf{Set}$ be the forgetful functor and F its left adjoint. There is a monadic functor $\mathcal{U}_n : \mathcal{C}^n\mathcal{G} \to \mathbf{Set}$ whose left adjoint $\mathcal{F}_n$ is given by $\mathcal{F}_0 = F$ and, for each $n \geq 1$ and any set X,*

$$\mathcal{F}_n X = \alpha_n(\overline{\mathcal{F}_{n-1}(X)}, \mathcal{F}_{n-1}(X) \amalg \mathcal{F}_{n-1}(X), i)$$

where i is the inclusion, $\amalg$ is the coproduct in $\mathcal{C}^{n-1}\mathcal{G}$, $\overline{\mathcal{F}_{n-1}(X)} = \ker p$, $p : \mathcal{F}_{n-1}(X) \amalg \mathcal{F}_{n-1}(X) \to \mathcal{F}_{n-1}(X)$, $pu_1 = 0$, $pu_2 = id$ and α_n is the equivalence of categories $\alpha_n : \mathbf{CM}(\mathcal{C}^{n-1}\mathcal{G}) \to \mathcal{C}^n\mathcal{G}$.

5. Modelling homotopy types. The category of n-fold categories internal to groups constitutes an algebraic model of connected $(n+1)$-types. In the case $n = 1$ this result was first proved by Mac Lane and Whitehead [42] using crossed modules. For $n > 1$, the first proof of this result, due to Loday [36], was formulated in the category $\mathcal{C}^n\mathcal{G}$. Two other proofs appeared later; one due to Bullejois, Cegarra and Duskin [12] was formulated in the category $\mathbf{Cat}^n(\mathbf{Gp})$ of n-fold categories internal to groups; one due to Porter [48] using the category of crossed n-cubes.

In this section we provide a concise exposition of the proof given in [12]. The algebraic modelling of homotopy types is realized there as follows.

There is a classifying space functor

$$B : \mathbf{Cat^n}\,(\mathbf{Gp}) \to \text{ connected } (n+1)\text{-types}$$

with a left adjoint, the fundamental cat^n-group functor

$$\Pi_n : \text{connected } (n+1)\text{-types } \to \mathbf{Cat^n}\,(\mathbf{Gp}),$$

inducing an equivalence between the localization of $\mathbf{Cat^n}\,(\mathbf{Gp})$ with respect to some suitably defined weak equivalences and the homotopy category of connected $(n+1)$-types:

$$\mathbf{Cat^n}\,(\mathbf{Gp})/\!\sim \; \simeq \; \mathcal{H}o\,(\text{connected } (n+1)\text{-types}). \qquad (10)$$

We start by describing the functor B. In analogy with the nerve of a small category, one defines the nerve of an internal category in $\mathcal{C}$, which is a simplicial object in $\mathcal{C}$. Hence, given an object of $\mathbf{Cat^n}\,(\mathbf{Gp}) \simeq \mathbf{Cat}(\mathbf{Cat^{n-1}}\,(\mathbf{Gp}))$, by applying the nerve we obtain a simplicial object in $\mathbf{Cat^{n-1}}\,(\mathbf{Gp})$ or, equivalently, an object of $\mathbf{Cat^{n-1}}\,(\mathbf{Simpl\ Gp}) \simeq \mathbf{Cat}\,(\mathbf{Cat^{n-2}}(\mathbf{Simpl\ Gp}))$. Iterating we obtain a multinerve functor

$$\mathcal{N} : \mathbf{Cat^n}\,(\mathbf{Gp}) \hookrightarrow \mathbf{Simpl^n}(\mathbf{Gp})$$

where $\mathbf{Simpl^n}(\mathbf{Gp})$ is the category of n-simplicial groups.

Let $\mathbf{Simpl\ Set_0}$ be the category of reduced simplicial sets. Recall that there is a functor $\overline{W} : \mathbf{Simpl\ Gp} \to \mathbf{Simpl\ Set_0}$ with a left adjoint $G : \mathbf{Simpl\ Set_0} \to \mathbf{Simpl\ Gp}$, called Kan loop group functor, such that the canonical maps $G\overline{W}H_* \to H_*$ and $X_* \to \overline{W}GX_*$ are weak equivalences for all simplicial groups H_* and reduced simplicial sets X_*. A detailed description of the functors $\overline{W}$ and G can be found, for instance, in [27] and [40]. The classifying space of an n-simplicial group can be defined by the composition

$$B : \mathbf{Simpl}^{\,n}\mathbf{Gp} \overset{diag}{\to} \mathbf{Simpl\ Gp} \overset{\overline{W}}{\to} \mathbf{Simpl\ Set_0} \overset{|\cdot|}{\to} \mathbf{Top_*}$$

where $diag$ is the diagonal functor and $|\cdot|$ is the geometric realization.

The *classifying space* of an object of $\mathbf{Cat^n}\,(\mathbf{Gp})$ is by definition the classifying space of its multinerve. It can be proved that this is an $(n+1)$-type.

PROPOSITION 5.1. [12] *Let $\mathcal{G}$ be an object of $\mathbf{Cat^n}\,(\mathbf{Gp})$. Then $\pi_i B\mathcal{G} = 0$ for all $i > (n+1)$.*

REMARK 5.1. The classifying space of a cat^n-group can be expressed, up to homotopy, in an alternative way, as follows. Recall that there is a functor $W_{AM} : \mathbf{Simpl}^2\mathbf{Set} \to \mathbf{Simpl\ Set}$ called Artin-Mazur codiagonal (see[1]) and, further, that for any bisimplicial set X_{**}, $|diag\,X_{**}|$ and $|W_{AM}X_{**}|$ are weakly homotopy equivalent (a detailed proof of this fact can be found in [17]). Any group can be considered as a category with just one object, so there is a nerve functor $Ner : \mathbf{Gp} \to \mathbf{Simpl\,Set}$, which induces a functor $Ner^* : \mathbf{Simpl\,Gp} \to \mathbf{Simpl}^2\mathbf{Set}$. It can be seen [17] that $W_{AM}Ner^* = \overline{W}$. It follows from above that, for any simplicial group H_*, there is a weak homotopy equivalence $|\overline{W}H_*| \simeq |diag\,Ner^*\,H_*|$. In particular, given an n-simplicial group Y, this implies

$$BY \simeq |diag\,Ner^*\,diag\,Y|.$$

On the other hand, we clearly have a commutative diagram

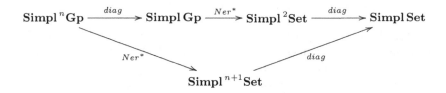

so that $BY \simeq |diag\,Ner^*\,Y|$. In particular, given a cat^n-group $\mathcal{G}$, we obtain a weak homotopy equivalence

$$BG \simeq |diag\,Ner^*\,\mathcal{N}\mathcal{G}|. \tag{11}$$

Thus (11) says that we can calculate the classifying space of a cat^n-group, up to homotopy, by first taking the multinerve, then taking the nerve of the group in each dimension and then taking the geometric realization of the diagonal of the resulting $(n+1)$-simplicial set. This is often useful when working with cat^n-groups.

A morphism $f : \mathcal{G} \to \mathcal{G}'$ in $\mathbf{Cat}^n\,(\mathbf{Gp})$ is a *weak equivalence* if the induced morphism $Bf : B\mathcal{G} \to B\mathcal{G}'$ is a weak homotopy equivalence of spaces.

Next we describe the fundamental cat^n-group functor Π_n. Since the classifying space functor B factors through the categories of n-simplicial groups as well as the category of simplicial groups, we expect the same to hold for its left adjoint Π_n. There are two main steps in identifying the functor Π_n: first find a left adjoint $\mathcal{P}_n$ to the multinerve $\mathcal{N}$, then find a functor $\mathcal{T}_n$ from simplicial groups to n-simplicial groups

$$\mathbf{Cat}^n\,(\mathbf{Gp}) \overset{\mathcal{P}_n}{\underset{\mathcal{N}}{\leftrightarrows}} \mathbf{Simpl}^n\mathbf{Gp} \underset{\mathcal{T}_n}{\leftarrow} \mathbf{Simpl\,Gp} \overset{G\mathcal{S}_0}{\underset{|\overline{W}|}{\leftrightarrows}} \mathrm{Top}_*$$

so that (10) holds with $\Pi_n = \mathcal{P}_n\mathcal{T}_nG\mathcal{S}_0$.

The left adjoint to $\mathcal{N}$ is easily identified by induction as follows. Let $\mathcal{C}$ be a category of groups with operations. Then the nerve functor

$$Ner : \mathbf{Cat}\,\mathcal{C} \hookrightarrow \mathbf{Simpl}\,\mathcal{C}$$

has a left adjoint $\mathcal{P}$, called the fundamental groupoid functor. For any object G_* of $\mathbf{Simpl}\,\mathcal{C}$, $\mathcal{P}(G_*)$ has arrows object $G_1/d_2(N_2G)$, where N_* denotes the Moore complex, objects object G_0, source, target and identity maps induced by face and degeneracy operators respectively. The simplicial identities easily imply that this is an object of $\mathbf{Cat}\,\mathcal{C}$. The unit of the adjunction $G_* \to \mathcal{N}er\mathcal{P}(G_*)$ induces an isomorphism of π_1 and π_2, as is easily checked.

Notice that, when restricting to the subcategory $\mathbf{Simpl}\,\mathcal{C}$ of simplicial objects in $\mathcal{C}$ of Moore complex of length 1 (namely the Moore complex is trivial in dimension greater than 1), then the adjunction $\mathcal{P} \dashv \mathcal{N}er$ becomes an equivalence of categories. Taking $\mathcal{C} = \mathbf{Gp}$ we deduce that crossed modules in groups are also equivalent to simplicial groups of Moore complex of length 1.

By induction, suppose we have constructed the left adjoint $\mathcal{P}_{n-1}$ to the multinerve

$$\mathcal{N} : \mathbf{Cat}^n\,(\mathbf{Gp}) \hookrightarrow \mathbf{Simpl}^{n-1}(\mathbf{Gp}).$$

This induces an adjunction $\mathcal{P}^*_{n-1} \dashv \mathcal{N}^*$ between the corresponding categories of simplicial objects; from above, taking $\mathcal{C} = \mathbf{Cat}^{n-1}(\mathbf{Gp})$, we obtain the commutative diagram

$$
\begin{array}{ccc}
\mathbf{Simpl\;Cat}^{n-1}(\mathbf{Gp}) & \underset{\mathcal{N}^*}{\overset{\mathcal{P}^*_{n-1}}{\longleftarrow\!\!\!\longrightarrow}} & \mathbf{Simpl}^n\;\mathbf{Gp} \\[2mm]
\Big\| & & \mathcal{P}_n \Big\downarrow\Big\uparrow \mathcal{N} \\[2mm]
\mathbf{Simpl\,Cat}^{n-1}(\mathbf{Gp}) & \underset{\mathcal{N}er}{\overset{\mathcal{P}}{\rightleftarrows}} & \mathbf{Cat}(\mathbf{Cat}^{n-1}(\mathbf{Gp})) \cong \mathbf{Cat}^n(\mathbf{Gp})
\end{array}
$$

where $\mathcal{P}_n = \mathcal{P}\mathcal{P}^*_{n-1}$. By construction, $\mathcal{P}_n$ is left adjoint to $\mathcal{N}$, and this concludes the inductive step.

To construct the fundamental cat^n-group functor in such a way that (10) holds we would need the unit of the adjunction $\mathcal{P} \dashv \mathcal{N}$ to induce a map of classifying spaces $B(G_*) \to B\mathcal{N}\mathcal{P}_n(G_*)$ which is an $(n+1)$-weak equivalence. Given any n-simplicial group G_*, this is in general false unless $n = 1$. The key observation of [12] is that under certain asphericity conditions on G_* the above is true; furthermore, there exists a functor $\mathcal{T}_n : \mathbf{Simpl\,Gp} \to \mathbf{Simpl}^n\mathbf{Gp}$ with the property that for any simplicial group H_* and $n \geq 2$, the n-simplicial group $\mathcal{T}_n(H_*)$ satisfies the required asphericity conditions and the natural transformation $B(H_*) \to B\mathcal{T}_n(H_*)$ is a weak homotopy equivalence.

The functor $\mathcal{T}_n$ is given by or^*, the functor induced by the ordinal sum $or : \Delta \times \cdots^n \times \Delta \to \Delta$.

In the case $n = 2$, the right adjoint to $\mathcal{T}_2$ is called in the literature the Artin-Mazur codiagonal functor [1]. Further, $\mathcal{T}_2(H_*)$ is given by "total

Dec;" this is obtained as cotriple resolution of H_* via the cotriple associated to the pair of adjoint functors

$$\textbf{Simpl Gp} \underset{Dec^1}{\overset{+}{\leftrightarrows}} Aug\ \textbf{Simpl Gp}.$$

Here the "shift functor" Dec^1 maps a simplicial group to an augmented simplicial group obtained by forgetting the last face operator; its left adjoint forgets the augmentation. Hence $\mathcal{T}_2(H_*)$ has the form

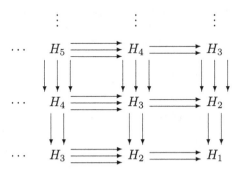

Thus, if X is a path-connected space and $H_* = GS_0(X)$, the fundamental cat^2-group of X has the form

$$
\begin{array}{ccc}
\dfrac{H_2 \times_{d_0} H_2}{\sim} & \rightrightarrows & H_1 \times_{d_0} H_1 \\
\downdownarrows & & \downdownarrows \\
H_1 \times_{d_0} H_1 & \rightrightarrows & H_1
\end{array}
$$

6. Recent developments. In this section we give a very broad outline of some recent developments that link strict higher dimensional algebra to higher category theory. As explained in Section 2, the area which we refer to as strict higher-dimensional algebra arose within algebraic topology and has as its main categorical input strict n-categories / ω-categories and n-fold categories internalized in a fixed algebraic category such as groups and algebras.

An important development has been taking place almost in parallel in category theory with the discovery of several notions of weak n-category. It is outside the scope of this paper to provide a survey of this rapidly expanding area: we refer the reader to existing surveys [18, 37] and references there.

The modelling of homotopy types provides an important link between weak n-categories and homotopy theory. Namely, it is understood that a good definition of weak n-category should provide a model of n-types in the weak n-groupoid case. This "conjecture-test" has been verified for some of the current notions of weak n-category for instance, for Tamsamani's model [54] and Batanin's model [19, 2]; in the weak ω-category case, it holds automatically for Street's [51] and Verity's [56], [57] definitions.

On the other hand, since strict higher-dimensional algebra does provide models of all n-types, it is entirely natural to ask how do these models compare with weak n-groupoids. One aspect that makes this a very intriguing question is that internal n-fold categorical structures seem to have a very different feature from weak n-groupoids: namely, in the first there is no apparent "weakness" in the definition (this is why we think of them as part of strict higher-dimensional algebra). Thus it ought to be interesting to directly construct, from a cat^n-group, a weak $(n+1)$-groupoid representing the same homotopy type. This comparison program has been carried out using Tamsamani's model of weak n-groupoids, for $n = 3$ in [47] and in [46] for general n. The main outcome of this comparison is a semistrictification result for Tamsamani's weak n-groupoids. This relates to "coherence theorems" for higher categories which, in the general case, are known so far in dimension 2 [41] and 3 [28].

At present a comparison between $\mathbf{Cat^n(Gp)}$ and weak $(n+1)$-groupoids, other than Tamsamani's, is not known. In view of the above such comparison could be interesting for the theory of weak n-categories itself.

We shall now give a very broad sketch of some ideas involved in the comparison between cat^n-groups and Tamsamani's model.

Suppose $\mathcal{C}$ is a category with finite limits and $\phi \in \mathbf{Simpl}\mathcal{C}$ is a simplicial object in $\mathcal{C}$; put $\phi_n = \phi[n]$ and let $\phi_1 \times_{\phi_0} \cdots \times_{\phi_0} \phi_1$ be the limit of the diagram

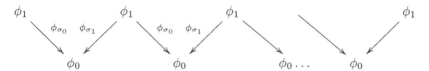

where $\sigma_0, \sigma_1 : [0] \to [1]$, $\sigma_0(0) = 0$, $\sigma_1(0) = 1$. It is immediate to see that, for each $n \geq 2$, there is a unique map $\eta_n : \phi_n \to \phi_1 \times_{\phi_0} \cdots \times_{\phi_0} \phi_1$ such that $\mathrm{pr}_j \eta_j = \phi \nu_j$, where $\nu_j : [1] \to [n]$, $\nu_j(0) = j - 1$, $\nu_j(1) = j$. The maps η_n are called *Segal maps* and they play a fundamental role in the cat^n-group and the Tamsamani model. In fact, a simplicial object in $\mathcal{C}$ is the nerve of an internal category in $\mathcal{C}$ precisely when all Segal maps are isomorphisms. Thus in view of Theorem 3.2 the definition of cat^n-groups could also be given, inductively, as follows:

i) $\mathbf{Cat^0(Gp)} = \mathbf{Gp}$.
ii) $\mathbf{Cat^n(Gp)}$ is the full subcategory of simplicial objects in   (12)
    $\mathbf{Cat^{n-1}(Gp)}$ such that all Segal maps are isomorphisms.

One of the key ideas of Tamsamani's model is to encode the weakness of associativity and identity laws in a weak n-category in the requirement that the Segal maps are not isomorphisms but some suitably defined n-equivalences. In the case $n = 2$ one obtains a notion strictly related to bicategories, see [35]. Thus the definition of Tamsamani's weak n-groupoids is given inductively by requiring:

i) A weak 1-groupoid is just a groupoid. An equivalence of weak 1-groupoids is a functor which is fully faithful and essentially surjective on objects.

ii) A weak n-groupoid is a simplicial object $\mathcal{G}$ in weak $(n-1)$-groupoids such that $\mathcal{G}_0$ is discrete and the Segal maps are $(n-1)$-equivalences.

$$(13)$$

Clearly, this definition is not complete until one specifies what is an n-equivalence, given inductively a notion of $(n-1)$-equivalence. The details of this important point can be found in [55] and [54]. Here we limit ourselves to comparing (12) and (13) on a qualitative basis. We can say (12) and (13) are similar in that they are both multi-simplicial structures: unravelling the inductive definitions (12) and (13) and using the nerve construction we see that $\mathbf{Cat}^n(\mathbf{Gp})$ can be viewed as a full subcategory of n-simplicial groups, Tamsamani's weak n-groupoids as a full subcategory of n-simplicial sets. Another similarity is that both definitions are given inductively on dimension.

There are however some important differences. In (13) we have a multi-simplicial structure which is discrete at level zero; this discreteness condition does not hold in (12). We say that (12) is a strict cubical structure, (13) is a weak globular one. This terminology refers, intuitively, to the shape of the 2-cells. A second difference is that (12) defines a structure internal to the category of groups, (13) defines a structure based on $\mathbf{Set}$. Thus in the comparison problem there are two main issues: a) How to pass from a strict cubical structure to a weak globular one without losing homotopical information. b) How to pass from an internal structure in groups to a set-based structure while preserving the homotopy type.

Point b) is the easiest to deal with, through the use of the nerve functor $\mathbf{Gp} \to [\Delta^{op}, \mathbf{Set}]$. Point a) is much more subtle and involves a discretization process which is made possible by the fact that every cat^n-group can be represented, up to homotopy, by one in which some of the 'faces' are weakly equivalent to discrete ones. For $n = 2$, for instance, it can be shown that every cat^2-group is weakly equivalent to one in which the object of objects is a cat^1-group weakly equivalent to a discrete one: see [47] for details. This property of cat^n-groups can be proved using the functor $\mathcal{F}_n : \mathbf{Set} \to \mathbf{Cat}^n(\mathbf{Gp})$ which was described in Theorem 4.1.

We conclude by mentioning that in recent years there has been a surge of interest in category theory in n-fold categories themselves, especially for $n = 2$. The theory of double categories is being developed in great detail: see for instance [20–22, 29]. Also in [25] double categories and their lax versions have been studied in relation to applications to conformal field theory.

These recent developments seem to indicate that n-fold categorical structures, although deceptively simple in their definition, encode much of the complexity and richness of higher category theory and could provide stimulating new developments in the years to come.

Acknowledgments. I am grateful to the organizers of the conference "n-Categories: Foundations and Applications" for the opportunity to speak on this topic. I also thank my colleagues at SUNY at Buffalo, where most of this paper was written, for providing a conducive working environment. I thank Ross Street for reading the final version and Peter May for helpful comments. I finally extend my gratitude to the many mathematicians I came in contact with from the time of my graduate studies until now, for enriching my understanding of the area of mathematics covered in this paper.

REFERENCES

[1]   M. ARTIN AND B. MAZUR, On the Van Kampen theorem, *Topology* **5**:179–189 (1966).

[2]   M.A. BATANIN, Monoidal globular categories as natural environment for the theory of weak n-categories, *Advances in Mathematics*, **136**:39–103 (1998).

[3]   C. BERGER, Double loop spaces, braided monoidal categories and algebraic 3-type of space, *Contemporary Mathematics*, **227**:46–66 (1999).

[4]   F. BORCEAUX, *Handbook of categorical algebra*, Encyclopedia of Mathematics and its applications, Cambridge University Press (1994).

[5]   D. BOURN, Moore normilization and Dold-Kan theorem for semi-abelian categories, to appear in *Proceeding of the conference 'Categories in Algebra, Geometry and Mathematical Physics,'* Macquarie University, July 2005.

[6]   L. BREEN, *Bitorsors et cohomologie non abélienne*, 1:401–476, Progr. Math. **86**, Birkhauser Boston, Boston MA (1990).

[7]   R. BROWN AND J.L. LODAY, Van Kampen theorems for diagrams of spaces, *Topology* **26**(3):311–335 (1987).

[8]   R. BROWN AND P. HIGGINS, The equivalence of ∞-groupoids and crossed complexes, *Cahiers Topologie Géom. Différentielle*, **22**:371–386 (1981).

[9]   R. BROWN, *Computing homotopy types using crossed n-cubes of groups*, London Math Soc., Lecture Note Ser., **175**, Cambridge University Press, (1992), pp. 187–210.

[10] R. BROWN AND J. HUEBSCHMANN, *Identities among relations.*, London Math Soc., Lecture Note Ser., **48**, Cambridge University Press, (1982), pp. 153–202.

[11] R. BROWN AND C.B. SPENCER, Double groupoids and crossed modules, *Cahiers Topologie Géom. Différentielle Categ.*, **17**(4):343–362 (1976).

[12] M. BULLEJOS, A.M. CEGARRA, AND J. DUSKIN, On cat^n-groups and homotopy types, *J. of Pure and Appl. Algebra*, **86**:134–154 (1993).

[13] J.M. CASAS, G. ELLIS, M. LADRA, AND T. PIRASHVILI, Derived functors and the homology of n-types, *Journal of Algebra*, **256**:583–598 (2002).

[14] J.M. CASAS, N. INASSARIDZE, E. KHMALADZE, AND M. LADRA, Homology of $(n+1)$-types and Hopf type formulas, *J. of Pure and Appl. Algebra*, **200**:267–280 (2005).

[15] P. CARRASCO, A.N. CEGARRA, AND A.R. GRANDJEAN, (Co)-homology of crossed modules, *J. of Pure and Appl. Algebra*, **168**:147–176 (2002).

[16] P. CARRASCO AND A.M. CEGARRA, Group-theoretic algebraic models for homotopy types, *J. of Pure and Appl. Algebra*, **75**(3):195–235 (1991).

[17] A.M. CEGARRA AND J. REMEDIOS, The relationship between the diagonal and the bar constructtions on a bisimplicial set, *Topology and its Applications*, **153**:21–51 (2005).

[18] E. CHENG, A. Lauda, Higher-Dimensional Categories: an illustrated guide book (2004).

[19] D.C. CISINSKI, Batanin higher operads and homotopy types, to appear in *Proceeding of the conference 'Categories in Algebra, Geometry and Mathematical Physics'*, Macquarie University, July 2005.

[20] R.J. MACG. DAWSON, R. PARÉ, AND D.A. PRONK, Free extensions of double categories, *Cahiers Topol. Géom. Différ., Catég.* **45**(1):35–80 (2004).

[21] R.J. MACG. DAWSON AND R. PARÉ, General associativity and general composition for double categories, *Cahiers Topol. Géom. Différ., Catég.* **34**(1):57–79 (1993).

[22] R.J. MACG. DAWSON AND R. PARÉ, What is a free double category like? *J. of Pure and Appl. Algebra*, **168**(1): 19–34 (2002).

[23] W.G. DWYER AND D. KAN, Homotopy theory and simplicial groupoids, *Proc. Konink. Nederl. Akad.*, **87**:379–389 (1984).

[24] G. ELLIS AND R. STEINER, Higher-dimensional crossed modules and the homotopy groups of $(n + 1)$-ads, *J. of Pure and Appl. Algebra*, **46**(2–3):117–136 (1987).

[25] T.M. FIORE, Pseudo algebras and pseudo double categories, *Journal of Homotopy and Related Structures*, to appear.

[26] A.R. GARZON AND J.G. MIRANDA, Models for homotopy n-types in diagram categories, *Appl. Categ. Structures*, **4**(2–3):213–225 (1996).

[27] P.G. GOERSS AND J.F. JARDINE, *Simplicial Homotopy Theory*, Progress in Mathematics, **174**, Birkhäuser Verlag, Basel (1999).

[28] R. GORDON, A.J. POWER, AND R. STREET, *Coherence for tricategories and indexed categories*, Memoirs of the American Mathematical Society, **117**(558) (1995).

[29] M. GRANDIS AND R. PARÉ, Limits in double categories,*Cahiers de Topologie et Géométrie Différentielle Catégorique*, **40**(3):162–220 (1999).

[30] A. GROTHENDIECK, Pursuing stacks, Manuscript (1984).

[31] D.F. HOLT, An interpretation of the cohomology groups $H^n(G, M)$, *Journal of Algebra*, **60**:307–329 (1979).

[32] J. HUEBSCHMANN, Crossed n-fold extensions of groups and cohomology, *Comment. Math. Helvetici*, **55**, (1980) 302–314.

[33] G. JANELIDZE, Internal crossed modules, *Georgian Math. Journal*, **10**(1): 99–114 (2003).

[34] S. LACK AND S. PAOLI, An operadic approach to internal structures, *Appl.Categ. Structures*, **13**(3):205–222 (2005).

[35] S. LACK AND S. PAOLI, 2-nerves for bicategories, *K-theory*, to appear. Available at math.CT/0607271.

[36] J.L. LODAY, Spaces with finitely many non-trivial homotopy groups, *J. of Pure and Appl. Algebra*, **24**:179–202 (1982).

[37] T. LEINSTER, A survey definition of n-category, *Theory and applications of Categories*, **10**(1):1–70 (2002).

[38] F.J. LINTON, Some aspects of equational categories, Proceedings of the conference on categorical algebra, La Jolla, Springer, berlin (1966), pp. 64–94.

[39] S. MACLANE, *Categories for the working mathematician*, Springer, Berlin (1971).

[40] J.P. MAY, *Simplicial objects in algebraic topology*, Chicaco Lectures in Mathematics, University of Chicago Press (1967).

[41] S. MAC LANE AND R. PARÉ, Coherence for bicategories and indexed categories, *J. of Pure and Appl. Algebra*, **37**(1):59–80 (1985).

[42] S. MACLANE AND J.H.C. WHITEHEAD, On the 3-type of a complex. *Proc. Nat. Acad. Sci. U.S.A.*, **30**:41–48 (1956).

[43] G. ORZECH, Obstruction theory in algebraic categories I and II, *J. of Pure and Appl. Algebra*, **2**:287–314; 315–340 (1972).

[44] S. PAOLI, (Co)homology of crossed modules with coefficients in a π_1-module, *Homology, Homotopy and Applications*, **5**(1):261–296 (2003).

[45] S. PAOLI, Developments in the (co)homology of crossed modules, PhD Thesis, University of Warwick, (2002).

[46] S. PAOLI, Semistrict Tamsamani n-groupoids and connected n-types, preprint (2007) math.AT/0701655v2.

[47] S. PAOLI, Semistrict models of connected 3-types and Tamsamani's weak 3-groupoids, *J. of Pure Appl. Algebra*, **211**:801–820 (2007).

[48] T. PORTER, N-types of simplicial groups and crossed N-cubes, *Topology*, **32**(1):5–24 (1993).

[49] T. PORTER, Extensions, crossed modules and internal categories in categories of groups with operations. *Proc. Edimburgh Math. Soc.* (1987) **30**:373–381.

[50] T. PORTER, Topological quantum field theories from homotopy n-types, *J. London Math. Soc.*, (2) **58**(3):723–732 (1998).

[51] R. STREET, Weak omega-categories, in 'Diagrammatic morphisms and applications', *Contemporary Mathematics*, **318**:207–213 (2003).

[52] M.J. VALE, Torsors and special extensions, *Cahiers de Topologie et Géométrie Différentielle Catégorique*, Vol. XXVI-1 (1985), pp. 63–90.

[53] J.H.C. WHITEHEAD, Combinatorial homotopy II, *Bull. Amer. Math. Soc.*, **55**: 453–496 (1949).

[54] Z. TAMSAMANI, Sur des notions de n-catégorie et n-groupoide non-strictes via des ensembles multi-simpliciaux, *K-theory*, **16**:51–99 (1999).

[55] B. TOEN, Tamsamani categories, Segal categories and applications, available at www.ima.umn.edu/talks/workshops/SP6.7-18.04/toen/bertrand-IMA.pdf.

[56] D. VERITY, Weak complicial sets, a simplicial weak omega-category theory. Part I: Basic homotopy theory, ArXiv preprint (2006) mathCT/0604414.

[57] D. VERITY, Weak complicial sets, a simplicial weak omega-category theory. Part II: Nerves of complicial Gray categories, ArXiv preprint (2006) mathCT/0604416.

A 2-CATEGORIES COMPANION

STEPHEN LACK*

Abstract. This paper is a rather informal guide to some of the basic theory of 2-categories and bicategories, including notions of limit and colimit, 2-dimensional universal algebra, formal category theory, and nerves of bicategories.

AMS(MOS) subject classifications. 18D05, 18C15, 18A30, 18G30, 18G55, 18C35.

1. Overview and basic examples. This paper is a rather informal guide to some of the basic theory of 2-categories and bicategories, including notions of limit and colimit, 2-dimensional universal algebra, formal category theory, and nerves of bicategories. As is the way of these things, the choice of topics is somewhat personal. No attempt is made at either rigour or completeness. Nor is it completely introductory: you will not find a definition of bicategory; but then nor will you really need one to read it. In keeping with the philosophy of category theory, the morphisms between bicategories play more of a role than the bicategories themselves.

1.1. The key players. There are bicategories, 2-categories, and **Cat**-categories. The latter two are exactly the same (except that strictly speaking a **Cat**-category should have small hom-categories, but that need not concern us here). The first two are nominally different — the 2-categories are the strict bicategories, and not every bicategory is strict — but every bicategory is *biequivalent* to a strict one, and biequivalence is the right general notion of equivalence for bicategories and for 2-categories. Nonetheless, the theories of bicategories, 2-categories, and **Cat**-categories have rather different flavours.

An enriched category is a category in which the hom-functors take their values not in **Set**, but in some other category $\mathcal{V}$. The theory of enriched categories is now very well developed, and **Cat**-category theory is the special case where $\mathcal{V} = \mathbf{Cat}$. In **Cat**-category theory one deals with higher-dimensional versions of the usual notions of functor, limit, monad, and so on, without any "weakening." The passage from category theory to **Cat**-category theory is well understood; unfortunately **Cat**-category theory is generally not what one wants to do — it is too strict, and fails to deal with the notions that arise in practice.

In bicategory theory all of these notions are weakened. One never says that arrows are equal, only isomorphic, or even sometimes only that there is a comparison 2-cell between them. If one wishes to generalize a

---

* School of Computing and Mathematics, University of Western Sydney, Locked Bag 1797 Penrith South DC NSW 1797, Australia (s.lack@uws.edu.au). The support of the Australian Research Council and DETYA is gratefully acknowledged.

J.C. Baez, J.P. May (eds.), *Towards Higher Categories*, The IMA Volumes in Mathematics and its Applications 152, DOI 10.1007/978-1-4419-1524-5_4,
© Springer Science+Business Media, LLC 2010

result about categories to bicategories, it is generally clear in principle what should be done, but the details can be technically very difficult.

2-category theory is a "middle way" between **Cat**-category theory and bicategory theory. It *uses* enriched category theory, but not in the simple minded way of **Cat**-category theory; and it cuts through some of the technical nightmares of bicategories. The prefix "2-," as in 2-functor or 2-limit, will always denote the strict notion; although often we will use it to describe or analyze non-strict phenomena.

There are also various other related notions, which will be less important in this companion. **SSet**-categories are categories enriched in simplicial sets; every 2-category induces an **SSet**-category, by taking nerves of the hom-categories. Double categories are internal categories in **Cat**. Once again every 2-category can be seen as a double category. A slight generalization of double categories allows bicategories to fit into this picture. Finally there are the internal categories in **SSet**; both **SSet**-categories and double categories can be seen as special cases of these.

1.2. Nomenclature and symbols. In keeping with our general policy, the word *2-functor* is understood in the strict sense: a 2-functor between 2-categories $\mathscr{A}$ and $\mathscr{B}$ assigns objects to objects, morphisms to morphisms, and 2-cells to 2-cells, preserving all of the 2-category structure strictly. We shall of course want to consider more general types of morphism between 2-categories later on.

If "widget" is the name of some particular categorical structure, then there are various systems of nomenclature for weak 2-widgets. Typically one speaks of *pseudo* widgets for the up-to-isomorphism notion, *lax* widgets for the up-to-not-necessarily-invertible comparison notion, and when the direction of the comparison is reversed, either *oplax* widget or *colax* widget, depending on the specific case. But there are also other conventions. In contexts where the pseudo notion is most important, this is called simply a widget, and then one speaks explicitly of *strict* widgets in the strict case. In contexts where the lax notion is most important (such as with monoidal functors), it is this which has no prefix; and one has *strict* widgets in the strict case or *strong* widgets in the pseudo.

As we move up to 2-categories and higher categories, there are various notions of sameness, having the following symbols:

- $=$ is equality
- $\cong$ is isomorphism (morphisms f and g with $gf = 1$, $fg = 1$)
- $\simeq$ is equivalence ($gf \cong 1$, $fg \cong 1$)
- $\sim$ is sometimes used for biequivalence.

In Sections 1.4 and 1.5 we look at various examples of 2-categories and bicategories. The separation between the 2-category examples and the bicategory examples is not really about strictness but about the sort of morphisms involved. The 2-category examples involve functions or functors of some sort; the bicategory examples (except the case of a monoidal category) involve more general types of morphism such as relations. These "non-functional" morphisms are often depicted using a slashed arrow ($\nrightarrow$) rather than an ordinary one ($\rightarrow$). Typically the functional morphisms can be seen as a special case of the non-functional ones. Sometimes it is also possible to characterize the non-functional ones as a special type of functional morphism (with different domain and/or codomain), and this can provide a concrete construction of a 2-category biequivalent to the given bicategory. The other special type of arrow often used is a "wobbly" one ($\rightsquigarrow$); this denotes a weak (pseudo, lax, etc.) morphism.

1.3. Contents. In the remainder of this section we look at examples of 2-categories and bicategories. In Section 2 we begin the study of formal category theory, including adjunctions, extensions, and monads, but stopping short of the full-blown formal theory of monads. In Section 3 we look at various types of morphism between bicategories or 2-categories: strict, pseudo, lax, partial; and see how these can be used to describe enriched and indexed categories. In Section 4 we begin the study of 2-dimensional universal algebra, with the basic definitions and the construction of weak morphism classifiers. This is continued in Section 5 on presentations for 2-monads, which demonstrates how various categorical structures can be described using 2-monads. Section 6 looks at various 2-categorical and bicategorical notions of limit and considers their existence in the 2-categories of algebras for 2-monads. Section 7 is about aspects of Quillen model structures related to 2-categories and to 2-monads. In Section 8 we return to the formal theory of monads, applying some of the earlier material on limits. Section 9 looks at the formal theory of *pseudomonads*, developed in a Gray-category. Section 10 looks at notions of nerve for bicategories. There are relatively few references throughout the text, but at the end of each section there is a brief commented bibliography.

One topic I was very disappointed not to cover is that of Yoneda structures [59], which later gave rise to the idea of *equipments* as developed by Wood and various collaborators.

1.4. Examples of 2-categories. Just as **Set** is the mother of all categories, so **Cat** is the mother of all 2-categories. From many points of view, it has all the best properties as a 2-category (but not as a category: for example colimits in **Cat** are not stable under pullback).

A small category involves a set of objects and a set of arrows, and also hom-sets between any two objects. One can generalize the notion of category in various ways by replacing various of these sets by objects of some other category.

(a) If $\mathscr{V}$ is a monoidal category one can consider the 2-category $\mathscr{V}$-**Cat** of categories enriched in $\mathscr{V}$; these have $\mathscr{V}$-valued hom-objects rather than hom-sets. The theory works best when $\mathscr{V}$ is symmetric monoidal closed, complete, and cocomplete. As for examples of enriched categories, one has ordinary categories ($\mathscr{V} = $ **Set**), additive categories

($\mathscr{V} = \mathbf{Ab}$), 2-categories ($\mathscr{V} = \mathbf{Cat}$), preorders ($\mathscr{V} = \mathbf{2}$, the "arrow category"), simplicially enriched categories ($\mathscr{V} = \mathbf{SSet}$), and DG-categories ($\mathscr{V}$ the category of chain complexes).

(b) More generally still, one can consider a bicategory $\mathscr{W}$ as a many-object version of a monoidal category; there is a corresponding notion of $\mathscr{W}$-enriched category: see [5] or Section 3.1. Sheaves on a site can be described as $\mathscr{W}$-categories for a suitable choice of $\mathscr{W}$.

(c) If $\mathbb{E}$ is a category with finite limits, one can consider the 2-category $\mathbf{Cat}(\mathbb{E})$ of categories internal to $\mathbb{E}$; these have an $\mathbb{E}$-object of objects and an $\mathbb{E}$-object of morphisms. The theory works better the better the category $\mathbb{E}$; the cases of a topos or an abelian category are particularly nice. This includes ordinary categories ($\mathbb{E} = \mathbf{Set}$), double categories ($\mathbb{E} = \mathbf{Cat}$), morphisms of abelian groups ($\mathbb{E} = \mathbf{Ab}$), and crossed modules ($\mathbb{E} = \mathbf{Grp}$).

There is another class of examples, in which the objects are "categories with structure." The structure could be something like

(d) category with finite products
(e) category with finite limits
(f) monoidal category
(g) topos
(h) category with finite products and coproducts and a distributive law.

For most of these there are also analogues involving enriched or internal categories with the relevant structure.

In each case you need to decide which morphisms to use. Normally you don't want the strictly algebraic ones (preserving the structure on the nose): although they can be technically useful, they are rare in nature. More common are the "pseudo" morphisms: these are functors preserving the structure "up to (suitably coherent) isomorphism." In (e), for example, this would correspond to the usual notion of finite-limit-preserving functor.

Sometimes, however, it's good to consider an even weaker notion of morphism, as in the 2-category $\mathbf{MonCat}$ of monoidal categories, monoidal functors, and monoidal natural transformations. Monoidal functors are the "lax" notion, involving maps $FA \otimes FB \to F(A \otimes B)$, coherent, but not necessarily invertible. Here are some reasons you might like this level of generality:

- Consider the monoidal categories $\mathbf{Ab}$ of abelian groups, with the usual tensor product, and $\mathbf{Set}$ of sets, with the cartesian product. The forgetful functor U from $\mathbf{Ab}$ to $\mathbf{Set}$ definitely does not preserve this structure, but we have the universal bilinear map $UG \times UH \to U(G \otimes H)$, and this makes U into a monoidal functor.
- A monoidal functor $\mathscr{V} \to \mathscr{W}$ sends monoids in $\mathscr{V}$ to monoids in $\mathscr{W}$, via the rule

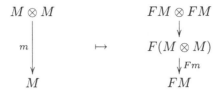

• Suppose $\mathcal{V}$ and $\mathcal{W}$ are monoidal categories and $F : \mathcal{V} \to \mathcal{W}$ is a left adjoint which does preserve the monoidal structure up to coherent isomorphism. There is no reason why the right adjoint U should do so, but there will be induced comparison maps $UA \otimes UB \to U(A \otimes B)$ making U a monoidal functor. (Think of the tensor product as a type of colimit, so the left adjoint preserves it, but the right adjoint doesn't necessarily.) In fact the monoidal functor $U : \mathbf{Ab} \to \mathbf{Set}$ arises in this way.

The case of monoidal categories is typical. Given an adjunction $F \dashv U$ between categories $\mathcal{A}$ and $\mathcal{B}$ with algebraic structure, to make the right adjoint U a colax morphism is equivalent to making the left adjoint F lax, while if the whole adjunction lives within the world of lax morphisms, then F is not just lax but pseudo. This phenomenon is called *doctrinal adjunction* [22].

For a further example, consider the structure of categories with finite coproducts. For a functor $F : \mathcal{A} \to \mathcal{B}$ between categories with finite coproducts there are canonical comparison maps $FA + FB \to F(A + B)$, and these make *every* such functor uniquely into a lax morphism; it is a pseudo morphism exactly when it preserves the coproducts in the usual sense. Thus in this case every adjunction between categories with finite coproducts lives in the lax world, and the fact that the left adjoint is actually pseudo reduces to the well known fact that left adjoints preserve coproducts.

In the case of categories with finite products or finite limits, however, the lax morphisms are the same as the pseudo morphisms; they are just the functors preserving the products or limits in the usual sense.

1.5. Examples of bicategories. Any monoidal category $\mathcal{V}$ determines a one-object bicategory $\Sigma\mathcal{V}$ whose morphisms are the objects of $\mathcal{V}$, and whose 2-cells are the morphisms of $\mathcal{V}$. The tensor product of $\mathcal{V}$ is the (horizontal) composition in $\Sigma\mathcal{V}$.

(i) **Rel** consists of sets and relations. The objects are sets and the morphisms $X \nrightarrow Y$ are the relations from X to Y; that is, the monomorphisms $R \rightarrowtail X \times Y$. This bicategory is 'locally posetal', in the sense that for any two parallel 1-cells, there is at most one 2-cell between them. There is a 2-cell from R to S if and only if R is contained in S as a subobject of $X \times Y$; in other words, if there is a morphism $R \to S$ making the triangles in

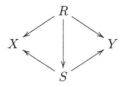

commute. As usual, xRy means that $(x, y) \in R$. The composite of $R \rightarrowtail X \times Y$ and $S \rightarrowtail Y \times Z$ is the relation $R \circ S$ defined by

$$x(R \circ S)z \iff (\exists y)xRySz.$$

We get a 2-category biequivalent to this one by identifying isomorphic 1-cells; this works for any locally posetal 2-category.

Another 2-category biequivalent to **Rel** has sets for objects, and as morphisms from X to Y the join-preserving maps from $\mathscr{P}X$ to $\mathscr{P}Y$, where $\mathscr{P}X$ denotes the set of all subsets of X. Here a relation R is represented by the function sending a subset $U \subset X$ to $\{y \in Y : (\exists x \in U)xRy\}$.

(j) **Par** consists of sets and partial functions. A partial function from X to Y is a diagram $X \leftarrowtail D \to Y$ in **Set**, where D is the domain of definition of the partial function; 2-cells and composition are defined as in **Rel**. Again, we get a biequivalent 2-category by identifying isomorphic 1-cells.

Alternatively, this is biequivalent to the 2-category of pointed sets and basepoint-preserving maps, with suitably defined (exercise!) 2-cells.

(k) **Span** consists of sets and "spans" $X \leftarrow E \to Y$ in **Set**, with composition by pullback, and with 2-cells given by diagrams such as

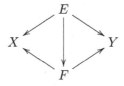

Unlike the previous two bicategories, this one is no longer locally posetal, so to get a biequivalent 2-category we need to do more than just identify isomorphic 1-cells. There are general results asserting that any bicategory is biequivalent to a 2-category, but in fact naturally occurring bicategories tend to be biequivalent to naturally occurring 2-categories. In this case, we can take the 2-category whose objects are sets and whose morphisms are the left adjoints $\mathbf{Set}/X \to \mathbf{Set}/Y$. Here the span

$$X \xleftarrow{u} E \xrightarrow{v} Y$$

is represented by the left adjoint

$$\mathbf{Set}/X \xrightarrow{u^*} \mathbf{Set}/E \xrightarrow{v_!} \mathbf{Set}/Y$$

given by pulling back along u then composing with v.

(l) **Mat** has sets as objects, $X \times Y$-indexed families ("matrices") of sets as morphisms from X to Y, and 2-cells are families of functions. Composition of 1-cells is given by matrix multiplication: if $A = (A_{xy})$ and $B = (B_{yz})$ then

$$(AB)_{xz} = \sum_y A_{xy} \times B_{yz}.$$

This is biequivalent to **Span**, but we'll see below that spans and matrices become different when we start to consider enrichment and internalization. A biequivalent 2-category consists of sets and left adjoints $\mathbf{Set}^X \to \mathbf{Set}^Y$. (Here $X \times Y \to \mathbf{Set}$ can be seen as a functor $X \to \mathbf{Set}^Y$, and so, since $\mathbf{Set}^X$ is the free cocompletion of X, as a left adjoint $\mathbf{Set}^X \to \mathbf{Set}^Y$.) This is really just the same as the construction given for **Span**, since $\mathbf{Set}/X \simeq \mathbf{Set}^X$; once again, though, when we start to enrich or internalize, the two pictures diverge.

(m) **Mod** has rings as objects, left R-, right S-modules as 1-cells $R \nrightarrow S$, and homomorphisms as 2-cells. The composite of modules $R \nrightarrow S$ and $S \nrightarrow T$ is given by tensoring over S. A biequivalent 2-category involves adjunctions $R\text{-}\mathbf{Mod} \rightleftarrows S\text{-}\mathbf{Mod}$.

A ring is the same thing as an **Ab**-category (a category enriched in abelian groups) with only one object. The underlying additive group of the ring is the single hom-object; the multiplication of the ring is the composition. If we identify rings with the corresponding one-object **Ab**-categories, then a module $R \nrightarrow S$ becomes an **Ab**-functor $R \to [S^{\mathrm{op}}, \mathbf{Ab}]$

But there is no reason to restrict ourselves to one-object categories, and there is a bicategory **Ab-Mod** whose objects are **Ab**-categories, and whose 1-cells are **Ab**-modules $\mathscr{A} \nrightarrow \mathscr{B}$; that is, **Ab**-functors $\mathscr{A} \to [\mathscr{B}^{\mathrm{op}}, \mathbf{Ab}]$.

More generally still, we can replace **Ab** by any monoidal category $\mathscr{V}$ with coequalizers which are preserved by tensoring on either side, and there is then a bicategory $\mathscr{V}$-**Mod** of $\mathscr{V}$-categories and $\mathscr{V}$-modules: once again, if $\mathscr{A}$ and $\mathscr{B}$ are $\mathscr{V}$-categories then a $\mathscr{V}$-module $\mathscr{A} \nrightarrow \mathscr{B}$ is a $\mathscr{V}$-functor $\mathscr{A} \to [\mathscr{B}^{\mathrm{op}}, \mathscr{V}]$, or equivalently a left adjoint $[\mathscr{A}^{\mathrm{op}}, \mathscr{V}] \to [\mathscr{B}^{\mathrm{op}}, \mathscr{V}]$, (and this last description gives a 2-category).

There's even, if you really want, a version with a bicategory $\mathscr{W}$ rather than a monoidal category $\mathscr{V}$.

Now let's internalize and enrich the other examples.

(n) If $\mathbb{E}$ is a *regular category*, meaning that any morphism factorizes as a strong epimorphism followed by a monomorphism, and the strong epimorphisms are stable under pullback, then we can form $\mathbf{Rel}(\mathbb{E})$ whose objects are those of $\mathbb{E}$ and whose morphisms $X \nrightarrow Y$ are monomorphisms $R \rightarrowtail X \times Y$. To compose $R : X \nrightarrow Y$ and $S : Y \nrightarrow Z$ we pullback over Y, but the resulting map into $X \times Z$ need not be monic, so we need the factorization system to define composition. It turns out that our assumption that strong epimorphisms are stable under pullback is precisely what is needed for this composition to be associative.

(o) Similarly, if $\mathscr{C}$ is a category and $\mathcal{M}$ is a class of monomorphisms in $\mathscr{C}$, then we can look at $\mathbf{Par}(\mathscr{C}, \mathcal{M})$, defined as above where the given monomorphism is in $\mathcal{M}$. There are conditions on $\mathcal{M}$ you need to make this work well: you want to be able to pullback an $\mathcal{M}$-map by an arbitrary map and obtain an $\mathcal{M}$-map, and you want $\mathcal{M}$ to be closed under composition and to contain the isomorphisms.

(p) If $\mathbb{E}$ has finite limits, we can look at $\mathbf{Span}(\mathbb{E})$ defined in an obvious way. You need the pullbacks for composition to work. You don't need any exactness properties to get a bicategory, but if you want to get a nice biequivalent 2-category, you'll need to start making more assumptions on $\mathbb{E}$. It turns out that $\mathbf{Span}(\mathbb{E})$ plays a crucial role in internal category: we shall see in Example 4 below that an internal category in $\mathbb{E}$ is the same thing as a monad in $\mathbf{Span}(\mathbb{E})$.

(q) $\mathbf{Mat}$, on the other hand, gets enriched rather than internalized. Then $\mathscr{V}$-$\mathbf{Mat}$ has *sets* as objects and $\mathscr{V}$-valued matrices $X \times Y \to \mathscr{V}$ as morphisms. $\mathscr{V}$-$\mathbf{Mat}$ stands in exactly the same relationship to $\mathscr{V}$-categories as $\mathbf{Span}(\mathbb{E})$ does to categories in $\mathscr{E}$. In the case $\mathscr{V} = \mathbf{Set}$ of course $\mathscr{V}$-$\mathbf{Mat}$ is just $\mathbf{Mat}$, but there is also another special case which we have already seen. Let $\mathscr{V}$ be the arrow-category $\mathbf{2}$, consisting of two objects 0 and 1, and a single non-identity arrow $0 \to 1$. This is cartesian closed (a $\mathscr{V}$-category in this case is just a preorder) and $\mathscr{V}$-$\mathbf{Mat}$ in this case is $\mathbf{Rel}$ (we identify a subject of $X \times Y$ with its characteristic function, seen as landing in $\mathbf{2}$).

1.6. Duality. A bicategory $\mathscr{B}$ has not one but three duals:

- $\mathscr{B}^{\mathrm{op}}$ is obtained by reversing the 1-cells
- $\mathscr{B}^{\mathrm{co}}$ is obtained by reversing the 2-cells
- $\mathscr{B}^{\mathrm{coop}}$ is obtained by reversing both

In the case of a monoidal category $\mathscr{V}$, we can form the monoidal category $\mathscr{V}^{\mathrm{op}}$ by reversing the sense of the morphisms; this reverses the 2-cells of the corresponding bicategory $\Sigma\mathscr{V}$, so $\Sigma(\mathscr{V}^{\mathrm{op}}) = (\Sigma\mathscr{V})^{\mathrm{co}}$. Reversing the 1-cells of $\Sigma\mathscr{V}$ corresponds to reversing the tensor of $\mathscr{V}$, denoted $\mathscr{V}^{\mathrm{rev}}$, so $\Sigma(\mathscr{V}^{\mathrm{rev}}) = (\Sigma\mathscr{V})^{\mathrm{op}}$.

1.7. References to the literature. The basic references for bicategories and 2-categories are [4], [17], [30], and [57]. The basic references for enriched categories are [14], [27], and [43]. For a good example of simplicially-enriched category theory that is very close to 2-category theory, see [10]. Both 2-categories and double categories were first defined by Ehresmann (see perhaps [13]); bicategories were first defined by Bénabou [4]. For (a generalization of) the fact that every bicategory is biequivalent to a 2-category, see [44].

Categories enriched in a bicategory were first defined by Walters to deal with the example of sheaves on a space (or site) [63, 64]. A good general reference is [5].

The importance of monoidal functors (not necessarily strong) was observed both by Eilenberg-Kelly [14] and by Bénabou [4].

For doctrinal adjunction see [22].

2. Formal category theory. One point of view is that a 2-category is a generalized category (add 2-cells). Another important one is that an *object of* a 2-category is a generalized category (since **Cat** is the primordial 2-category). This is "formal category theory": think of a 2-category as a collection of category-like things.

You don't capture all of $\mathscr{V}$-category theory by thinking of $\mathscr{V}$-categories as objects of $\mathscr{V}$-**Cat**, just as you don't capture all of group theory by thinking of groups as objects of **Grp**, but many things do work out well when we take this "element-free" approach. In formal category theory you tend to avoid talking about objects of a category, instead talking about morphisms (functors) into the category. Thus morphisms become generalized objects (of their codomain) in exactly the same way that morphisms in categories are generalized elements.

One of the starting points of formal category theory was Street's beautiful work on the "formal theory of monads." This was motivated by the desire to develop a uniform approach to universal algebra for enriched and internal categories. It uses all four dualities to incredible effect.

2.1. Adjunctions and equivalences. We start here with the notion of adjunction in a 2-category (in other words, adjunction between objects of a 2-category — this is not to be confused with adjunctions between 2-categories). In ordinary category theory there are two main ways to say that a functor $f : A \to B$ is left adjoint to $u : B \to A$. First there is the local approach, consisting of a bijection between hom-sets

$$B(fa, b) \cong A(a, ub)$$

for each object $a \in A$ and $b \in B$, natural in both a and b. Alternatively, there is the global approach, involving natural transformations $\eta : 1_A \to uf$ and $\varepsilon : fu \to 1_B$ satisfying the usual triangle equations. Each can be generalized to the 2-categorical setting.

Let $\mathscr{K}$ be a 2-category. Everything I'm going to say works for bicategories, but let's keep things simple; of course you can always replace a bicategory by a biequivalent 2-category anyway.

An *adjunction* in $\mathscr{K}$ consists of 1-cells $f : A \to B$ and $u : B \to A$, and 2-cells $\eta : 1_A \to uf$ and $\varepsilon : fu \to 1_B$ satisfying the triangle equations. This is exactly the global approach to ordinary adjunctions, with functors replaced by 1-cells, and natural transformations by 2-cells. In a lot of 2-categories, this is a good thing to study. We mentioned above the case

MonCat. The study of adjunctions in **Mod** is called *Morita theory*: it involves adjunctions and equivalences between categories of the form R-**Mod** for a ring R.

In the case where η and ε are invertible, we have not just an adjunction but an *adjoint equivalence*.

The local approach to adjunctions also works well here, provided that one uses generalized objects rather than objects. For any 1-cells $a : X \to A$ and $b : X \to B$, there is a bijection between 2-cells $fa \to b$ and 2-cells $a \to ub$. One now has naturality with respect to both 1-cells $x : Y \to X$, and 2-cells $a \to a'$ or $b \to b'$. This local-global correspondence can be proved more or less as in the usual case, or it can be deduced from the usual case using a suitable version of the Yoneda lemma. In fact the global-to-local part follows from the easy fact that *2-functors preserve adjunctions*, so that the representable 2-functors $\mathscr{K}(X, -)$ send the adjunction $f \dashv u$ in $\mathscr{K}$ to an adjunction $\mathscr{K}(X, f) \dashv \mathscr{K}(X, u)$ in **Cat**, between $\mathscr{K}(X, A)$ and $\mathscr{K}(X, B)$, and so the usual properties of adjunctions give the correspondence between $fa = \mathscr{K}(X, f)a \to b$ and $a \to \mathscr{K}(X, u)b = ub$.

The contravariant representable functors

$$\mathscr{K}(-, X) \colon \mathscr{K}^{\mathrm{op}} \to \mathbf{Cat}$$

also preserve adjunctions. This prepares you for:

EXERCISE 2.1. *f is a left adjoint in $\mathscr{K}$ if and only if it is a right adjoint in $\mathscr{K}^{\mathrm{co}}$ if and only if it is a right adjoint in $\mathscr{K}^{\mathrm{op}}$.*

EXERCISE 2.2. *A morphism $f : A \to B$ in a 2-category $\mathscr{K}$ is said to be an equivalence if there exist a morphism $g : B \to A$ and isomorphisms $gf \cong 1_A$ and $fg \cong 1_B$. Show that for any equivalence f these data can be chosen so as to give an adjoint equivalence. Hint: you can keep the same f and g; you'll need to change at most one of the isomorphisms.*

Considering an adjunction $f \dashv u$ in $\mathscr{K}$ as an adjunction in $\mathscr{K}^{\mathrm{op}}$, and using the local approach, we see that to give a 2-cell $s \to tf$ is the same as to give a 2-cell $su \to t$. Even in the case $\mathscr{K} = \mathbf{Cat}$ this is not as well known as it should be.

More generally, given a pair of adjunctions $f \dashv u$ and $f' \dashv u'$, we have bijections between 2-cells $f'a \to bf$, 2-cells $a \to u'bf$, and 2-cells $af' \to u'b$: squares

$$
\begin{array}{ccc}
A & \xrightarrow{\ f\ } & B \\
{\scriptstyle a}\downarrow & \overset{\beta}{\Rightarrow} & \downarrow{\scriptstyle b} \\
A' & \xrightarrow[\ f'\ ]{} & B'
\end{array}
$$

correspond to squares

$$
\begin{array}{ccc}
A & \xleftarrow{\ u\ } & B \\
a \downarrow & \overset{\alpha}{\Rightarrow} & \downarrow b \\
A' & \xleftarrow[u']{} & B'
\end{array}
$$

These pairs of 2-cells are called *mates*. To pass between α and β one pastes with the unit and counit:

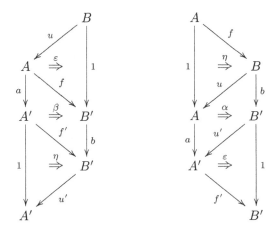

2.2. Extensions. Extensions generalize Kan extensions. They provide limit and colimit notions for *objects of a 2-category*, generalizing the usual notions for categories.

Let $\mathcal{K}$ be a 2-category. What is the universal solution to extending f along j?

Such a universal solution is denoted $\mathrm{lan}_j f$; by universal we mean that it induces a bijection

$$
\frac{f \longrightarrow gj}{\mathrm{lan}_j f \longrightarrow g}
$$

for any $g : B \to C$. When such a $\mathrm{lan}_j f$ exists in $\mathcal{K}$, it is called a *left extension* of f along j.

A colimit is called *absolute* if it is preserved by any functor; similarly we say that the left extension $\mathrm{lan}_j f$ is absolute if composing with any $h : C \to D$ gives another extension, so that $h\,\mathrm{lan}_j f = \mathrm{lan}_j(hf)$.

Consider the case $\mathscr{K} = \mathbf{Cat}$. There would be such a bijection if $\mathrm{lan}_j f$ were the left Kan extension $\mathrm{Lan}_j f$ of f along j, as indeed the notation is supposed to suggest. In the case of (pointwise) left Kan extensions, we have a coend formula

$$(\mathrm{lan}_j f) b = \int^a B(ja, b) \cdot fa.$$

Alternatively the right hand side can be expressed using colimits: given b we can form the comma category j/b, with pairs $(a \in A, ja \to b)$ as objects, and the canonical functor $d : j/b \to A$, then the coend on the right hand side is (canonically isomorphic to) the colimit of $fd : j/b \to C$.

Kan extensions which are not 'pointwise' — in other words, which don't satisfy this formula — can exist if C is not cocomplete, but should be regarded as somewhat pathological.

How might we express this formula so that it makes sense in an arbitrary 2-category? Once again, the answer will involve generalized objects. Staying for a moment in the case of $\mathbf{Cat}$, consider an object $b \in B$ as a morphism $b : 1 \to B$, and then consider the diagram

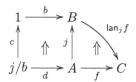

in which j/b is the comma category. The coend $\int^a B(ja, b) \cdot fa$ is isomorphic to the colimit of fd, as we saw, but the colimit of fd is itself isomorphic to the left Kan extension of fd along the unique map $j/b \to 1$. A careful calculation of the isomorphisms involved reveals that the coend formula amounts to the assertion that the diagram above is a left extension.

This motivates the definition of pointwise extension in a general 2-category $\mathscr{K}$ with comma objects. We say that the left extension $\mathrm{lan}_j f$ is *pointwise* if, for any $b : X \to B$, when we form the comma object the 2-cell

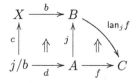

exhibits $(\mathrm{lan}_j f) b$ as $\mathrm{lan}_c(fd)$.

This agrees with the usual definition in the case $\mathscr{K} = \mathbf{Cat}$, works perfectly in the case of $\mathbf{Cat}(\mathbb{E})$, and captures many but not all features in $\mathscr{V}\text{-}\mathbf{Cat}$. The problem is that for $\mathscr{V}$-categories A and B, the $(\mathscr{V}\text{-})$functor category $[A, B]$ should really be regarded as a $\mathscr{V}$-category, but the 2-category

$\mathscr{V}$-**Cat** can't see this extra structure. There are ways around this if B is sufficiently complete or cocomplete.

Let's leave the pointwise aspect aside and go back to extensions.

- A left extension in $\mathscr{K}^{co}$ (reverse the 2-cells) is called a *right extension*.
- A left extension in $\mathscr{K}^{op}$ (reverse the 1-cells) is called a *left lifting*.
- A left extension in $\mathscr{K}^{coop}$ (reverse both) is called a *right lifting*.

The right lifting $r : X \to A$ of $b : X \to B$ through $f : A \to B$ is characterized by a bijection

$$\frac{fa \longrightarrow b}{a \longrightarrow r}$$

which is a sort of internal-hom; indeed, in the one-object case, where the composite fa is given by tensoring, it really is an internal hom. Some people use the notation $r = f \setminus b$ for this lifting.

A special case is adjunctions. Given $f \dashv u : B \to A$, we have a bijection

$$\frac{fa \longrightarrow b}{a \longrightarrow ub}$$

and so $ub = f \setminus b$ is the right lifting of b through f. In particular, u is the right lifting of the identity 1_B through f. Conversely, a right lifting u of the identity through f is a right adjoint if and only if it is absolute; in other words, if ub is the right lifting of b through f for all $b : X \to B$; in symbols $f \setminus b = (f \setminus 1)b$.

Dually, given an adjunction $f \dashv u : B \to A$ we have a bijection

$$\frac{xu \longrightarrow y}{x \longrightarrow yf}$$

and so $yf = \mathrm{ran}_u y$ and $f = \mathrm{ran}_u 1_B$; while in general a right extension $f = \mathrm{ran}_u 1_B$ of the identity is a left adjoint of u if and only if it is absolute.

A bicategory is said to be *closed* if it has right extensions and right liftings. In the one-object case, this means that the endofunctors $- \otimes c$ and $c \otimes -$ of the monoidal category have right adjoints for any object c.

We saw that pointwise left extensions in **Cat** are given by colimits. Thus the existence of left extensions is some kind of internal cocompleteness condition. So in 2-categories like **Cat**($\mathbb{E}$) or $\mathscr{V}$-**Cat** they will exist only in some cases. In bicategories like $\mathscr{V}$-**Mod**, on the other hand, all extensions exist (provided that $\mathscr{V}$ is itself complete and cocomplete).

Let me point out a little lemma which everyone knows for **Cat**, but which is true for 2-categories basically because everything is representable. A morphism $f : A \to B$ in a 2-category $\mathscr{K}$ is said to be *representably fully faithful* if $\mathscr{K}(X, f) : \mathscr{K}(X, A) \to \mathscr{K}(X, B)$ is a fully faithful functor for all objects X of $\mathscr{K}$. For $\mathscr{K} = $ **Cat** this is equivalent to f being fully faithful.

LEMMA 2.1. *Let $f \dashv u$ be an adjunction in a 2-category $\mathscr{K}$ for which the unit $\eta : 1 \to uf$ is invertible. Then f is representably fully faithful.*

Similarly, under the same hypotheses, u will be (representably) "cofully-faithful," in the sense that each $\mathscr{K}(u, X) : \mathscr{K}(B, X) \to \mathscr{K}(A, X)$ is fully faithful.

2.3. Monads. Just as in ordinary category theory, an adjunction $f \dashv u : B \to A$ in a 2-category induces a 1-cell $t = uf$, with 2-cells $\eta : 1 \to uf = t$, given by the unit of the adjunction, and a multiplication $\mu = u\varepsilon f : t^2 = ufuf \to uf = t$, where $\varepsilon : fu \to 1$ is the counit. This η and μ make t into a monoid in the monoidal category $\mathscr{K}(A, A)$.

More generally, a monad in a 2-category $\mathscr{K}$ on an object $A \in \mathscr{K}$ consists of a 1-cell $t : A \to A$ equipped with 2-cells $\eta : 1 \to t$ and $\mu : t^2 \to t$ satisfying the usual (associative and identity) equations; the situation of the previous paragraph is a special case. One often speaks simply of a monad (A, t), when η and μ are understood.

The case $\mathscr{K} = \mathbf{Cat}$ is just the usual notion of monad on a category A. (This is sometimes called a monad *in* A, but this usage is to be avoided: it is *in* $\mathscr{K}$ and *on* A.)

EXAMPLE 1. *Monads in* **Cat** *are the usual monads. Monads in $\mathscr{V}$-**Cat** or **Cat**($\mathbb{E}$) correspond to the obvious notion of enriched or internal monad. Monads in* **MonCat** *are called monoidal monads. Monads in the 2-category* **OpMonCat** *of monoidal categories, opmonoidal functors, and opmonoidal natural transformations are called opmonoidal monads, or sometimes Hopf monads (see [48]).*

EXAMPLE 2. *Monads in the one-object 2-category $\Sigma\mathscr{V}$ are monoids in the strict monoidal category $\mathscr{V}$. Conversely, a monad in an arbitrary 2-category $\mathscr{K}$, on an object X of $\mathscr{K}$, is a monoid in the (strict) monoidal category $\mathscr{K}(X, X)$. There are analogous facts for bicategories and (not necessarily strict) monoidal categories.*

EXAMPLE 3. *Monads in* **Rel**. *We have a set E_0 and a relation $t : E_0 \nrightarrow E_0$, in the form of a subset R of $E_0 \times E_0$; the "identity" $1 \to t$ amounts to the assertion that the relation R is reflexive, and the multiplication to the fact that R is transitive. The associative and unit laws are automatic.*

EXAMPLE 4. *Monads in* **Span**($\mathbb{E}$). *We have an object E_0, a 1-cell $t : E_0 \nrightarrow E_0$, as in*

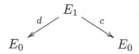

(a directed graph in $\mathbb{E}$), with a multiplication

$$\mu : E_1 \times_{E_0} E_1 \to E_1$$

from the object of composable pairs to the object of morphisms, giving a composite; associativity of the monad multiplication is precisely associativity of the composition. Similarly the unit $1 \xrightarrow{\eta} t$ gives $E_0 \to E_1$ since the identity span is

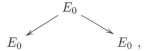

and the unit laws for the monad are precisely the identity laws for the internal category. Thus a monad in **Span**(𝔼) is the same as an internal category in 𝔼 .

This is one of the main reasons for considering the span construction.

EXAMPLE 5. *Monads in $\mathscr{V}$-Mat. We have an object X, which is just a set, a 1-cell $X \rightarrow X$, in the form of a matrix $X \times X \to \mathscr{V}$, which we think of as sending (x, y) to a hom-object $\mathscr{C}(x, y)$. The multiplication map goes from the matrix product, as in*

$$\sum_y \mathscr{C}(y, z) \otimes \mathscr{C}(x, y) \longrightarrow \mathscr{C}(x, z)$$

and gives a composition map. Once again the associative and identity laws for the composition are precisely the associative and unit laws for the monad, and we see that a monad in $\mathscr{V}$-**Mat** is the same as a category enriched in $\mathscr{V}$.

In the special case $\mathscr{V} = 2$ we have $\mathscr{V}$-Mat = Rel, and so we recover the observation, made in Example (q) above, that a category enriched in 2 is just a preorder (a reflexive and transitive relation).

A morphism of monads from (A, t) to (B, s) consists of a 1-cell $f : A \to B$ equipped with a 2-cell $\varphi : sf \to ft$, satisfying two conditions: see [54] or Section 8 below. A morphism of monads in **Span**(𝔼) is *not* an internal functor, since it would involve a 1-cell (a span) $E_0 \rightarrow F_0$ between the objects of objects, rather than a morphism in 𝔼. In order to get internal functors, we need to consider not **Span**(𝔼) itself, but rather **Span**(𝔼) equipped with the class of "special" 1-cells consisting of those spans whose left leg is the identity; these can of course be identified with the 1-cells in 𝔼. An internal functor will turn out to be a monad morphism, for which the span $E_0 \rightarrow F_0$ is "special."

The case of enriched functors is similar: one needs to keep track of which 1-cells in $\mathscr{V}$-**Mat** are really just functions.

To get (enriched or internal) natural transformations, you do not use the obvious notion of monad 2-cells as in [54], but rather those of [40]; once again see Section 8 below.

2.4. References to the literature. For adjunctions in 2-categories and the calculus of mates see [30] or [17]. For monads in 2-categories see the classic [54]. For extensions and liftings see [59]. The idea that categories can be seen as monads in **Span** comes from [4].

3. Morphisms between bicategories.

3.1. Lax morphisms. We have seen the importance of *monoidal functors* between monoidal categories. The corresponding morphisms between bicategories are the *lax functors* (originally just called morphisms of bicategories by Bénabou). A lax functor $\mathscr{A} \to \mathscr{B}$ sends objects $A \in \mathscr{A}$ to objects $FA \in \mathscr{B}$, has functors $F\colon \mathscr{A}(A, B) \to \mathscr{B}(FA, FB)$ (thus preserving 2-cell composition in a strict way), and has comparison maps $\varphi\colon Fg \cdot Ff \to F(gf)$ and $\varphi_0\colon 1_{FA} \to F(1_A)$ and some coherence conditions, which are formally identical to those for monoidal functors.

All the good things that happen for monoidal functors happen for lax functors. For example, monoidal functors take monoids to monoids, and lax functors take monads to monads. (Recall that a monad in $\mathscr{B}$ on an object X is the same as a monoid in the monoidal category $\mathscr{B}(X, X)$.)

As a very special case, consider the terminal 2-category 1. This has a unique object $*$, and a unique monad on $*$ (the identity monad). Then for any lax functor $1 \to \mathscr{B}$, the object $*$ gets sent to $F* = A$, the identity 1 is sent to $F1 = t$, the comparison maps become $\mu\colon tt \to t$ and $\eta\colon 1 \to t$, and the coherence conditions make this precisely a monad. In fact, *monads in $\mathscr{B}$ are the same as lax functors $1 \to \mathscr{B}$*. For Bénabou, this was a key reason to consider lax morphisms of bicategories, rather than the stronger version.

In particular, $\mathscr{V}$-categories are the same as monads in $\mathscr{V}$-**Mat**, and so also the same as lax functors $1 \to \mathscr{V}$-**Mat**. This is the same as a set X together with a lax functor

$$X_{\mathrm{ch}} \longrightarrow \Sigma \mathscr{V}$$

where X_{ch} is X made into a chaotic bicategory (also called indiscrete: every hom-category $X_{\mathrm{ch}}(x, y)$ is trivial). Why? We send each x to $*$, we have a functor

$$1 = X_{\mathrm{ch}}(x, y) \to \Sigma \mathscr{V}(*, *) = \mathscr{V}$$

picking out the hom-object $\mathscr{C}(x, y) \in \mathscr{V}$, and the lax comparison maps φ become the composition and identity maps.

If we replace $\Sigma \mathscr{V}$ by an arbitrary bicategory $\mathscr{W}$, we get the notion of a $\mathscr{W}$-enriched category: a set X with a lax functor

$$X_{\mathrm{ch}} \longrightarrow \mathscr{W}.$$

Another way to think about X_{ch}, as a bicategory, is to say that the unique map $X \to 1$ is fully faithful. But we can also consider, more generally, a pair of bicategories with a partial map

where the hooked arrow $\hookrightarrow$ denotes a fully faithful strict morphism, and the wobbly map $\rightsquigarrow$ denotes a lax functor. This partial map is called a *2-sided enrichment* or a *category enriched from $\mathscr{A}$ to $\mathscr{B}$* . If $\mathscr{A}$ is 1, it's just a category enriched over $\mathscr{B}$.

Using the notion of composition for these things is very helpful in analyzing the change of base between different bicategories. For example, a $\mathscr{B}$-category is a partial map from 1 to $\mathscr{B}$; this can be composed with a partial map from $\mathscr{B}$ to $\mathscr{C}$ to get a $\mathscr{C}$-category. As a (better-known) special case, lax functors from $\mathscr{B}$ to $\mathscr{C}$ send $\mathscr{B}$-categories to $\mathscr{C}$-categories; as a still more special case, monoidal functors from $\mathscr{V}$ to $\mathscr{W}$ send $\mathscr{V}$-categories to $\mathscr{W}$-categories.

3.2. Pseudofunctors and 2-functors. A pseudofunctor (or homomorphism of bicategories) is a lax functor for which φ and φ_0 are invertible.

EXAMPLE 6. *For a bicategory $\mathscr{B}$, the representables*

$$\mathscr{B} \xrightarrow{\mathscr{B}(B,-)} \mathbf{Cat}$$

are pseudofunctors, not strict in general.

EXAMPLE 7 (Indexed categories). *A pseudofunctor $\mathscr{B}^{\mathrm{op}} \to \mathbf{Cat}$ is sometimes called a $\mathscr{B}$-indexed category. Often $\mathscr{B}$ itself will just be a category (no non-identity 2-cells), in which case such a pseudofunctor corresponds to a fibration $\mathscr{E} \to \mathscr{B}$ in the Grothendieck picture.*

An important property of pseudofunctors not shared by lax functors is that they preserve adjunctions. Consider a pseudofunctor $F : \mathscr{A} \to \mathscr{B}$, and an adjunction $f \dashv u : B \to A$ in $\mathscr{A}$, with unit $\eta : 1_A \to uf$ and $\varepsilon : fu \to 1_B$. We may apply F to f and u to get $Ff : FA \to FB$ and $Fu : FB \to FA$, and now the composite 2-cells

$$Ff.Fu \xrightarrow{\varphi} F(fu) \xrightarrow{F\varepsilon} F1_B \xrightarrow{\varphi_0^{-1}} 1_{FB}$$

$$1_{FA} \xrightarrow{\varphi_0} F1_A \xrightarrow{F\eta} F(uf) \xrightarrow{\varphi^{-1}} Fu.Ff$$

provide the unit and counit for an adjunction $Ff \dashv Fu$. This fails for a general lax functor F.

If φ and φ_0 are not just invertible, but in fact identities, then one speaks of a *strict homomorphism*; or, in the case of 2-categories, of a *2-functor*. Note that in the bicategory case the associativity and identity constraints must still be preserved: this is the content of the coherence condition for φ and φ_0.

2-functors are much nicer to work with, but often it is the pseudo-functors which arise in nature. One reason you might prefer 2-functors is so as not to have to worry about coherence. Furthermore, 2-functors have better properties than pseudofunctors: for example, the category **2-Cat** of 2-categories and 2-functors has limits and colimits, but the category **2-Cat**$_{\mathrm{ps}}$ of 2-categories and pseudofunctors does not. For example the diagram

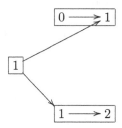

has no pushout: such a pushout would have to have morphisms $0 \to 1 \to 2$ and a composite, but in some other cocone we have no way to decide where to send the composite. If, however, we made **2-Cat**$_{\mathrm{ps}}$ into a tricategory, then it would have trilimits (the relevant "weak" notion of limits for tri-categories).

On the other hand, even if you start in the world of 2-categories and 2-functors, you may be forced out of it. A 2-functor $\mathscr{A} \xrightarrow{F} \mathscr{B}$ is a *biequiv-alence* if $\mathscr{A}(A, B) \to \mathscr{B}(FA, FB)$ are equivalences and it is "bi-essentially surjective," in the sense that for all $X \in \mathscr{B}$, there exists an $A \in \mathscr{A}$ and an equivalence $FA \simeq X$ in $\mathscr{B}$. This is the "right notion" of equivalence for 2-functors.

The point is that you'd like something going back the other way from $\mathscr{B}$ to $\mathscr{A}$. Well you do have *something*, but it's just not a 2-functor in general. Given $X \in \mathscr{B}$, pick $A \in \mathscr{A}$ and $FA \simeq X$ and let $GX = A$. Given $X \xrightarrow{x} Y$, we can bring it across the equivalences $FA \simeq X$ and $FB \simeq Y$ to get $\overline{x} \colon FA \to FB$, and since F is locally an equivalence, $\overline{x} \cong Fa$ for some $a \colon A \to B$; let $Gx = a$. This all works, but since everything is only defined up to isomorphism, there's no way you can possibly hope for G to preserve things strictly.

There is a Quillen model structure on **2-Cat** — see Section 7.5 below — for which the weak equivalences are the biequivalences, and clearly get-ting a 2-functor $\mathscr{B} \to \mathscr{A}$ is going to have something to do with $\mathscr{B}$ being cofibrant.

3.3. Higher structure. As well as lax (and other) morphisms be-tween bicategories, there is higher structure. Given morphisms $F, G : \mathscr{A} \to \mathscr{B}$, one can consider families $\alpha A : FA \to GA$ of morphisms in $\mathscr{B}$ indexed

by the objects of $\mathscr{A}$, and subject to (lax, oplax, pseudo, or strict) naturality conditions. There is even a further level of structure, consisting of
morphisms between such transformations: these are called modifications.

3.4. References to the literature. The importance of lax functors,
especially lax functors with domain 1, was observed by Bénabou [4].

Categories enriched in a bicategory were first defined by Walters to
deal with the example of sheaves (on a space or site) [63, 64]. A good
general reference is [5]. Two-sided enrichments (although not from the
point of view of partial morphisms of bicategories) were defined in [31], in
order to deal with change of base issues.

4. 2-dimensional universal algebra. There are various categorical
approaches to universal algebra: theories, operads, sketches, and others,
but I'll mostly talk about monads, although you may see parallels with
operads and with theories if you know about those.

The ordinary universal algebra picture you might have in mind is
monoids (or groups, rings, etc.) living over sets. But our algebras don't
have to be single-sorted; they could live over some power of sets. Ab-
stractly, of course, we could be living over almost everything. A good
many-sorted example to have in mind is the functor category $[\mathscr{C}, \mathbf{Set}]$ liv-
ing over $[\mathrm{ob}\mathscr{C}, \mathbf{Set}]$, for a small category $\mathscr{C}$. If $\mathscr{C}$ has one object, then we
may identify $\mathscr{C}$ with the monoid M of its arrows, and the functor category
is then the category of M-sets.

When we come to 2-categories, we might generalize monoids over sets
to monoidal categories over categories; or (also living over categories) categories with finite products, or with finite coproducts, or with both, or with
finite products and finite coproducts and a distributive law.

For an example of the many-sorted case, let $\mathscr{B}$ be a small bicategory.
There is a 2-category $\mathbf{Hom}(\mathscr{B}, \mathbf{Cat})$ of homomorphisms from $\mathscr{B}$ to $\mathbf{Cat}$
($\mathscr{B}$-indexed categories), whose morphisms and 2-cells are the pseudonatural
transformations and modifications. The domain of the forgetful 2-functor

is an example of the sort of algebraic structure we have in mind.

In the next two sections there is a lot of interplay between 2-category
theory and $\mathbf{Cat}$-category theory. Since I don't want to assume enriched
category theory, I'll tend to describe the ordinary (unenriched) setting, take
it for granted that one can modify this to get a $\mathbf{Cat}$-enriched version, and
concentrate more on how to modify this to do the proper 2-categorical one.

4.1. 2-monads. We continue to follow the convention that the prefix
2- indicates a strict notion. Thus a 2-monad consists of a 2-category $\mathscr{K}$

equipped with a 2-functor $T : \mathscr{K} \to \mathscr{K}$, and 2-natural transformations $m : T^2 \to T$ and $i : 1 \to T$, satisfying the usual equations for a monad. In other words, this is a monad in the (large) 2-category of 2-categories, 2-functors, and 2-natural transformations. (This could be made into a 3-category, but we don't need to do so for this observation.)

There is a good theory of enriched monads — this was one of the motivations of the formal theory of monads — and 2-monads are just $\mathscr{V}$-monads in the case $\mathscr{V} = \mathbf{Cat}$.

A (strict) T-*algebra* is the usual thing, an object $A \in \mathscr{K}$ with a morphism $a : TA \to A$ satisfying the usual equations, written (A, a). Once again, this is the strict (or $\mathbf{Cat}$-enriched) notion.

REMARK 4.1. There are pseudo and lax notions of monad and of algebra, but they seem to be less important in universal algebra than the strict ones. The main reason for this is that the actual structures one wants to describe using 2-dimensional monads are the strict algebras for strict monads in a fairly straightforward way — an example is given below — whereas identifying the structures of interest with pseudoalgebras is rather more work. A secondary reason is that in reasonable cases a pseudomonad T can be replaced by a strict monad T' whose strict algebras are the pseudoalgebras of T.

It is when we come to the *morphisms* of algebras that we are forced to depart from the strict setting. A *lax T-morphism* $(A, a) \to (B, b)$ is a morphism $f : A \to B$ in $\mathscr{K}$, equipped with a 2-cell

$$
\begin{array}{ccc}
TA & \xrightarrow{\ Tf\ } & TB \\
{\scriptstyle a}\downarrow & \Downarrow \bar{f} & \downarrow{\scriptstyle b} \\
A & \xrightarrow[\ f\ ]{} & B
\end{array}
$$

satisfying two coherence conditions:

$$
\begin{array}{ccc}
T^2A & \xrightarrow{\ T^2f\ } & T^2B \\
{\scriptstyle Ta}\downarrow & \Downarrow T\bar{f} & \downarrow{\scriptstyle Tb} \\
TA & \xrightarrow{\ Tf\ } & TB \\
{\scriptstyle a}\downarrow & \Downarrow \bar{f} & \downarrow{\scriptstyle b} \\
A & \xrightarrow[\ f\ ]{} & B
\end{array}
\quad = \quad
\begin{array}{ccc}
T^2A & \xrightarrow{\ T^2f\ } & T^2B \\
{\scriptstyle mA}\downarrow & & \downarrow{\scriptstyle mB} \\
TA & \xrightarrow{\ Tf\ } & TB \\
{\scriptstyle a}\downarrow & \Downarrow \bar{f} & \downarrow{\scriptstyle b} \\
A & \xrightarrow[\ f\ ]{} & B
\end{array}
$$

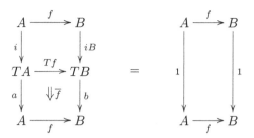

Note that the outer 1-cells are the same (I wouldn't write this equation down if they weren't), and that empty regions commute, and are deemed to contain the relevant identity 2-cell.

Let's do a baby example: $\mathcal{K} = \mathbf{Cat}$ and $TA = \sum_n A^n$ the usual free monoid construction. The T-algebras are strict monoidal categories, and a lax morphism involves 2-cells

$$
\begin{array}{ccc}
\sum_n A^n & \longrightarrow & \sum_n B^n \\
\otimes \downarrow & \Downarrow \overline{f} & \downarrow \otimes \\
A & \longrightarrow & B
\end{array}
$$

so we have transformations

$$ f(a_1) \otimes \ldots \otimes f(a_n) \longrightarrow f(a_1 \otimes \ldots \otimes a_n) $$

for each n. The definition of monoidal functor only mentions the cases $n = 0$ and $n = 2$, but all the others can be built up from these in an obvious way; the coherence conditions for lax T-morphisms say that you *do* build them up in this sensible way, and that the coherence conditions for monoidal functors are satisfied.

So for this T, the lax morphisms are precisely the monoidal functors. This provides a practical motivation for the definition of lax T-morphism. Here's a theoretical one. There's a 2-category $\mathbf{Lax}(\mathbf{2}, \mathcal{K})$ where $\mathbf{2}$ is the arrow category. In detail:

- An object is an arrow $a : A' \to A$ in $\mathcal{K}$
- A 1-cell is a square

$$
\begin{array}{ccc}
A' & \xrightarrow{f'} & B' \\
a \downarrow & \Downarrow \varphi & \downarrow b \\
A & \xrightarrow{f} & B
\end{array}
$$

- A 2-cell $(f, \varphi, f') \to (g, \psi, g')$ consists of 2-cells $\alpha : f \to g$ and $\alpha' : f' \to g'$ satisfying the equation

$$
\begin{array}{ccc}
A' \xrightarrow{f'} B' & = & A' \xrightarrow{f'} B' \\
\Downarrow\alpha' & & \Downarrow\varphi \\
a \downarrow \quad \Downarrow\psi \quad \downarrow b & & a \downarrow \quad f \quad \downarrow b \\
A \xrightarrow{g} B & & A \Downarrow\alpha B \\
& & \qquad g
\end{array}
$$

Since this is functorial in $\mathscr{K}$, the 2-monad T induces a 2-monad $\mathbf{Lax}(2, T)$ on $\mathbf{Lax}(2, \mathscr{K})$. Then a (strict) $\mathbf{Lax}(2, T)$-algebra is precisely a lax T-morphism. The coherence conditions for lax morphisms become the usual axioms for algebras.

Similarly, a T-*transformation* between lax T-morphisms

$$(f, \overline{f}), (g, \overline{g}) \colon (A, a) \to (B, b)$$

is a 2-cell $\rho \colon f \to g$ in $\mathscr{K}$ such that

$$
\begin{array}{ccc}
TA \xrightarrow{\quad} TB & = & TA \xrightarrow{f} TB \\
\Downarrow T\rho & & \\
a \downarrow \quad \Downarrow\overline{g} \quad \downarrow b & & a \downarrow \quad \Downarrow\overline{f} \quad \downarrow b \\
A \xrightarrow{g} B & & A \Downarrow\rho B
\end{array}
$$

In the baby example, for $n = 2$ this says that

$$
\begin{array}{ccc}
fa_1 \otimes fa_2 & \longrightarrow & f(a_1 \otimes a_2) \\
\rho a_1 \otimes \rho a_2 \downarrow & & \downarrow \rho \\
ga_1 \otimes ga_2 & \longrightarrow & g(a_1 \otimes a_2)
\end{array}
$$

which is exactly the condition for $\rho : f \to g$ to be a monoidal natural transformation.

EXERCISE 4.1. *Play the* $\mathbf{Lax}(2, \mathscr{K})$ *game with T-transformations: find a 2-category $\mathscr{K}'$ and a 2-monad T' on $\mathscr{K}'$ whose algebras are the T-transformations.*

There is a 2-category $T\text{-Alg}_\ell$ of T-algebras, lax T-morphisms, and T-transformations, and a forgetful 2-functor

$$T\text{-Alg}_\ell \xrightarrow{U_\ell} \mathscr{K}$$

and in some cases, such as that of monoidal categories, this is the 2-category of primary interest, but often the pseudo case is more important (and of course strong monoidal functors are themselves important). If $\overline{f}$ is invertible, we say that $(f, \overline{f})$ is a *pseudo T-morphism* or just a T-*morphism* (privileging these over the strict or the lax). These are the morphisms of the 2-category T-Alg of T-algebras, pseudo T-morphisms, and T-transformations; it has a forgetful 2-functor

$$T\text{-Alg} \xrightarrow{U} \mathcal{K}.$$

When $\overline{f}$ is an identity we have a *strict* T-morphism. Of course this just means that the square commutes, and we have a morphism in the usual unenriched sense, but it is still useful to think of the identity 2-cell as being "an $\overline{f}$," since it is used in the condition on 2-cells. The T-algebras, strict T-morphisms, and T-transformations form a 2-category T-Alg$_s$ with a 2-functor

$$T\text{-Alg}_s \xrightarrow{U_s} \mathcal{K}.$$

Each of these 2-categories has the same objects, and we have a diagram

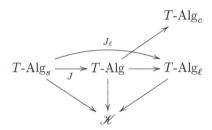

of 2-categories and 2-functors, where T-Alg$_c$ is the 2-category of *colax* morphisms, defined like lax morphisms except that the direction of the 2-cell is reversed. I won't worry too much about them since they can be treated as the lax morphisms for an associated 2-monad on $\mathcal{K}^{co}$.

At this point we need to start making some assumptions. To start with, suppose that $\mathcal{K}$ is cocomplete, and that T has a *rank*, which means that $T \colon \mathcal{K} \to \mathcal{K}$ preserves α-filtered colimits for some α. For ordinary monads on categories, it says that we can describe the structure in terms of operations which may not be finitary, but are at least α-ary for some regular cardinal α. The famous example of a monad on **Set** which is not α-filtered for any α is the covariant power set monad.

Under these conditions

$$T\text{-Alg}_s \xrightarrow{J} T\text{-Alg}$$

$$T\text{-Alg}_s \xrightarrow{J_\ell} T\text{-Alg}_\ell$$

have left adjoints. What does this mean? Among other things it means that for each algebra A there is an algebra A' and bijections

$$\frac{A \rightsquigarrow B}{A' \to B}$$

where the wobbly arrow denotes a weak morphism and the normal arrow a strict one. Here "weak" might mean either pseudo or lax, depending on the context; of course there will be a different A' depending on whether we consider the pseudo or the lax case.

These are 2-adjunctions, so these bijections are just part of isomorphisms of categories

$$T\text{-Alg}_s(A', B) \cong T\text{-Alg}(A, JB)$$

2-natural in A and B. We usually omit writing the J, since it is the identity on objects. We say that such an A' *classifies weak morphisms out of* A. From this we get a unit

$$p \colon A \rightsquigarrow A'$$

and counit

$$q \colon A' \to A$$

and one of the triangle equations tells you that $qp = 1$. An unfortunate consequence of the (otherwise reasonable) notation A', is that the left adjoint to $J : T\text{-Alg}_s \to T\text{-Alg}$ is sometimes saddled with the rather embarrassing name $(\ )'$; I shall call it Q instead.

4.2. Sketch proof of the existence of A'.

Step 1. $T\text{-Alg}_s$ **is cocomplete.**

This part is entirely "strict": it is really an enriched category phenomenon, and not really any harder than the corresponding fact for ordinary categories. It is here that you use the assumptions on $\mathscr{K}$ and T.

Colimits of algebras, as we know, are generally hard. The problem is essentially that algebras are a "quadratic" notion, involving $a : TA \to A$ with two copies of A. We "linearize" and it becomes easy. What does that mean?

Take the T-algebra $(A, a : TA \to A)$, forget the axioms, and also forget that the two A's are the same, so consider it only as a map $a : TA \to A_1$. This defines the objects of a new category, whose morphisms are squares of the form

$$
\begin{array}{ccc}
TA & \xrightarrow{Tf} & TB \\
\downarrow{\scriptstyle a} & & \downarrow{\scriptstyle b} \\
A_1 & \xrightarrow{f_1} & B_1
\end{array}
$$

With the obvious notion of 2-cell this becomes a 2-category; in fact it is just the comma 2-category $T/\mathscr{K}$. The point is that we have a full embedding

$$T\text{-Alg}_s \hookrightarrow T/\mathscr{K}$$

since by the unit condition for algebras, any morphism in $T/\mathscr{K}$ between algebras must have $f = f_1$ and so be a strict T-morphism. It is this $T/\mathscr{K}$ which is the "linearization" of $T\text{-Alg}_s$, and colimits in it are easy. Say we have a diagram of things $TA_i \to B_i$. Take the colimits in $\mathscr{K}$ and take the pushout

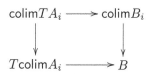

$$
\begin{array}{ccc}
\operatorname{colim} TA_i & \longrightarrow & \operatorname{colim} B_i \\
\downarrow & & \downarrow \\
T\operatorname{colim} A_i & \longrightarrow & B
\end{array}
$$

to get the colimits in $T/\mathscr{K}$.

The hard bit, which I'll leave out, is the construction of a reflection $T/\mathscr{K} \to T\text{-Alg}_s$ (a left adjoint to the inclusion). This is where we use the assumption on T. There are some transfinite calculations, as you might expect given the condition on α-filtered colimits.

Note, however, that should T preserve all colimits, then this Step 1 becomes easy: the colimits are constructed pointwise. In particular, this is true in the case of categories of diagrams ($T\text{-Alg}_s = [\mathscr{B}, \mathbf{Cat}]$).

In fact when we come to step 2, we'll see that only certain (finite) colimits in $T\text{-Alg}_s$ are actually needed, and if T should preserve these colimits, as does sometimes happen, then once again the proof simplifies.

Step 2. Let (A, a) be an algebra; we want to construct the pseudomorphism classifier A' using colimits in $T\text{-Alg}_s$. A lax T-morphism $(A, a) \to (B, b)$ consists of various data in $\mathscr{K}$, and we want to translate all these data into $T\text{-Alg}_s$.

A lax T-morphism $A \to B$ consists of
- A morphism $f : A \to B$ in $\mathscr{K}$; equivalently a morphism $g : TA \to B$ in $T\text{-Alg}_s$, where $g = b \cdot Tf$.
- A 2-cell

$$
\begin{array}{ccc}
TA & \xrightarrow{Tf} & TB \\
a \downarrow & \Downarrow \bar{f} & \downarrow b \\
A & \xrightarrow{f} & B
\end{array}
$$

in $\mathscr{K}$, which becomes a 2-cell

$$
\begin{array}{ccc}
T^2A & \xrightarrow{mA} & TA \\
Ta \downarrow & \Downarrow \zeta & \downarrow g \\
TA & \xrightarrow{g} & B
\end{array}
$$

in T-Alg$_s$, since

$$b.T(fa) = b.Tf.Ta = g.Ta$$
$$b.T(b.Tf) = b.Tb.T^2f = \cdots = g.mA.$$

- The condition $\overline{f}.iA = $ id corresponds to saying that $\zeta.TiA = $ id
- The other condition becomes

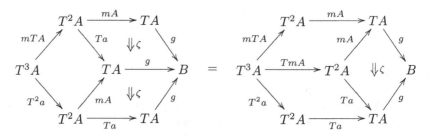

We have a truncated simplicial object:

$$T^3A \underset{\underset{mTA}{\overset{TmA}{\longrightarrow}}}{\overset{T^2a}{\longrightarrow}} T^2A \underset{\overset{mA}{\underset{\longrightarrow}{\longleftarrow}}}{\overset{Ta}{\underset{TiA}{\longrightarrow}}} A \ .$$

We now form a 2-categorical colimit, called the *codescent object*, of this truncated simplicial object, and the result is the desired A'. Alternatively, we can break this up into bite-sized pieces. We first construct the *coinserter* of mA and Ta: this is the universal $p : TA \to A_1$ equipped with a 2-cell $\rho : p.mA \to p.Ta$. To give a map $A_1 \to B$ in T-Alg$_s$ is equivalently to give a map $f : A \to B$ in $\mathscr{K}$ and a 2-cell $\overline{f} : b.Tf \to fa$, without any coherence conditions. To capture the coherence conditions, we have to perform a special sort of quotient, called a *coequifier*, which universally makes equal a parallel pair of 2-cells. We'll talk about 2-categorical limits and colimits later.

If we used the "pseudo" version of weak morphisms, then we'd use a *co-isoinserter* instead of an coinserter, which is the obvious analogue in which ρ is invertible.

4.3. Consequences of the pseudomorphism classifier. Recall that we have

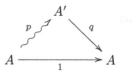

with $qp = 1$. It's also true that $pq \cong 1$, so that this is an equivalence, and thus $A \simeq A'$ in T-Alg, although generally not in T-Alg$_s$. If however q has a section s in T-Alg$_s$, so that $qs = 1$, then $s \cong p$, so $sq \cong 1$, and q is an equivalence in T-Alg$_s$. When q does have such a section, the algebra A is said to be *flexible*.

We'll see in Section 7.3 that there is a model structure on T-Alg$_s$ for which A' is a cofibrant replacement of A. The weak equivalences are the strict morphisms which become equivalences in T-Alg, or equivalently in $\mathscr{K}$; the cofibrant objects are precisely the flexible algebras.

EXERCISE 4.2. *If A is flexible, then any pseudo $A \rightsquigarrow B$ is isomorphic to a strict $A \to B$.*

The equivalence $A \simeq A'$ is a kind of coherence result for morphisms. There are also coherence results for algebras. Consider the composite

$$T\text{-Alg}_s \to T\text{-Alg} \to \text{Ps-}T\text{-Alg}.$$

To give a left adjoint is to construct a pseudo morphism classifier $A' \in T$-Alg$_s$ not just for each strict T-algebra, but also for pseudo-T-algebras. This can still be done; rather than a truncated simplicial object one has a truncated pseudosimplicial object (some of the simplicial identities are satisfied only up to isomorphism), but we can still form the codescent object A' and obtain an isomorphism of categories

$$T\text{-Alg}_s(A', B) \cong \text{Ps-}T\text{-Alg}(A, B)$$

for any strict algebra B, natural in B with respect to strict maps. This time we have a counit $q : B' \to B$ only when B is strict, and a unit $A \rightsquigarrow A'$ for any pseudo algebra A. For a general pseudo algebra A, there seems no way to construct a map from A' back to A, and so no way to show that p is an equivalence. In some cases, however p is an equivalence. In particular it is so if T preserves the relevant codescent objects, since then one can construct the codescent object in $\mathscr{K}$, and get the inverse-equivalence down there. There are various other sufficient conditions for this to work.

The existence of A' for each pseudoalgebra A, along with the fact that the unit $A \rightsquigarrow A'$ is an equivalence is sometimes called the "full coherence result."

There are 2-monads for which not every pseudoalgebra is equivalent to a strict one, but the only examples I know involve horrible 2-categories $\mathscr{K}$. I don't know of an example satisfying the assumptions made in this section ($\mathscr{K}$ cocomplete and T preserving α-filtered colimits).

4.4. References to the literature. 2-monads were first considered in [42]. The basic reference is [7], although many of the key ideas go all the way back to Kelly's [21], including the constructions $\langle A, A \rangle$ and $\{f, f\}$ which allow one to describe algebras in terms of monad morphisms. For the latter, see also [28]. The accessibility issues in [7] were treated in Blackwell's (unpublished) thesis, and later in the monumental (and somewhat impenetrable) [24], which builds on many earlier papers, in particular [2].

Anyway, [7] contains the results about limits and (bi)colimits in T-Alg, biadjoints to algebraic 2-functors, and the left adjoint to T-Alg$_s$ → T-Alg. The proof given here for the existence of this left adjoint follows [34]. See [34] and the references therein for a discussion of proofs that pseudo algebras are equivalent to strict ones. (But I can't omit explicit mention of one those references: the short and beautiful paper [50] of Power.) In [15], analogues to some of the results of [7] are proved in the context of a notion of theory which is rather less expressive than that of 2-monad: see [51] for more on the relationship between monads and theories. Kelly's work [23] on clubs was not at first appreciated, but it has been influential recently in monad-theoretic approaches to higher categories.

5. Presentations for 2-monads. Presentations involve *free* gadgets and *colimits*. Both are defined in terms of a universal property involving maps *out of* the constructed gadget. Why are these important in the case of 2-monads (or monads)? It turns out that one can use colimits to build up 2-monads out of free ones exactly as one builds up algebraic structure using basic operations, derived operations, and equations. Both the colimits and the freeness will involve the world of strict morphisms of monads. Exactly what this world might be is discussed below, but to start with we indicate why (strict) maps out of a given monad are important.

5.1. Endomorphism monads. Let T be a monad on a complete category $\mathcal{K}$. Everything works without change for 2-categories, or indeed for $\mathcal{V}$-categories. For objects $A, B \in \mathcal{K}$, the right Kan extension

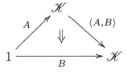

can be computed as

$$\langle A, B \rangle C = \mathcal{K}(C, A) \pitchfork B$$

where $\pitchfork$ means the cotensor, defined by

$$\mathcal{K}(D, X \pitchfork B) \cong \mathbf{Cat}(X, \mathcal{K}(D, B))$$

for a set (or category or object of $\mathcal{V}$, as the case may be) X, and objects B and D of $\mathcal{K}$. The universal property of the right Kan extension implies in particular that we have bijections of natural transformations

$$\frac{T \longrightarrow \langle A, B \rangle}{TA \longrightarrow B}.$$

(This is starting to look like something you might want to do if T is a monad.)

We have natural "composition" and "identity" maps

$$\langle B, C \rangle \langle A, B \rangle \longrightarrow \langle A, C \rangle$$

$$1 \to \langle A, A \rangle$$

which provide $\mathscr{K}$ with an enrichment over $[\mathscr{K}, \mathscr{K}]$ with internal-hom $\langle A, B \rangle$. (Writing down where the composition and identity come from is a good exercise.) Thus $\langle A, A \rangle$ becomes a monoid in $[\mathscr{K}, \mathscr{K}]$; that is, a monad. This can be regarded as the monoid of endomorphisms of A in the $[\mathscr{K}, \mathscr{K}]$-category $\mathscr{K}$.

The important thing about this monad is that the bijection

$$\frac{T \longrightarrow \langle A, A \rangle}{TA \longrightarrow B}$$

restricts to a bijection

$$\frac{T \overset{\text{monad}}{\longrightarrow} \langle A, A \rangle}{TA \overset{\text{alg. str.}}{\longrightarrow} A}$$

between monad maps into $\langle A, A \rangle$ and algebra structures on A.

This is exactly like the endomorphism operad of an object, except that instead of an object of n-ary operations for each $n \in \mathbb{N}$, we have an object "C-ary operations"

$$\langle A, A \rangle C = \mathscr{K}(C, A) \pitchfork A$$

for each object $C \in \mathscr{K}$.

Thus colimits of monads are interesting. For simple example, algebras for $S + T$ (coproduct as monads) are objects with an algebra structure for S and an algebra structure for T, with no particular relationship between the two.

We can play the same game with morphisms. First observe that $\langle A, B \rangle$ is functorial (covariant in B, contravariant in A), and so for any $f : A \to B$ we can form the solid part of

$$
\begin{array}{ccc}
T & \overset{\beta}{\dashrightarrow} & \langle B, B \rangle \\
{\scriptstyle \alpha} \big\downarrow & & \big\downarrow {\scriptstyle \langle f, B \rangle} \\
\langle A, A \rangle & \underset{\langle A, f \rangle}{\longrightarrow} & \langle A, B \rangle
\end{array}
$$

and now if we have monad maps $\alpha : T \to \langle A, A \rangle$ and $\beta : T \to \langle B, B \rangle$, then the square commutes if and only if f is a strict map between the corresponding algebras.

It is at this point that we want to make things 2-categorical, and allow for pseudo or lax morphisms. So suppose that $\mathscr{K}$ is a (complete) 2-category, and that T is a 2-monad on $\mathscr{K}$. To give a 2-cell

$$
\begin{array}{ccc}
TA & \xrightarrow{Tf} & TB \\
a \downarrow & \Downarrow \bar{f} & \downarrow b \\
A & \xrightarrow{f} & B
\end{array}
$$

is equivalently to give a 2-cell

$$
\begin{array}{ccc}
T & \xrightarrow{\beta} & \langle B, B \rangle \\
\alpha \downarrow & \Downarrow \tilde{f} & \downarrow \langle f, B \rangle \\
\langle A, A \rangle & \xrightarrow[\langle A, f \rangle]{} & \langle A, B \rangle.
\end{array}
$$

The *comma object*

$$
\begin{array}{ccc}
\{f, f\}_\ell & \xrightarrow{d} & \langle B, B \rangle \\
c \downarrow & \Downarrow & \downarrow \langle f, B \rangle \\
\langle A, A \rangle & \xrightarrow[\langle A, f \rangle]{} & \langle A, B \rangle
\end{array}
$$

is the universal diagram of this shape, so to give $\tilde{f}$ as above is equivalent to giving a 1-cell $\varphi : T \to \{f, f\}_\ell$ with $d\varphi = \beta$ and $c\varphi = \alpha$.

Now $\{f, f\}_\ell$ becomes a monad: this can be seen via a routine argument using pasting diagrams; or one can get more sophisticated, and show that $\mathbf{Lax}(2, \mathscr{K})$ is enriched over $[\mathscr{K}, \mathscr{K}]$, and now regard $\{f, f\}_\ell$ as the endomorphism monoid. The important thing is that

$$
\begin{array}{c}
\{f, f\}_\ell \xrightarrow{d} \langle B, B \rangle \\
c \downarrow \\
\langle A, A \rangle
\end{array}
$$

are monad maps (although $\langle A, B \rangle$ is not a monad), and that a morphism $T \to \{f, f\}_\ell$ is a monad map if and only if the corresponding $(f, \bar{f})$ is a lax T-morphism.

Of course there is also a pseudo version of this: use the *iso-comma object* $\{f, f\}$ rather than the comma object $\{f, f\}_\ell$; this is the evident analogue in which the 2-cell is required to be invertible.

Thus we can work out the algebras and the (strict, pseudo, or lax) morphisms for a monad just by looking at monad morphisms out of T, and *that* shows why free monads and colimits of monads should be important.

EXERCISE 5.1. *Describe the T-transformations in this way.*

5.2. Pseudomorphisms of monads. In addition to strict monad maps, where the good colimits live, there are also pseudo maps of monads. A *pseudomorphism* of 2-monads on $\mathcal{K}$ is a 2-natural transformation which preserves the multiplication and unit in exactly the sense that strong monoidal functors preserve the tensor product and unit of monoidal categories. Thus there are invertible 2-cells

$$
\begin{array}{ccc}
1 \xrightarrow{\;i\;} T & \xleftarrow{\;m\;} & T^2 \\
 \searrow{\scriptstyle\cong} \;\; \downarrow f & {\scriptstyle\cong} & \downarrow f^2 \\
{\scriptstyle j}\quad T & \xleftarrow{\;n\;} & S
\end{array}
$$

satisfying the same usual coherence conditions.

We have seen that to give an arbitrary map $\alpha : T \to \langle A, A \rangle$ is equivalent to giving $a : TA \to A$ in $\mathcal{K}$, and that α is a strict map of monads if and only if a makes A into a strict algebra; it turns out that to make α into a pseudomorphism of monads

$$
\alpha : T \rightsquigarrow \langle A, A \rangle
$$

is precisely equivalent to making

$$
a : TA \to A
$$

into a pseudoalgebra.

5.3. Locally finitely presentable 2-categories. If the 2-category $\mathcal{K}$ is large, there are all sorts of problems with the 2-category $\mathbf{Mnd}(\mathcal{K})$ of 2-monads on $\mathcal{K}$: its hom-categories are large, it is not cocomplete, and free monads don't exist. We shall therefore pass to a smaller 2-category of 2-monads.

Assume that $\mathcal{K}$ is a locally finitely presentable 2-category. If you know what a locally finitely presentable category is then this is just the obvious 2-categorical analogue. If not, then here are some ways you could think about them:

- The formal definition (which you don't need to know because I'm not going to prove anything): a cocomplete 2-category with a small full subcategory which is a strong generator and consists of finitely presentable objects.
- A 2-category which is complete and cocomplete and in which transfinite arguments are more inclined to work then is usually the case.
- A 2-category of all finite-limit-preserving 2-functors from $\mathscr{C}$ to **Cat**, where $\mathscr{C}$ is a small 2-category with finite limits; you can take $\mathscr{C}$ to be $\mathcal{K}_f^{\mathrm{op}}$ where $\mathcal{K}_f$ is the full subcategory of finitely presentable objects.
- Full reflective sub-2-categories of presheaf 2-categories which are closed under filtered colimits.

- A 2-category which is complete and cocomplete, and is the free cocompletion under filtered colimits of some small 2-category (an Ind-completion). In fact you don't need to suppose both completeness and cocompleteness: for an Ind-completion, either implies the other.

Examples include the presheaf 2-category $[\mathscr{A}, \mathbf{Cat}]$ for any small 2-category $\mathscr{A}$, or $\mathbf{Cat}^X$ for any set X. The 2-category of groupoids is another example.

Once again, this is really an enriched categorical notion: there is a notion of locally finitely presentable $\mathscr{V}$-category, provided that $\mathscr{V}$ itself has a good notion of finitely presentable object: more precisely, provided that $\mathscr{V}$ is a locally finitely presentable category and the full subcategory of finitely presentable objects is closed under the monoidal structure.

Because $\mathscr{K}$ is the free completion of $\mathscr{K}_f$ under filtered colimits, to give an arbitrary 2-functor $\mathscr{K}_f \to \mathscr{K}$ is equivalent to giving a finitary (that is, filtered-colimit-preserving) 2-functor $\mathscr{K} \to \mathscr{K}$. We write $\mathbf{End}_f(\mathscr{K})$ for the monoidal 2-category of finitary endo(-2-)functors on $\mathscr{K}$. Unlike $[\mathscr{K}, \mathscr{K}]$ this is locally small, since $\mathscr{K}_f$ is small.

A 2-monad is said to be finitary if its endo-2-functor part is so. Then the 2-category $\mathbf{Mnd}_f(\mathscr{K})$ of finitary 2-monads on $\mathscr{K}$ is the 2-category of monoids in $\mathbf{End}_f(\mathscr{K})$, and the forgetful 2-functor U from $\mathbf{Mnd}_f(\mathscr{K})$ to $\mathbf{End}_f(\mathscr{K})$ does indeed have a left adjoint, so in this world we do have free monads.

Moreover, the adjunction is monadic; there is a 2-monad on $\mathbf{End}_f(\mathscr{K})$ for which $\mathbf{Mnd}_f(\mathscr{K})$ is the strict algebras and strict morphisms. We can drop down even further to get

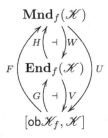

and go back up (along G) by left Kan extension along the inclusion of $\mathrm{ob}\mathscr{K}_f$ in $\mathscr{K}$. The lower adjunction is also monadic, as indeed is the composite, although this does not follow from monadicity of the two other adjunctions.

Thus $\mathbf{Mnd}_f(\mathscr{K})$ is monadic both over $\mathbf{End}_f(\mathscr{K})$ and over $[\mathrm{ob}\mathscr{K}_f, \mathscr{K}]$, and the choice of which base 2-category to work over affects what the pseudomorphisms and pseudoalgebras will be. Dropping down one level, the transformations are 2-natural ones, as in Section 5.2; while if we drop down the whole way, they will be only pseudonatural.

The induced monads on $[\mathrm{ob}\mathscr{K}_f, \mathscr{K}]$ are finitary, and so it follows that $\mathbf{End}_f(\mathscr{K})$ and $\mathbf{Mnd}_f(\mathscr{K})$ are themselves locally finitely presentable, and in particular are complete and cocomplete. In fact slightly more is true, since the inclusion of $\mathbf{Mnd}_f(\mathscr{K})$ in $\mathbf{Mnd}(\mathscr{K})$ has a right adjoint, and so preserves colimits. $\mathbf{Mnd}(\mathscr{K})$ does not have colimits in general, but it does have colimits of finitary monads, and these are finitary. Free monads on arbitrary endo-2-functors may not exist, but free monads on finitary endo-2-functors do, and they are themselves finitary. This is useful since the $\langle A, A \rangle$ are *not* finitary, although we can use the coreflection of $\mathbf{Mnd}(\mathscr{K})$ into $\mathbf{Mnd}_f(\mathscr{K})$ to obtain a finitary analogue.

Everything in this section remains true if you replace "finite" by some regular cardinal α.

5.4. Presentations. The most primitive generator for a 2-monad is an object of $[\mathrm{ob}.\mathscr{K}_f, \mathscr{K}]$: a family $(X_c)_{c \in \mathrm{ob}.\mathscr{K}_f}$ of objects of $\mathscr{K}$, indexed by the objects of $\mathscr{K}_f$. This then generates a free 2-monad FX. What is an FX-algebra? A monad map

$$FX \to \langle A, A \rangle$$

which is the same as

$$X \to U \langle A, A \rangle.$$

This just means that for each c, we have

$$Xc \to \langle A, A \rangle c$$

which unravels to a functor

$$\mathscr{K}(c, A) \to \mathscr{K}(Xc, A)$$

between hom-categories. Since $\mathscr{K}$ is cocomplete, this is the same as a map

$$\sum_c \mathscr{K}(c, A) \cdot Xc \longrightarrow A$$

where $\mathscr{K}(c, A) \cdot Xc$ denotes the tensor of $Xc \in \mathscr{K}$ by the category $\mathscr{K}(c, A)$. Thus we can think of Xc as the "object of all c-ary operations."

EXAMPLE 8. *Let $\mathscr{K} = \mathbf{Cat}$, so $\mathscr{K}_f$ is the finitely presentable categories, and X assigns to every such c a category Xc of c-ary operations. We take*

$$Xc = \begin{cases} 1 & c = 0, 2 \\ 0 & otherwise \end{cases}$$

where 2 is the discrete category $1 + 1$. Thus we have one binary operation and one nullary operation. An FX-algebra is then a category A with maps as above. If Xc is empty, then $\mathscr{K}(Xc, A)$ is terminal, so there's nothing to do. In the other cases, we get maps

$$A^2 \to A$$

when c = 2 and

$$A^0 = 1 \to A$$

when $c = 0$. This is the first step along the path of building up the 2-monad for monoidal categories. The pseudo (or lax) morphisms can be determined using $\{f, f\}$ or $\{f, f\}_\ell$: they will preserve $\otimes$ and I in the pseudo or the lax sense, as the case may be, but without coherence conditions.

EXAMPLE 9. *Again let $\mathscr{K} = $ **Cat**, and let*

$$Xc = \begin{cases} 2 & c = 1 \\ 0 & otherwise. \end{cases}$$

Then an FX algebra is a category with a map

$$A \to A^2$$

in other words, a pair of maps with a natural transformation

$$A \overset{\Downarrow}{\underset{}{\rightrightarrows}} A \,.$$

This is an example in which Xc is not discrete. We say that X specifies a "basic operation of arity 1 (unary) and type arrow."

In the case of monoidal categories, you can use operations of "type arrow" to provide the associativity and unit isomorphisms, but I'll take a different approach.

5.5. Monoidal categories. Actually, let's forget about the units, just worry about the binary operation. Then Xc is 1 if $c = 2$ and 0 otherwise, so an FX-algebra is a category with a single binary operation. Then we have a (non-commutative) diagram of 2-categories and 2-functors

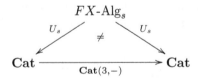

which act on an FX-algebra $(C, \otimes)$ by

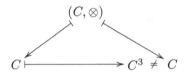

and now we have the two maps

$$C^3 \underset{\otimes(1\otimes)}{\overset{\otimes(\otimes1)}{\rightrightarrows}} C$$

which are natural in $(C, \otimes)$, and so induce two natural transformations $\mathbf{Cat}(3, -)U_s \to U_s$ in the previous triangle. We can take their mates under the adjunction $F_s \dashv U_s$ to get 2-cells in

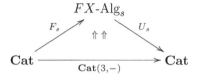

with two 2-cells in the middle. Note that $U_s F_s = FX$ is the monad, so that we have two natural transformations

$$\mathbf{Cat}(3, -) \rightrightarrows FX,$$

which are morphisms of endofunctors. We can now construct the free 2-monad $H\mathbf{Cat}(3, -)$ on $\mathbf{Cat}(3, -)$ and the induced monad morphisms

$$\kappa_1, \kappa_2 \colon H\mathbf{Cat}(3, -) \rightrightarrows FX.$$

Consider now an FX-algebra $(C, \otimes)$, and the corresponding monad map $\gamma : FX \to \langle C, C \rangle$. Then $(C, \otimes)$ is strictly associative if and only if $\gamma\kappa_1 = \gamma\kappa_2$, while to give an isomorphism $\otimes(1\otimes) \cong \otimes(\otimes1)$ is equivalent to giving an isomorphism $\gamma\kappa_1 \cong \gamma\kappa_2$. In the 2-category $\mathbf{Mnd}_f(\mathscr{K})$ construct the universal map $\rho : FX \to S$ equipped with an isomorphism $\rho\kappa_1 \cong \rho\kappa_2$: this is called a *co-iso-inserter*, and it's a (completely strict) 2-categorical colimit, which we'll meet later on.

Now, an S-algebra is a category C equipped with a monad map $S \to \langle C, C \rangle$, or equivalently a monad map $\rho : FX \to \langle C, C \rangle$ and an isomorphism $\rho\kappa_1 \cong \rho\kappa_2$, or equivalently with a functor $\otimes : C^2 \to C$ and a natural isomorphism $\alpha : \otimes(1\otimes) \cong \otimes(\otimes1)$. You can also write down what it means to be a pseudo or lax morphism of such algebras, and it's what you want it to be: the tensor-preserving isomorphisms must be compatible with the associativity constraints.

The coherence condition states that a pair of 2-cells

$$C^4 \underset{\otimes(1\otimes)(11\otimes)}{\overset{\otimes(\otimes1)(\otimes11)}{\rightrightarrows}} C$$

built up using α are equal. Much as before, these are all natural in $(C, \otimes, \alpha)$, so $\otimes(\otimes1)(\otimes11)$ corresponds to one monad map $\lambda_1 : H\mathbf{Cat}(4, -) \to S$, and $\otimes(1\otimes)(11\otimes)$ to another $\lambda_2 : H\mathbf{Cat}(4, -) \to S$. Furthermore the two isomorphisms $\otimes(\otimes 1)(\otimes 11) \cong \otimes(1\otimes)(11\otimes)$ correspond to invertible monad 2-cells

$$H\mathbf{Cat}(4, -) \underset{\lambda_2}{\overset{\lambda_1}{\rightleftarrows}} \,\,{}_{\Lambda_1\Downarrow \,\Downarrow \Lambda_2}\,\, S.$$

Now we form the *coequifier* $q\colon S \to T$, in the category of monads, of these two 2-cells Λ_1 and Λ_2: this is the universal map q with the property that $q\Lambda_1 = q\Lambda_2$.

Then the 2-category T-Alg is the 2-category of "semigroupoidal categories" and strong morphisms (we can get the strict and lax morphisms in the obvious way too). All this follows from the universal property of the monad T.

Often, as here, we build up structure in a particular order, starting with the operations of type object, then those of type arrow or isomorphism, and finally impose equations on these arrows or isomorphisms.

5.6. Terminal objects. Consider the structure of *category with terminal object*. This is a baby example, but you can do any limits you like once you understand this example.

How do you say algebraically that a category A has a terminal object? You give an object

$$1 \xrightarrow{\ t\ } A$$

with a natural transformation

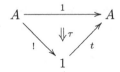

such that the component

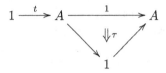

of τ at t is the identity. This last condition plus naturality of τ guarantees that $\tau a : a \to t$ is the only map from a to t, and so that t is terminal.

Let's give a presentation for it. First we have the nullary operation t, which takes the form

$$\mathbf{Cat}(0, A) \to \mathbf{Cat}(1, A)$$

or equivalently

$$\mathbf{Cat}(0, A) \cdot X0 \to A$$

where $X0 = 1$, or equivalently

$$\sum_c \mathbf{Cat}(c, A) \cdot Xc \to A$$

where now Xc is 0 unless $c = 0$. Thus an object A with nullary operation $t : 1 \to A$ is precisely an FX-algebra, where

$$Xc = \begin{cases} 1 & \text{if } c = 0 \\ 0 & \text{otherwise.} \end{cases}$$

For any FX-algebra (A, t), there are two canonical maps from $A \to A$, given by

and these are clearly natural in (A, t); in other words, they define a pair of natural transformations from $U_s : FX\text{-Alg}_s \to \mathbf{Cat}$ to itself. Taking mates under the adjunction $F_s \dashv U_s$ gives a pair of natural transformations $1_{\mathbf{Cat}} \to U_s F_s$. Now $U_s F_s$ is just FX, so forming the free monad $H1$ on the identity $1_{\mathbf{Cat}}$, we get a pair of monad maps $\kappa_1, \kappa_2 : H1 \to FX$. We now form the *coinserter* $\rho : FX \to S$ of κ_1 and κ_2. This is another 2-categorical colimit; it is the universal ρ equipped with a 2-cell $\rho\kappa_1 \to \rho\kappa_2$. An S-algebra is now an A equipped with an object $t : 1 \to A$, and a natural transformation $\tau : 1_A \to t{\circ}!$, as in our earlier description of terminal objects. Finally one can construct a suitable coequifier $q : S \to T$ to obtain the 2-monad T for categories with terminal objects.

Here's a different presentation: it starts as before by putting in a nullary operation

$$\mathbf{Cat}(0, A) \xrightarrow{\;t\;} \mathbf{Cat}(1, A)$$

but then adds a unary operation of type arrow:

$$\mathbf{Cat}(1, A) \xrightarrow{\;\tau\;} \mathbf{Cat}(2, A)$$

which specifies two endomorphisms of A and a natural transformation between them:

$$A \underset{g}{\overset{f}{\Rightarrow}}{}_{\tau} A$$

Later we'll introduce equations $f = 1$, $g = t\circ!$, and $\tau t = \mathrm{id}$.

To specify t and τ, define

$$
Xc = \begin{cases} 1 & \text{if } c = 0 \\ 2 & \text{if } c = 1 \\ 0 & \text{otherwise} \end{cases}
$$

so that FX-algebra structure an a category A amounts to

$$
\sum_c \mathbf{Cat}(c, A) \cdot Xc \to A
$$

or equivalently $t : 1 \to A$ and $\tau : f \to g : A \to A$.

Now we turn to the equations. Consider the (non-commuting) diagram

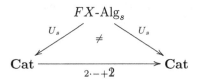

which acts on an FX-algebra (A, t, τ) by

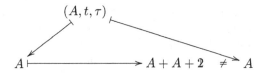

There is a map $\alpha_{(A,t,\tau)} : A + A + 2 \to A$ whose components are $f : A \to A$, $g : A \to A$, and the functor $2 \to A$ corresponding to $\tau \circ t$. This is natural in (A, t, τ).

There is another map $\beta_{(A,t,\tau)} : A + A + 2 \to A$ whose components are $1 : A \to A$, $t\circ! : A \to A$, and the functor $2 \to A$ corresponding to the identity natural transformation on t. Once again this is natural in (A, t, τ).

A category with terminal object is precisely an FX-algebra (A, t, τ) for which $\alpha_{(A,t,\tau)} = \beta_{(A,t,\tau)}$.

Now α and β live in the diagram

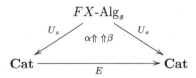

where $EC = C + C + 2$, and we can take their mates under the adjunction $F_s \dashv U_s$ to obtain natural transformations

$$
\alpha', \beta' : E \to U_s F_s
$$

and now $U_s F_s$ is the monad FX, so there are induced monad maps

$$\overline{\alpha}, \overline{\beta} : HE \to FX$$

from the free monad HE on E, and the required 2-monad T for categories-with-terminal object is obtained as the coequalizer

$$HE \underset{\overline{\beta}}{\overset{\overline{\alpha}}{\rightrightarrows}} FX \overset{q}{\longrightarrow} T.$$

REMARK 5.1. Whichever approach we take, the algebras will be the categories with a *chosen* terminal object. This may seem strange, but is not really a problem. The strict morphisms preserve the chosen terminal object strictly, which is probably not what we really want, but the pseudo morphisms preserve it in the usual sense.

5.7. Bicategories. There are two reasons for including this example: first of all it's a fairly easy case with $\mathcal{K} \neq \mathbf{Cat}$, and second it's important for 2-nerves. I won't give all the details.

Let $\mathcal{K} = \mathbf{Cat}$-Grph, the 2-category of category-enriched graphs. A $\mathbf{Cat}$-graph consists of objects $G, H, \ldots$, and hom-categories like $\mathscr{G}(G, H)$. (Of course one could do this for any $\mathscr{V}$ in place of $\mathbf{Cat}$.) A morphism is a function $G \mapsto FG$ on objects, along with functors $\mathscr{G}(G, H) \to \mathscr{H}(FG, FH)$ between hom-categories. One might hope that the 2-cells would be some sort of natural transformations, but since $\mathbf{Cat}$-graphs have no composition law, there is no way to assert that a square in a $\mathbf{Cat}$-graph commutes, and so no way to state naturality. Instead, we use a special sort of lax naturality. We only allow 2-cells

$$\mathscr{G} \overset{F}{\underset{F'}{\Downarrow}} \mathscr{H}$$

to exist when F and F' agree on objects, and then the 2-cell consists of natural transformations

$$\mathscr{G}(G, H) \overset{F}{\underset{F'}{\Downarrow}} \mathscr{H}(FG, FH)$$

on all hom-categories.

Now, given a $\mathbf{Cat}$-graph, what do you need to do to turn it into a bicategory? To start with, you have to give compositions

$$\mathscr{G}(H, K) \times \mathscr{G}(G, H) \longrightarrow \mathscr{G}(G, K).$$

Let comp and arr be the **Cat**-graphs $\cdot \to \cdot \to \cdot$ and $\cdot \to \cdot$ (no 2-cells). Then

$$\mathcal{K}(\text{comp}, \mathcal{G}) = \sum_{G,H,K} \mathcal{G}(H,K) \times \mathcal{G}(G,H)$$

$$\mathcal{K}(\text{arr}, \mathcal{G}) = \sum_{G,K} \mathcal{G}(G,K)$$

so if we define

$$Xc = \begin{cases} \text{arr} & \text{if } c = \text{comp} \\ 0 & \text{otherwise} \end{cases}$$

then an FX-algebra structure on $\mathcal{G}$ amounts to a map

$$\sum_c \mathcal{K}(c, \mathcal{G}) \cdot Xc \to \mathcal{G}$$

and so to a map

$$M : \sum_{G,H,K} \mathcal{G}(H,K) \times \mathcal{G}(G,H) \to \sum_{G,K} \mathcal{G}(G,K).$$

We need to make sure that the restriction

$$M_{G,H,K} : \mathcal{G}(H,K) \times \mathcal{G}(G,H) \to \sum_{G,K} \mathcal{G}(G,K)$$

to the (G,H,K)-component lands in the (G,K)-component: this can be done by constructing a quotient of FX.

Define

$$Yc = \begin{cases} \text{ob} & \text{if } c = \text{comp} \\ 0 & \text{otherwise} \end{cases}$$

where ob denotes the **Cat**-graph $\cdot$ (no 1-cells or 2-cells). An FY-algebra structure on a **Cat**-graph $\mathcal{G}$ is a map

$$\sum_{G,H,K} \mathcal{G}(H,K) \times \mathcal{G}(G,H) \to \sum_G 1$$

where 1 denotes the terminal category.

Suppose now that $(\mathcal{G}, M)$ is an FX-algebra. There are many induced FY-algebra structures on $\mathcal{G}$; in particular, there are the following two:

$$\sum_{G,H,K} \mathcal{G}(H,K) \times \mathcal{G}(G,H) \xrightarrow{M} \sum_G \sum_K \mathcal{G}(G,K) \xrightarrow{\sum_G !} \sum_G 1$$

$$\sum_{G,H,K} \mathcal{G}(H,K) \times \mathcal{G}(G,H) \xrightarrow{\quad ! \quad} 1 \xrightarrow{\text{inj}_G} \sum_G 1.$$

Each is functorial, and so each induces a monad map $FY \to FX$; we form their coequalizer $q_1 : FX \to S_1$, and now an S_1-algebra is a **Cat**-graph $\mathscr{G}$ equipped with a composition M such that $M_{G,H,K}$ lands in $\sum_K \mathscr{G}(G, K)$. A further quotient forces $M_{G,H,K}$ to land in $\mathscr{G}(G, K)$ as desired.

One now introduces an associativity isomorphism. This has the form of a map

$$\mathscr{K}(\mathsf{triple}, \mathscr{G}) \to \mathscr{K}(\mathsf{iso}, \mathscr{G})$$

where triple is the **Cat**-graph $\cdot \to \cdot \to \cdot \to \cdot$ and iso is $\cdot \overset{\cong}{\underset{}{\rightleftarrows}} \cdot$. There are also left and right identity isomorphisms, and various coherence conditions to be encoded, but I'll leave all that as an exercise. The result of the exercise is:

- An algebra is a bicategory.
- A lax morphism is a lax functor.
- A pseudo morphism is a pseudo functor.
- A strict morphism is a strict functor.
- A 2-cell is an *icon*. This is an oplax natural transformation (which we haven't officially met yet) for which the 1-cell components are identities. ICON stands for "Identity Component Oplax Natural-transformation." An icon $F \to G$ can exist only if F and G agree on objects, in which case it consists of a 2-cell

$$
\begin{array}{ccc}
FA & =\!=\!= & GA \\
{\scriptstyle Ff}\downarrow & \Rightarrow & \downarrow{\scriptstyle Gf} \\
FB & =\!=\!= & GB
\end{array}
$$

for each $f : A \to B$ in $\mathscr{G}$, subject to conditions expressing compatibility with respect to composition of 1-cells and identities, and naturality in f with respect to 2-cells. In the case of one-object bicategories these are precisely the monoidal natural transformations.

These icons are just nice enough to give us a 2-category of bicategories. In general, lax natural transformations between lax functors can't even be whiskered by lax functors — the composite

isn't well-defined. In the pseudo case it is defined, but not associative, and so we are led into the world of tricategories. But with just icons, we do get a 2-category, which is moreover the category of algebras for the 2-monad just described.

For example, in this 2-category, it's true that every bicategory is equivalent (in the 2-category) to a 2-category; this works because in replacing a bicategory by a biequivalent 2-category you don't have to change the objects of the bicategory. The 2-category **NHom** of bicategories, normal homomorphisms, and icons, is a full sub-2-category of the the 2-category $[\Delta^{\mathrm{op}}, \mathbf{Cat}]$ of simplicial objects in **Cat**, via a "2-nerve" construction. In order to deal with normal homomorphisms (which preserve identities strictly) rather than general ones, it's convenient to start with *reflexive* **Cat**-graphs rather than **Cat**-graphs.

The choice of direction of the 2-cell in lax transformations and oplax transformations goes back to Bénabou. It seems that the oplax transformations are generally more important than the lax ones.

5.8. Cartesian closed categories. The comments in this section apply equally to monoidal closed categories, symmetric monoidal closed categories, and toposes.

There is no problem constructing a monad for categories with finite products, similarly to the constructions given above. When we come to the closed structure, however, things are not so straightforward. The internal hom is a functor

$$A^{\mathrm{op}} \times A \to A$$

and we're not allowed to talk about A^{op} the way we're doing things: our operations are supposed to be of the form $A^c \to A$. How can we deal with this?

In fact, it's a theorem that cartesian closed categories *don't* have the form T-Alg for a 2-monad T on **Cat**. What you can do, however, is change the base 2-category $\mathscr{K}$ to the 2-category $\mathbf{Cat}_g$ of categories, functors, and natural *isomorphisms*. Recall that $\mathbf{Cat}(2, A)$ is just $A \times A$, but in $\mathbf{Cat}_g(2, A)$ we have only $A_{\mathrm{iso}} \times A_{\mathrm{iso}}$, where A_{iso} is the subcategory of A consisting of the isomorphisms. The internal-hom *does* give us a functor

$$A_{\mathrm{iso}} \times A_{\mathrm{iso}} \longrightarrow A_{\mathrm{iso}}$$

$$(a, b) \longmapsto [a, b]$$

which has the form

$$\mathbf{Cat}_g(2, A) \longrightarrow \mathbf{Cat}_g(1, A)$$

since we can turn around an isomorphism in the first variable to make everything covariant. This gives a new problem; the product is now only given as a functor $A_{\mathrm{iso}} \times A_{\mathrm{iso}} \to A_{\mathrm{iso}}$, we have to put in the rest of the functoriality separately "by hand," using an operation

$$\mathbf{Cat}_g(2 + 2, A) \longrightarrow \mathbf{Cat}_g(2, A)$$

$$(f : a \to a', g : b \to b') \longmapsto (f \times g : a \times b \to a' \times b')$$

subject to various equations. You also have to relate the product to the internal hom.

Any 2-monad on **Cat** induces monads on $\mathbf{Cat}_g$ and on the 1-category $\mathbf{Cat}_0$ (since things are stable under change of base monoidal category: categories to groupoids to sets). But at each stage, to present the same structure becomes harder. In the groupoid enriched stage we can still talk about pseudomorphisms, although at this stage every lax morphism is pseudo; by the time we get to the **Set**-enriched stage there is no longer any genuine pseudo notion at all — everything is strict.

5.9. Diagram 2-categories. The first version of this is not really an example of a presentation at all, since the 2-monad pops out for free. Let $\mathscr{C}$ be a small 2-category, and consider the 2-category $[\mathscr{C}, \mathbf{Cat}]$ of (strict) 2-functors, 2-natural transformations, and modifications. This is the **Cat**-enriched functor category. The forgetful 2-functor has both adjoints

$$
\begin{array}{c}
[\mathscr{C}, \mathbf{Cat}] \\
\left(\nearrow \;\; \downarrow \;\; \nwarrow \right) \\
\dashv \quad \dashv \\
[\mathrm{ob}\mathscr{C}, \mathbf{Cat}]
\end{array}
$$

given by left and right Kan extension. The existence of the right adjoint tells us that the forgetful functor preserves all colimits. In this case U_s is strictly monadic as is easily proved using the enriched version of Beck's theorem. The induced monad T then preserves *all* colimits, and we can write, using the Kan extension formula,

$$
(TX)c = \sum_d \mathscr{C}(d, c) \cdot Xd.
$$

It's now a long, but essentially routine, exercise to check that
- pseudo T-algebras are pseudo-functors,
- lax algebras are lax functors,
- pseudo morphisms are pseudo-natural transformations,

and so on. When you write down the coherence conditions for a lax morphism it will tell you more than is in the *definition* of a lax functor: it will also include a whole lot of consequences of the definition. Notice, by the way, that this is a 2-monad for which every pseudoalgebra is equivalent to a strict one, and so every pseudofunctor from $\mathscr{C}$ to **Cat** is equivalent to a strict one.

Now let $\mathscr{C}$ be a bicategory. If we tried the same game, we wouldn't get a 2-monad, since the associativity of the multiplication for the monad corresponds to the associativity of composition in $\mathscr{C}$, so we'd just get a pseudo-monad. We could just go ahead and do this, but we've been avoiding pseudo-monads, and there is an alternative. One can give a presentation for a 2-monad T on $[\mathrm{ob}\mathscr{C}, \mathbf{Cat}]$ whose

- (strict) algebras are pseudofunctors $\mathscr{C} \to \mathbf{Cat}$,
- pseudomorphisms of algebras are pseudonatural transformations,

and so on. You start with a family $(X_c)_{c \in \mathrm{ob}\mathscr{C}}$, then introduce operations

$$\mathscr{C}(c,d) \times X_c \to X_d$$

and so on. The target doesn't really need to be $\mathbf{Cat}$, although it would need to be cocomplete.

5.10. References to the literature. For locally finitely presentable categories see [16] or [1]; and for the enriched version see [25]. A more formal approach to presentations for 2-monads can be found in [29], which in turn builds upon ideas of [12] See [32] for the monadicity of $\mathbf{Mnd}_f(\mathscr{K})$ over $[\mathrm{ob}\mathscr{K}_f, \mathscr{K}]$.

6. Limits. We'll begin with some concrete examples of limits, looking in particular at limits in T-Alg, for a finitary 2-monad T on a complete and cocomplete 2-category $\mathscr{K}$ (you could get by with much less for most of this). Recall that T-Alg is the 2-category of strict algebras and pseudomorphisms. A good example to bear in mind would have $\mathscr{K} = \mathbf{Cat}$, and T-Alg the 2-category of categories with chosen limits of some particular type, and functors which preserve these limits in the usual, up-to-isomorphism, sense.

6.1. Terminal objects. Let's start with something really easy: terminal objects. Let 1 be terminal in $\mathscr{K}$; we have a unique map $T1 \to 1$, making 1 a T-algebra, and then for any T-algebra $(A, a : TA \to A)$ we have a unique $! : A \to 1$, and

commutes strictly, so there's a unique *strict* algebra morphism $A \to 1$. Moreover, by the 2-universal property of 1, there's a unique isomorphism in the above square, which happens to be an identity; thus there is only one pseudo morphism as well (which happens to be strict). A similar argument works for endomorphisms of this morphism; thus

$$T\text{-Alg}((A,a),(1,!)) \cong 1$$

so $(1,!)$ is a terminal object in T-Alg.

6.2. Products. Similarly for products: given T-algebras (A,a) and (B,b), and a product $A \times B$ in $\mathscr{K}$, there is an obvious map $\langle a, b \rangle$ as in

$$T(A \times B) \to TA \times TB \to A \times B$$

which makes $A \times B$ into a T-algebra (exactly as for ordinary monads: nothing 2-categorical going on here). The point is that if we have *pseudo* morphisms

$$
\begin{array}{ccc}
TC & \longrightarrow & TA \\
c \downarrow & \cong & \downarrow a \\
C & \longrightarrow & A
\end{array}
\qquad\qquad
\begin{array}{ccc}
TC & \longrightarrow & TB \\
c \downarrow & \cong & \downarrow b \\
C & \longrightarrow & B
\end{array}
$$

we get a unique induced pseudo morphism

$$
\begin{array}{ccc}
TC & \longrightarrow & T(A \times B) \\
c \downarrow & \cong & \downarrow \langle a,b \rangle \\
C & \longrightarrow & A \times B
\end{array}
$$

and indeed there is a natural isomorphism (of categories)

$$
T\text{-Alg}(C, A \times B) \cong T\text{-Alg}(C, A) \times T\text{-Alg}(C, B).
$$

Thus $A \times B$ is a product in T-Alg in the strict **Cat**-enriched sense.

Note that the projections $A \times B \to A$ and $A \times B \to B$ are actually strict maps, by construction. Moreover, they jointly "detect strictness": a map into $A \times B$ is strict if and only if its composites into A and B are strict. This is a useful technical property.

Actually, we didn't really need to check anything, since we've already seen that $T\text{-Alg}_s \hookrightarrow T\text{-Alg}$ has a left adjoint, hence preserves all limits, and in the case of terminal objects and products the diagram of which we are taking the limit consists only of objects, so already exists in the strict world. (On the other hand, the explicit argument works for any 2-monad on any 2-category with the relevant products, whereas the adjunction needs a transfinite argument, and much stronger assumptions on T and $\mathscr{K}$.)

6.3. Equalizers. Now let's look at equalizers. Here it's different, because the morphisms whose equalizer we seek may not be strict. If they *are*, then the equalizer exists in $T\text{-Alg}_s$ and is preserved, but if they aren't, the adjunction doesn't help. In fact, in general equalizers of pseudo morphisms need *not* exist.

For example, let T be the 2-monad on **Cat** for categories with a terminal object. Let 1 be the terminal category and let $\mathscr{I}$ be the free-living isomorphism, consisting of two objects and a single isomorphism between them. Clearly both categories have a terminal object, and both inclusions are pseudo morphisms. But any functor which equalizes them has to have empty domain, and no category with an empty domain has a terminal object.

Thus $T\text{-Alg}$ is not complete, but we can look at some of the limits that it does have.

6.4. Equifiers. Consider a parallel pair of 1-cells in T-Alg with a parallel pair of 2-cells between them:

$$A \underset{g}{\overset{f}{\rightrightarrows}} B \qquad \alpha\Downarrow \ \Downarrow\beta$$

The *equifier* of these 2-cells, is the universal 1-cell $k : C \to A$ with $\alpha k = \beta k$. Here universality means that $\mathscr{K}(D, C)$ is *isomorphic* (not just equivalent) to the category of morphisms $D \xrightarrow{h} A$ with $\alpha h = \beta h$. Equifiers do lift from $\mathscr{K}$ to T-Alg: if (A, a) and (B, b) are T-algebras, $(f, \bar{f})$ and $(g, \bar{g})$ are T-morphisms, and α and β are T-transformations, then the composites

$$TC \xrightarrow{Tk} TA \underset{Tg}{\overset{Tf}{\rightrightarrows}} TB \xrightarrow{b} B \ \Downarrow T\alpha \quad = \quad TC \xrightarrow{Tk} TA \underset{Tg}{\overset{Tf}{\rightrightarrows}} TB \xrightarrow{b} B \ \Downarrow T\beta$$

are equal. Paste the isomorphism $\bar{g} : b.Tg \cong ga$ on the bottom of each side and the isomorphism $\bar{f} : fa \cong b.Tf$ on the top, and use the T-transformation condition for α and β to get the equation

$$TC \xrightarrow{Tk} TA \xrightarrow{a} A \underset{g}{\overset{f}{\rightrightarrows}} B \ \Downarrow\alpha \quad = \quad TC \xrightarrow{Tk} TA \xrightarrow{a} A \underset{g}{\overset{f}{\rightrightarrows}} B \ \Downarrow\beta$$

and now by the universal property of the equifier C there is a unique $c : TC \to C$ satisfying $kc = a.Tk$. Two applications of the universal property show that c makes C into a T-algebra, and so clearly k becomes a strict T-morphism $(C, c) \to (A, a)$. Further judicious use of the universal property shows that $k : (C, c) \to (A, a)$ is indeed the equifier in T-Alg.

Observe that once again, the projection map k of the limit is actually a strict map, and detects strictness of incoming maps.

Why does the analogous argument for equalizers fail? Given pseudo morphisms $(f, \bar{f})$ and $(g, \bar{g})$ from (A, a) to (B, b), we could form the equalizer $k : C \to A$ of f and g, and then hope to make C into a T-algebra using the universal property of C, but we'd need to show that $fa.Tk = ga.Tk$. All we actually know is that $c.Tf.Tk = c.Tg.Tk$ while $c.Tf.Tk \cong fa.Tk$ and $c.Tg.Tk \cong ga.Tk$, which just isn't good enough.

The moral is that in forming limits in T-Alg, we can ask for existence or invertibility of 2-cells, and equations between them, but we can't generally force equations between 1-cells.

6.5. Inserters. There is a sort of lax version of an equalizer, called an inserter. Rather than making 1-cells equal, you put a 2-cell in between them. The *inserter* of a parallel pair of arrows $f, g : A \to B$ is the universal $k : C \to A$ equipped with a 2-cell $\kappa : fk \to gk$. More precisely, the universal property states that $\mathscr{K}(D,C)$ should be isomorphic to the category whose objects are morphisms $\ell : D \to A$ equipped with a 2-cell $\lambda : f\ell \to g\ell$, and whose morphisms $(\ell,\lambda) \to (m,\mu)$ are 2-cells

$$D \underset{m}{\overset{\ell}{\Downarrow \alpha}} A$$

such that

$$
\begin{array}{ccc}
f\ell & \xrightarrow{\lambda} & g\ell \\
{\scriptstyle f\alpha}\downarrow & & \downarrow{\scriptstyle g\alpha} \\
fm & \xrightarrow{\kappa} & gm
\end{array}
$$

commutes. Thus for every pair $(l : D \to A, \lambda : fl \to gl)$, there is a unique $l' : D \to C$ with $kl' = l$ and $\kappa l' = \lambda$. Furthermore, given $l, \lambda, m, \mu, \alpha$ as above, there is a unique $\alpha' : l' \to m'$ with $k\alpha' = \alpha$.

Once again, inserters in $\mathscr{K}$ lift to T-Alg, where they have strict projections and detect strictness. Given a pair

$$(A,a) \underset{(g,\bar{g})}{\overset{(f,\bar{f})}{\rightrightarrows}} (B,b)$$

of pseudo morphisms, we construct the inserter $(k : C \to A, \kappa : fk \to gk)$ of f and g in $\mathscr{K}$, and want to make it an algebra. We need a 2-cell $f.a.Tk \to g.a.Tk$ to induce $c : TC \to C$, so we follow our nose:

$$f.a.Tk \xrightarrow{\bar{f}^{-1}.Tk} b.Tf.Tk \xrightarrow{b.T\kappa} b.Tg.Tk \xrightarrow{\bar{g}.Tk} g.a.Tk.$$

This composite must be κc for a unique c, by the universal property of the inserter in $\mathscr{K}$. Now check that c makes C into an algebra, and so on; everything goes through just as before.

Observe that an inserter in a (2-)category with no non-identity 2-cells is just an equalizer.

6.6. PIE-limits. Thus T-Alg has Products, Inserters, and Equifiers, and many important types of limit can be constructed out of these. A limit which can be so constructed is called a *PIE-limit*, so clearly T-Alg has all PIE-limits, and equally clearly equalizers are not PIE-limits. Some other examples of PIE-limits are:

- *iso-inserters*, which are inserters where we ask the 2-cell to be invertible. *Insert* 2-cells in each direction, then *equify* their composites with identities. (Of course you can't go the other way: iso-inserters don't suffice to construct inserters.)

- *inverters*, where we start with a 2-cell α and make it invertible: we want the universal k such that αk is invertible. *Insert* something going back the other way, then *equify* composites with the identities.
- *cotensors* by categories. Cotensors by discrete categories can be constructed using *products*. Any category can be constructed from discrete ones using coinserters (to add morphisms) and coequifiers (to specify composites). So cotensors by arbitrary categories can be constructed from cotensors by discrete categories using *inserters* and *equifiers*.

The dual (colimit) notions of *coinserter*, *coequifier*, and *co-iso-inserter* were important in giving presentations of monads. The dual of inverter is the *coinverter*. The coinverter of a 2-cell $\alpha : f \to g : A \to B$ is the universal $q : B \to C$ with $q\alpha$ invertible. In **Cat**, this is just the category of fractions $B[\Sigma^{-1}]$, where Σ consists of all arrows in B which appear as components of α. Of course the dual of cotensor is tensor, not cocotensor!

6.7. Weighted Limits. In this section we briefly review the general notion of weighted limit, before turning in the next section to the case $\mathscr{V} = \mathbf{Cat}$, where we shall see how the various examples of the previous section arise.

Let $S : \mathscr{C} \to \mathscr{K}$ be a functor between, say, ordinary categories. The limit is supposed to be defined by the fact that

$$\mathscr{K}(A, \lim S) \cong \mathrm{Cone}(A, S)$$

where the right hand side is the set of cones under S with vertex A. This is typically defined as the hom-set $[\mathscr{C}, \mathscr{K}](\Delta A, S)$, where ΔA denotes the constant functor at A, but it can also be expressed as $[\mathscr{C}, \mathbf{Set}](\Delta 1, \mathscr{K}(A, S))$. It is this last description of cones which forms the basis for the generalization to weighted limits; we're going to replace $\Delta 1$ by some more general functor $\mathscr{C} \to \mathbf{Set}$.

EXAMPLE 10. *No one really uses this in practice, but it's useful to think about, and motivates the name "weighted" in "weighted limit." Let $\mathscr{C} = 2$ have two objects, so a functor $S : \mathscr{C} \to \mathscr{K}$ is a pair of objects B and C, and a weight is a functor $J : \mathscr{C} \to \mathbf{Set}$, say it sends one to 2 and the other to 3. Then*

$$[\mathscr{C}, \mathbf{Set}](J, \mathscr{K}(A, S))$$

consists of functions $2 \to \mathscr{K}(A, B)$ and $3 \to \mathscr{K}(A, C)$, or equivalently two arrows $A \to B$ and three arrows $A \to C$, so that the "weighted product" is $B^2 \times C^3$.

For general $\mathscr{V}$, we start with $\mathscr{V}$-functors $S : \mathscr{C} \to \mathscr{K}$ and $J : \mathscr{C} \to \mathscr{V}$ and consider

$$[\mathscr{C}, \mathscr{V}](J, \mathscr{K}(A, S)).$$

If this is representable as a functor of A, the representing object is called the *J-weighted limit* of S and written $\{J, S\}$. Thus we have a natural isomorphism

$$\mathscr{K}(A, \{J, S\}) \cong [\mathscr{C}, \mathscr{V}](J, \mathscr{K}(A, S)).$$

which defines the limit.

EXERCISE 6.1. *If $\mathscr{K} = \mathscr{V}$, then $\{J, S\}$ is just the $\mathscr{V}$-valued hom $[\mathscr{C}, \mathscr{V}](J, S)$.*

When $\mathscr{V} = \mathbf{Set}$, weighted limits don't give you any *new* limits: if $\mathscr{K}$ is an ordinary category which is complete in the usual sense of having all conical limits ($J = \Delta 1$), then it also has all weighted limits. More precisely, for any weight $J : \mathscr{C} \to \mathbf{Set}$ and any diagram $S : \mathscr{C} \to \mathscr{K}$, there is a category $\mathscr{D}$ and a diagram $R : \mathscr{D} \to \mathscr{K}$, such that the universal property of $\{J, S\}$ is precisely the universal property of the usual limit of R.

But the weighted ones are more expressive, so it's still useful to think about them. In particular, you might want to talk about all limits indexed by a particular weight $J : \mathscr{C} \to \mathbf{Set}$; this class is not so easy to express using only conical limits.

When $\mathscr{V} \neq \mathbf{Set}$ it's not longer true that all limits can be reduced to conical ones. But if you have all conical limits *and* cotensors, you can construct all weighted limits.

REMARK 6.1. There is a slight subtlety here. In the case $\mathscr{V} = \mathbf{Set}$, the conical limit of a functor $S : \mathscr{C} \to \mathscr{K}$ is just the limit of S weighted by $\Delta 1 : \mathscr{C} \to \mathbf{Set}$. But for a $\mathscr{V}$-category $\mathscr{C}$, the "constant functor at 1" (from $\mathscr{C}$ to $\mathscr{V}$) is usually not what you want to look at, and indeed may fail to exist. What you really want, to get the right universal property, is the constant functor ΔI at the unit object I of $\mathscr{V}$. But even this may not exist, unless $\mathscr{C}$ is the free $\mathscr{V}$-category on an ordinary category $\mathscr{B}$. So this is the right general context for conical limits in enriched category theory.

That's all I want to say about general $\mathscr{V}$.

6.8. Cat-weighted limits. Here I describe the weights for some of the limit notions introduced earlier.

EXAMPLE 11 (Inserters). *Let $\mathscr{C}$ be the 2-category $\cdot \rightrightarrows \cdot$, so S is determined by a parallel pair of arrows $A \rightrightarrows B$. The weight $J : \mathscr{C} \to \mathbf{Cat}$ has image $(1 \rightrightarrows 2)$. Then a natural transformation $J \to \mathscr{K}(C, S)$ gives has two components. The first is a functor $1 \to \mathscr{K}(C, A)$, or equivalently a morphism $h : C \to A$, while the second is a functor $2 \to \mathscr{K}(C, B)$, or equivalently a 2-cell*

$$C \underset{v}{\overset{u}{\Rrightarrow}} \Downarrow\beta \; B$$

Naturality of these components means precisely that $u = fh$ and $v = gh$, so the data consists of a 1-cell $h : C \to A$ and a 2-cell $\beta : fh \to gh$. To give h and β is just to give a map from C into the inserter.

This is the 1-dimensional aspect of the universal property, which characterizes the 1-cells into C; there is also a 2-dimensional aspect characterizing the 2-cells, since the limit is defined in terms of an isomorphism of categories, not just a bijection between sets. In general, this 2-dimensional aspect must be checked, but if the 2-category $\mathscr{K}$ should admit *tensors*, the 2-dimensional aspect follows from the 1-dimensional one. Similar comments apply to all the examples.

EXAMPLE 12 (Equifiers). *Here, our 2-category $\mathscr{C}$ is*

and our weight is

in which α and β get mapped to the same 2-cell in **Cat**.

EXAMPLE 13 (Comma objects). *$\mathscr{C}$ is the same shape as for pullbacks*

and J is

There is no 2-cell in $\mathscr{C}$, since we don't start *with a 2-cell, we only add one universally.*

EXAMPLE 14 (Inverters). *Recall, this is where we start with a 2-cell and universally make it invertible. Then $\mathscr{C}$ is*

and J is

$$1 \quad \overset{\Downarrow}{\underset{}{\rightrightarrows}} \quad \mathscr{I}$$

where $\mathscr{I}$ is the "free-living isomorphism" $\cdot \rightleftarrows \cdot$.

6.9. Colimits. Colimits in $\mathcal{K}$ are limits in $\mathcal{K}^{\mathrm{op}}$. That's really all you have to say, but I should show you the notation. As usual, we rewrite things so as to refer to $\mathcal{K}$. In this case, it's also convenient to replace $\mathscr{C}$ by $\mathscr{C}^{\mathrm{op}}$, so that we start with

$$S : \mathscr{C} \to \mathcal{K}$$
$$J : \mathscr{C}^{\mathrm{op}} \to \mathcal{V}$$

and now the weighted colimit is written $J \star S$ and defined by a natural isomorphism

$$\mathcal{K}(J \star S, A) \cong [\mathscr{C}^{\mathrm{op}}, \mathcal{V}](J, \mathcal{K}(S, A)).$$

One form of the *Yoneda lemma* says that

$$J \cong J \star Y$$

where $Y : \mathscr{C} \to [\mathscr{C}^{\mathrm{op}}, \mathcal{V}]$ is the Yoneda embedding and $J : \mathscr{C}^{\mathrm{op}} \to \mathcal{V}$ is arbitrary.

Here's an application. Suppose you have some "limit-notion" which you know in advance is a weighted limit, but you don't know what the weight is. Thus you know $\{J, S\}$ given S, but you don't know J itself. Consider the version of the Yoneda embedding $Y : \mathscr{C} \to [\mathscr{C}, \mathcal{V}]^{\mathrm{op}}$ and take its "limit," for the notion of limit we're interested in; equivalently, take the relevant *colimit* of $Y : \mathscr{C} \to [\mathscr{C}^{\mathrm{op}}, \mathcal{V}]$. This is $J \star Y$ for our as yet unknown J; but by the Yoneda lemma this $J \star Y$ is itself the desired weight. This can be used to calculate the weights for all the concrete examples of **Cat**-weighted limits discussed here.

6.10. Pseudolimits. The pseudolimit of a 2-functor $S : \mathscr{C} \to \mathcal{K}$ is defined an object $\mathrm{pslim}S$ of $\mathcal{K}$ equipped with an isomorphism

$$\mathcal{K}(A, \mathrm{pslim}S) \cong \mathrm{Ps}(\mathscr{C}, \mathbf{Cat})(\Delta 1, \mathcal{K}(A, S))$$

of categories natural in A, where $\mathrm{Ps}(\mathscr{A}, \mathscr{B})$ is the 2-category of 2-functors, pseudonaturals, and modifications from $\mathscr{A}$ to $\mathscr{B}$. The right side is what we mean by a *pseudo-cone*. Note that this is still an *isomorphism* of categories, not an equivalence, so such pseudolimits are determined up to isomorphism not just equivalence.

EXAMPLE 15 (Pseudopullbacks). *Again we take $\mathscr{C}$ to be*

A pseudo-cone then consists of

with isomorphisms in each triangle. We have made the cones commute only up to isomorphisms, but the universal property and factorizations are still strict. Note that the pseudopullback is equivalent (not isomorphic) to the iso-comma object (assuming both exist). In the latter, we specify $fa \cong gb$ without specifying the middle diagonal arrow. Of course, we can take it to be fa, or gb, so we get ways of going back and forth.

The pseudopullback is not *in general equivalent to the pullback, although it is possible to characterize when they are [19]. This situation is entirely analogous to homotopy pullbacks, and indeed it can be regarded as a special case, via the "categorical" Quillen model structure on* **Cat** *(see Section 7).*

Again, given a weight $J \colon \mathscr{C} \to \mathbf{Cat}$, the *weighted pseudolimit* is defined by

$$\mathscr{K}(C, \{J, S\}_{ps}) \cong \mathsf{Ps}(\mathscr{C}, \mathbf{Cat})(J, \mathscr{K}(C, S)).$$

I don't really want to do any examples of this one, I want to do some general nonsense instead.

Recall that $\mathsf{Ps}(\mathscr{C}, \mathbf{Cat}) = T\text{-Alg}$ for a 2-monad T on $[\mathrm{ob}\mathscr{C}, \mathbf{Cat}]$, while $T\text{-Alg}_s = [\mathscr{C}, \mathbf{Cat}]$, so the inclusion

$$[\mathscr{C}, \mathbf{Cat}] \to \mathsf{Ps}(\mathscr{C}, \mathbf{Cat})$$

has a left adjoint Q, with $QJ = J'$. Thus

$$\mathsf{Ps}(\mathscr{C}, \mathbf{Cat})(J, \mathscr{K}(C, S)) \cong [\mathscr{C}, \mathbf{Cat}](J', \mathscr{K}(C, S))$$

which just defines the universal property for the J'-weighted limit. In other words, *pseudolimits are not some more general thing, but a special case of ordinary (weighted) limits.* Thus we say that a weight "is" a pseudolimit if it has the form J' for some J.

REMARK 6.2. This sort of phenomenon is common. Recall, for example, that pseudo-algebras for monads are strict algebras over a cofibrant replacement monad. Thus talking about things of the form Ps-T-Alg is actually *less* general than things of the form T-Alg, since everything of the former form has the latter form, but not conversely.

6.11. PIE-limits again. Recall that PIE-limits are the limits constructible from products, inserters, and equifiers. We can now make this more precise. A weight $J : \mathscr{C} \to \mathbf{Cat}$ is a (weight for a) PIE-limit if and only if the following conditions hold:

- any 2-category $\mathscr{K}$ with products, inserters, and equalizers has J-weighted-limits;
- any 2-functor $F : \mathscr{K} \to \mathscr{L}$ which preserves products, inserters, and equalizers (and for which $\mathscr{K}$ has these limits) also preserves J-weighted limits.

There is a characterization of such weights. Given a 2-functor J : $\mathscr{C} \to \mathbf{Cat}$, first consider the underlying ordinary functor $J_0 : \mathscr{C}_0 \to \mathbf{Cat}_0$ obtained by throwing away all 2-cells. Now compose this with the functor ob : $\mathbf{Cat}_0 \to \mathbf{Set}$ which throws away the arrows of a category, leaving just the set of objects. This gives a functor $j : \mathscr{C}_0 \to \mathbf{Set}$. Then J is a PIE-weight if and only if j is a coproduct of representables; and j will be a coproduct of representables if and only if each connected component of the category of elements of j has an initial object.

Pseudolimits are also PIE-limits, as we shall now see. For a general T-algebra A, the pseudomorphism classifier A' was constructed from free algebras using coinserters and coequifiers. Thus for a general weight $J : \mathscr{C}^{\mathrm{op}} \to \mathbf{Cat}$ we can construct J' from "free weights," using coinserters and coequifiers. Free weights, in this context, are coproducts of representables, thus J' can be constructed from representables using coproducts, coinserters, and coequifiers. It will follow that pseudolimits can indeed be constructed using products, inserters, and equifiers, and so that they are PIE-limits.

Now a limit weighted by a representable $\mathscr{C}(C, -)$ is given by evaluation at C; a limit weighted by a coproduct of representables is given by the product of the evaluations; a limit weighted by a coinserter of coproducts of representables is given by an inserter of products of evaluations, and so on. This is part of a general result, not needed here, that colimits of weights give iterated limits, as in the formula

$$\{J \star H, S\} \cong \{J, \{H, S\}\}$$

reminiscent of a tensor-hom situation. In other words, the 2-functor

$$\{-, S\} : [\mathscr{C}, \mathbf{Cat}]^{\mathrm{op}} \to \mathscr{K},$$

sending a weight J to the J-weighted limit of S, sends colimits in $[\mathscr{C}, \mathbf{Cat}]$ to limits in $\mathscr{K}$. In any case we can conclude in the current context that J'-weighted limits can be constructed using products, inserters, and equifiers, and so we have:

PROPOSITION 6.1. *Pseudolimits are PIE-limits.*

The converse is false: for example inserters are PIE-limits but are not pseudolimits. Neither are iso-comma objects, although they're pretty close (as we saw above).

Remember that T-Alg has all PIE-limits. It therefore has all pseudolimits as well. But consider the class of all limits (weights) which are equivalent (in $[\mathscr{C}, \mathbf{Cat}]$, so that the equivalences are 2-natural) to pseudolimits. It is not the case that T-Alg has all of those limits. So equivalence of limits is not always totally trivial.

For example, consider the apparently rather benign act of splitting idempotent equivalences. If we split an idempotent equivalence

$$
\begin{array}{ccc}
TA & \xrightarrow{\;Te\;} & TA \\
{\scriptstyle a}\big\downarrow & \cong & \big\downarrow{\scriptstyle a} \\
T & \xrightarrow[\;e\;]{} & T
\end{array}
$$

in T-Alg, we won't necessarily get a (strict) T-algebra back, only a pseudo-algebra.

As an example, let $\mathscr{C}$ be a non-strict monoidal category, and $\mathscr{C}_{\mathrm{st}}$ its strictification. Then there is an idempotent equivalence on $\mathscr{C}_{\mathrm{st}}$, which when split, gives $\mathscr{C}$. This shows that T-Alg does not have splittings of idempotents when T is the 2-monad for strict monoidal categories.

6.12. Bilimits. I'm going to write down all the same symbols, but they'll just mean different things! So now $\mathscr{C}$ and $\mathscr{K}$ are bicategories, while $S : \mathscr{C} \to \mathscr{K}$ and $J : \mathscr{C} \to \mathbf{Cat}$ are now homomorphisms (pseudofunctors). The *weighted bilimit* is defined by an *equivalence*

$$
\mathscr{K}(C, \{J, S\}_b) \simeq \mathbf{Hom}(\mathscr{C}, \mathbf{Cat})(J, \mathscr{K}(C, S)).
$$

Now our limits are determined only up to equivalence, instead of up to isomorphism.

In the case when $\mathscr{C}$ and $\mathscr{K}$ are 2-categories and J and S are 2-functors, then the right hand side is equal to the right hand side for pseudolimits, just by definition (since $\mathsf{Ps}(\mathscr{C}, \mathbf{Cat}) \hookrightarrow \mathbf{Hom}(\mathscr{C}, \mathbf{Cat})$ is locally an isomorphism). Thus *every pseudolimit is a bilimit*.

On the other hand, if just $\mathscr{K}$ is a 2-category, then you can replace $\mathscr{C}$ by a 2-category $\mathscr{C}'$ such that homomorphisms out of $\mathscr{C}$ are the same as 2-functors out of $\mathscr{C}'$. Now for any $A \in \mathscr{K}$ we have

$$
\mathbf{Hom}(\mathscr{C}^{\mathrm{op}}, \mathbf{Cat})(J, \mathscr{K}(S, A)) \simeq \mathsf{Ps}(\mathscr{C}^{\mathrm{op}}, \mathbf{Cat})(\tilde{J}, \mathscr{K}(\tilde{S}, A))
$$

where $\tilde{J}$ and $\tilde{S}$ are the 2-functors corresponding to J and S. Thus *a 2-category with all pseudolimits also has all bilimits*.

As we shall see in the following section, the converse is false: there are 2-categories with bilimits which do not have pseudolimits, so the definition of pseudolimit is logically harder to *satisfy* than that of bilimit. On the other hand in concrete examples it is often much easier to *verify* the definition using pseudolimits than bilimits. This is certainly the case for pseudolimits in T-Alg. It's also the case for the opposite of the 2-category of Grothendieck toposes.

6.13. Bilimits and bicolimits in T-Alg . Suppose once more that $\mathscr{K}$ is locally finitely presentable and T is finitary, and consider the 2-category T-Alg. It has PIE-limits, as we saw, and so has pseudo-limits, and so has bilimits. So from a bicategorical perspective, we have all the limits we might want.

T-Alg also has *bicolimits*, although not in general pseudocolimits or PIE-colimits. Thus T-Alg^{op} is an example of a 2-category with bilimits but not pseudolimits. Just as in the case of ordinary monads, the (bi)colimits in T-Alg are not generally constructed as in $\mathscr{K}$.

The existence of bicolimits follows from:

THEOREM 6.1. *Suppose we have a 2-functor* $G : T\text{-Alg} \to \mathscr{L}$ *such that the composite* GJ *in*

$$T\text{-Alg}_s \xrightarrow{\ J\ } T\text{-Alg} \xrightarrow{\ G\ } \mathscr{L}$$

has a left adjoint F. *Then* JF *is left biadjoint to* G.

We start with a left 2-adjoint F to GJ but end up with only a left biadjoint to G. Here's the idea of the proof. The biadjunction amounts to a (pseudonatural) equivalence

$$T\text{-Alg}(JFL, A) \simeq \mathscr{L}(L, GA).$$

Since T-Alg$_s$ and T-Alg have the same objects, we may write A as JA. Now the adjunction $F \dashv GJ$ gives an isomorphism of categories

$$T\text{-Alg}_s(FL, A) \cong \mathscr{L}(L, JGA)$$

so it suffices to show that

$$T\text{-Alg}_s(FL, A) \simeq T\text{-Alg}(JFL, JA)$$

which in turn amounts to the fact that every pseudomorphism from FL to A is isomorphic to a strict one. This will hold if we know that $(FL)' \simeq FL$. Writing Q for the left adjoint to $J : T\text{-Alg}_s \to T\text{-Alg}$, we have a pseudomorphism $p : JFL \rightsquigarrow JQJFL$ (unit of $Q \dashv J$), and a map $n : L \to GJFL$ (unit of $F \dashv GJ$), so we can form the composite

$$L \xrightarrow{\ n\ } GJFL \xrightarrow{\ Gp\ } GJQJFL$$

and the corresponding *strict* map

$$FL \xrightarrow{\ r\ } QJFL$$

under the adjunction $F \dashv GJ$ provides the desired inverse-equivalence to $q : QJFL \to FL$ (the counit of $Q \dashv J$).

COROLLARY 6.1.
  (i) T-Alg *has bicolimits;*
 (ii) *for any monad morphism* $f : S \to T$, *the induced 2-functor* $f^* : T\text{-Alg} \to S\text{-Alg}$ *has a left biadjoint.*

Part (ii) is easier: we have a commutative diagram of 2-functors

$$
\begin{array}{ccc}
T\text{-Alg}_s & \xrightarrow{\ J\ } & T\text{-Alg} \\
{\scriptstyle f_s^*}\downarrow & & \downarrow{\scriptstyle f^*} \\
S\text{-Alg}_s & \xrightarrow[\ J\ ]{} & S\text{-Alg}
\end{array}
$$

in which the left hand map has a left adjoint, by a general enriched-category-theoretic fact (no harder than the corresponding fact for ordinary categories), and the bottom map has a left adjoint (the pseudomorphism classifier for S-algebras). Thus the composite has a left adjoint, and so f^* has a left biadjoint. (The argument as stated uses the pseudomorphism classifier for S-algebras, and so requires S to have rank, but this can be avoided.)

What about part (i)? For any $S : \mathscr{C} \to T\text{-Alg}$, we can form the diagram

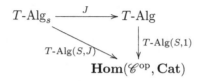

and now the existence of bicolimits in T-Alg amounts to the existence of left biadjoints for all such $T\text{-Alg}(S, 1)$. So it will suffice to show that the composite $T\text{-Alg}(S, J) : T\text{-Alg}_s \to \mathbf{Hom}(\mathscr{C}^{\mathrm{op}}, \mathbf{Cat})$ has a left adjoint. But $T\text{-Alg}(S, J) \cong T\text{-Alg}_s(QS, 1)$, where $Q \dashv J$, and $T\text{-Alg}_s(QS, 1)$ has a left adjoint provided that $T\text{-Alg}_s$ has pseudocolimits. Finally since T is finitary, $T\text{-Alg}_s$ is cocomplete (by a general enriched-category-theoretic fact no harder than the corresponding fact for ordinary categories) and so in particular has pseudocolimits.

A *direct* proof that T-Alg has bicolimits would be a nightmare, but using pseudocolimits it becomes manageable.

6.14. References to the literature. Many people came up with some notion of weighted limit at about the same time. But I guess the main reference for general $\mathcal{V}$ is now just [27]. On the other hand, for various limit notions for 2-categories, [26] is very readable. Once you've got through that, you should turn to [6], [56], and [17]. For the beautiful theory of PIE-limits, see [52]. For the connection between pullbacks and pseudopullbacks, see [19]. Section 6.13 is based on [7]. For relationships between pseudolimits and bilimits, see [49].

7. Model categories, 2-categories, and 2-monads. This section involves Quillen model categories, henceforth called model categories, or model structures on categories. There are various connections between model categories, 2-categories, and 2-monads which I'll discuss.

(i) Model structures *on* 2-categories: a model 2-category is a category with a model structure and an enrichment over **Cat**, with suitable compatibility conditions between these structures. Any 2-category with finite limits and colimits has a "trivial" such structure, in which the weak equivalences are the categorical equivalences. These trivial model 2-categories are not so interesting in themselves, but can be used to generate other more interesting model 2-categories.

(ii) Model categories *for* 2-categories: There's a model structure on the category of 2-categories and 2-functors, and one for bicategories too.

(iii) Model structures *induced by* 2-monads. If T is a 2-monad on a 2-category $\mathscr{K}$, we can lift the trivial model 2-category structure on $\mathscr{K}$ coming from the 2-category structure to get a model structure on $T\text{-Alg}_s$.

(iv) Model structures *for* 2-monads: the 2-category $\mathbf{Mnd}_f(\mathscr{K})$ of finitary 2-monads on $\mathscr{K}$ is also a model 2-category.

One thing which I won't discuss, but deserves further study:

(v) "Many-object monoidal model categories." A *monoidal model category* is a monoidal category with a model structure, suitably compatible with the tensor product. The many-object version of this would involve a bicategory (or 2-category) with a model structure on each hom-category, subject to certain conditions (somewhat more complicated than those for monoidal model categories).

7.1. Model 2-categories. There's a model structure on the category $\mathbf{Cat}_0$ of categories and functors in which the weak equivalences are the equivalences of categories, and the fibrations are the functors $f : A \to B$ such that for any object $a \in A$ and any isomorphism $\beta : b \cong fa$ in B, there is an isomorphism $\alpha : a' \cong a$ in A with $fa' = b$ and $f\alpha = \beta$. This is sometimes called the "categorical model structure" or "folklore model structure." (There are other model structures on $\mathbf{Cat}_0$, in particular the famous one due to Thomason that gives you a homotopy theory equivalent to simplicial sets.)

As mentioned above, a category with a monoidal structure and a model structure satisfying certain compatibility conditions is called a monoidal model category. The cartesian product makes $\mathbf{Cat}_0$ into a monoidal model category.

If we now consider a category that has both a model structure and an enrichment over **Cat**, there is a notion of compatibility between these structures, which can be expressed in terms of the monoidal model structure on $\mathbf{Cat}_0$. We call this notion a *model 2-category*.

First of all the 2-category $\mathscr{K}$ is required to have finite limits and colimits in the 2-categorical sense; if the underlying ordinary category $\mathscr{K}_0$ already has finite limits and colimits, then the extra requirement is that $\mathscr{K}$ have tensors and cotensors with 2. A model structure on $\mathscr{K}_0$ makes $\mathscr{K}$ into a model 2-category if two new axioms hold for any cofibration $i : A \to B$ and fibration $p : C \to D$:

(a) Given morphisms $x : A \to C$, $y : B \to C$, and $z : B \to D$, with $px = zi$, and invertible 2-cells $\alpha : x \cong yi$ and $\beta : z \cong py$ with $p\alpha = \beta i$, there exist a morphism $y' : B \to C$ and an isomorphism $\gamma : y' \cong y$ with $p\gamma = \beta$ and $\gamma i = \alpha$;

(b) If either i or p is trivial, then for any morphisms $x, y : B \to C$ and any 2-cells $\alpha : xi \to yi$ and $\beta : px \to py$ with $\beta i = p\alpha$, there exists a unique 2-cell $\gamma : x \to y$ with $p\gamma = \beta$ and $\gamma i = \alpha$.

It follows that every equivalence is a weak equivalence, and that any morphism isomorphic to a weak equivalence is itself a weak equivalence.

7.2. Trivial model 2-categories. Let $\mathscr{K}$ be a 2-category with finite limits and colimits. The most important limit here will be the *pseudolimit of an arrow* $f : A \to B$. Ordinarily we don't talk about the limit of an arrow, since the ordinary limit of an arrow is just its domain, but the pseudolimit is only equivalent to the domain, not equal. It's the universal diagram

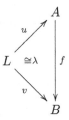

such that given $a : X \to A$, $b : X \to B$, and $\beta : fa \cong b$, there is a unique $c : X \to L$ with $\lambda c = \beta$ (and so also $uc = a$ and $vc = b$). In this case, u is an equivalence, because id $: f1 \cong f$ factors through by a $d : A \to L$ with $ud = 1$, and one can also check that $du \cong 1$. The technique of Section 6.9 can be used to calculate the weight for pseudolimits of arrows.

The model structure on $\mathscr{K}$ is:

- The weak equivalences are the equivalences;
- The fibrations are the *isofibrations*, the maps such that each invertible 2-cell

lifts to an invertible 2-cell

- The cofibrations have the left lifting property with respect to the trivial fibrations.
- It follows that the trivial fibrations are the *surjective equivalences*: the p for which there exists an s with $ps = 1$ and $sp \cong 1$.

We call such a model 2-category $\mathcal{K}$ a trivial model 2-category. When $\mathcal{K}$ is **Cat** this is just the folklore structure. When $\mathcal{K}$ has no non-identity 2-cells, then the equivalences are the isomorphisms, and all maps are isofibrations, so this agrees with the usual notion of trivial model category. For a general 2-category $\mathcal{K}$, however, there will be weak equivalences which are not invertible.

The pseudolimit of f gives us, for any f, a factorization $f = vd$ where v is a fibration (which follows from the universal property of the pseudolimit) and d is an equivalence. In the case of **Cat**, you could stop there and d would already be a trivial cofibration, but in general there's more work to do, although we have reduced the problem to factorizing an equivalence.

The way you do that is also the way you get the other factorization: use the dual construction. Form the pseudo*co*limit of the arrow f, as in the diagram below, and let e be the unique map with $ei = f$, $ej = 1$, and $e\varphi = \mathsf{id}$.

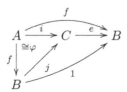

This time i is a cofibration and e is a trivial fibration, and if f itself is an equivalence, then i has the left lifting property with respect to the fibrations (so it's what's going to become a trivial cofibration).

That's all I'll say about the proof. There is, of course, a dual model structure in which the cofibrations are characterized and the fibrations are defined by a right lifting property. For **Cat**, these coincide, in general they don't.

When $\mathcal{K}$ is arbitrary, there is no reason why the model structure should be cofibrantly generated. Certainly for **Cat** it is, but even for such a simple 2-category as $\mathbf{Cat}^2$ it is not. From the homotopical point of view the trivial model structure is trivial in several ways, including:

- All objects are cofibrant and fibrant;
- The morphisms in the homotopy category $\mathrm{Ho}(\mathcal{K}_0)$ are the isomorphism classes of 1-cells in $\mathcal{K}$.

In the case $\mathscr{K} = \mathbf{Cat}(\mathbb{E})$, one typically considers different model struc- tures: an internal functor $F \colon \mathbb{A} \to \mathbb{B}$ is usually called a weak equivalence if it's full and faithful and essentially surjective in an internal sense. For $\mathbf{Cat}$ this is equivalent to the usual notion (by the axiom of choice), but in general it won't be. It's the weak equivalences in this sense that people tend to use as their weak equivalences for $\mathbf{Cat}(\mathbb{E})$. When $\mathbb{E}$ is a topos, this was studied by Joyal and Tierney, and there's been recent work on other cases, when $\mathbb{E}$ is groups (so that internal categories are crossed modules) or abelian groups.

7.3. Model structures for T-algebras. Now let T be a (finitary) 2-monad on (a locally finitely presentable) 2-category $\mathscr{K}$, and $T\text{-Alg}_s$ the 2-category of strict algebras and strict morphisms. In the usual way, one can lift the model structure on $\mathscr{K}$ to get one on $T\text{-Alg}_s$: a strict T-morphism $f : (A, a) \to (B, b)$ is a weak equivalence or fibration if and only if $U_s f :$ $A \to B$ is one in $\mathscr{K}$; the cofibrations are then defined via a left lifting property.

Now the lifted model 2-category structure on $T\text{-Alg}_s$ is *not* trivial. In general, if $(f, \overline{f}) : (A, a) \to (B, b)$ is a pseudomorphism of T-algebras and $f : A \to B$ is an equivalence, then any inverse-equivalence $g : B \to A$ naturally becomes an equivalence upstairs in $T\text{-Alg}$. This is a 2-categorical analogue of the fact that if an algebra morphism is a bijection, its inverse also preserves the algebra structure. But if f is strict ($\overline{f}$ is an identity), there is no reason why its inverse equivalence should also be strict, and thus no reason why f should be an equivalence in $T\text{-Alg}_s$. For example, a strict monoidal functor which is an equivalence of categories has an inverse which is strong monoidal, but which need not be strict.

Recall the adjoint to the inclusion

$$T\text{-Alg}_s \underset{J}{\overset{Q}{\underset{\longrightarrow}{\overset{\longleftarrow}{\bot}}}} T\text{-Alg}$$

where $QA = A'$, so that we have a bijection

$$\frac{A \rightsquigarrow B}{A' \to B}.$$

This fits into the model category framework very nicely. The counit of this adjunction

$$A' \xrightarrow{q} A$$

is a cofibrant replacement: a trivial fibration with A' being cofibrant. So we see that $T\text{-Alg}$, which is the thing we're more interested in, is starting to come out of the picture: a weak morphism out of A is *the same thing as* a strict morphism out of the "special cofibrant replacement" A' of A. This is much tighter than the general philosophy that "we should think of maps in the homotopy category as maps out of a cofibrant replacement."

An algebra turns out to be cofibrant if and only if $q : A' \to A$ has a section *in* T-Alg$_s$. (There's always a section in T-Alg, using a pseudomorphism.) Since q is a trivial fibration, there will certainly be a section if A is cofibrant. Conversely, if there is a section, then A is a retract of A'; but A' is always cofibrant, and so then A must be cofibrant too. In 2-categorical algebra, the word *flexible* is used in place of cofibrant.

7.4. Model structures for 2-monads. Recall now that we have adjunctions

$$\mathbf{Mnd}_f(\mathscr{K})$$

$$H \left(\dashv \right) W$$

$$\mathbf{End}_f(\mathscr{K})$$

$$G \left(\dashv \right) V$$

$$[\mathrm{ob}\mathscr{K}_f, \mathscr{K}]$$

both of which are monadic, as is the composite. Thus $\mathbf{Mnd}_f(\mathscr{K})$ is both M-Alg$_s$ and N-Alg$_s$ where M is the induced monad on $\mathbf{End}_f(\mathscr{K})$ and N is the induced monad on $[\mathrm{ob}\mathscr{K}_f, \mathscr{K}]$.

Thus $\mathbf{Mnd}_f(\mathscr{K})$ has *two* lifted model structures, coming from the trivial structures on $\mathbf{End}_f(\mathscr{K})$ and on $[\mathrm{ob}\mathscr{K}_f, \mathscr{K}]$. They're not the same, since something can be an equivalence all the way downstairs without being one in $\mathbf{End}_f(\mathscr{K})$ (which is itself the 2-category of algebras for another induced monad on $[\mathrm{ob}\mathscr{K}_f, \mathscr{K}]$).

A monad map $S \xrightarrow{f} T$ is a 2-natural transformation compatible with the unit and multiplication. If the 2-natural transformation is an equivalence in $\mathbf{End}_f(\mathscr{K})$, it is a weak equivalence for the M-model structure; if the *components* of the 2-natural transformation are equivalences, it is a weak equivalence for the N-model structure.

It's the M-model structure (the one lifted from $\mathbf{End}_f(\mathscr{K})$) which seems to be more important, and we'll only consider that one here. The corresponding prime construction classifies pseudomorphisms of monads. These are precisely the things that arise when talking about pseudoalgebras: recall that a pseudo-T-algebra was an object A with a pseudomorphism

$$T \rightsquigarrow \langle A, A \rangle$$

into the "endomorphism 2-monad" of A, corresponding to maps $TA \xrightarrow{a} A$ which are associative and unital up to coherent isomorphism.

This corresponds to a strict map $T' \to \langle A, A \rangle$, so that T'-Alg is Ps-T-Alg. (This is the part of the justification for working with strict algebras that people tend to understand first, but it's the less important one: see Remark 4.1 above.)

If $q : T' \to T$ has a section in M-Alg$_s$ = $\mathbf{Mnd}_f(\mathscr{K})$, then T is said to be *flexible* (=*cofibrant*). This was the context in which the notion of flexibility was first introduced. *Any monad that you can give a presentation for without having to use equations between objects is always flexible.* For example, the monad for monoidal categories is flexible, but the monad for strict monoidal categories, which involves the equation $(X \otimes Y) \otimes Z = X \otimes (Y \otimes Z)$, is not.

Flexible monads have the property that every pseudo-algebra is (not just equivalent but) *isomorphic* to a strict one; in fact isomorphic via a pseudomorphism whose underlying $\mathscr{K}$-morphism is an identity! Remember that the importance of pseudo-algebras is *not* for describing concrete things, but for the theoretical side, since various constructions don't preserve strictness of algebras. For particular structures like monoidal categories, you're better off choosing the "right" monad to start with: the one for which monoidal categories are the strict algebras.

7.5. Model structures on 2-Cat. There is a category **2-Cat** of 2-categories and 2-functors which underlies a 3-category, and a 2-category, and perhaps more importantly a Gray-category. But we want to describe a model structure on the mere category **2-Cat**, analogous to the one above for **Cat**.

The weak equivalences will be the *biequivalences*. Recall that $F \colon \mathscr{A} \to \mathscr{B}$ is a biequivalence if

- each $F : \mathscr{A}(A, B) \to \mathscr{B}(FA, FB)$ is an equivalence of categories; and
- F is "biessentially surjective" on objects: if $C \in \mathscr{B}$, there exists an $A \in \mathscr{A}$ with $FA \simeq C$ in $\mathscr{B}$.

Every equivalence has an inverse-equivalence, going back the other way. For a biequivalence $F : \mathscr{A} \to \mathscr{B}$ you can build a thing $G \colon \mathscr{B} \to \mathscr{A}$ with $GF \simeq 1$ and $FG \simeq 1$. You can make G a pseudofunctor, but generally not a 2-functor, even when F is one. That's somehow the whole point of the model structure. Similarly the equivalences $FG \simeq 1$ and $GF \simeq 1$ will generally only be pseudonatural.

Clearly biequivalence is the right notion of "sameness" for bicategories, or 2-categories, but there is this stability (under biequivalence-inverses) problem, if you want to work entirely within **2-Cat**. If you allow pseudofunctors, and so move to **2-Cat**$_{\text{ps}}$, then as we have seen, you lose completeness and cocompleteness.

The fibrations are similar to the case of categories. Fibrations for the model structure on **Cat** involved lifting invertible 2-cells; here we lift equivalences: a 2-functor $F \colon \mathscr{A} \to \mathscr{B}$ is a fibration if

- given an object A upstairs and equivalence downstairs, we have a lift as in

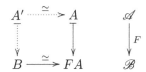

- given a 1-cell f upstairs and an invertible 2-cell downstairs, we have a lift as in

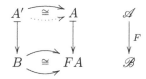

Equivalently, each of the functors $\mathscr{A}(A_1, A_2) \to \mathscr{B}(FA_1, FA_2)$ is an (iso)fibration in **Cat**.

Note that the notion of biequivalence is *not* internal to any reasonable 3-category or Gray-category of 2-categories and 2-functors, which speaks against the existence of a general model structure on an arbitrary 3-category or Gray-category which would reduce to this one.

There's an equivalent way of characterizing the fibrations which is useful. Keep the iso-2-cell lifting property as is, but modify the equivalence-lifting to deal with adjoint equivalences rather than equivalences. Here it is not just the 1-cell, but also the equivalence-inverse, and the invertible unit and counit which must be lifted.

In the presence of the iso-2-cell lifting property, these two types of equivalence-liftings are equivalent: clearly the lifting of adjoint equivalences implies the lifting of equivalences, since we can complete any equivalence to an adjoint equivalence, but the converse is also true provided that we can lift 2-cell isomorphisms.

This is related to a mistake I made in my first paper on this topic, where I used a condition like this on lifting equivalences that aren't necessarily adjoint equivalences. Regard "being an equivalence" as a property, and "an adjoint equivalence" as a structure, but be wary of regarding "a not-necessarily-adjoint equivalence" as a structure. Adjoint equivalences are now completely algebraic, classified by maps out of "the free-living adjoint equivalence," which is biequivalent to the terminal 2-category 1. A "free-living not-necessarily-adjoint equivalence" would *not* be biequivalent to 1.

The trivial fibrations, which are the things which are both fibrations and weak equivalences, can be characterized as the 2-functors that

- are surjective on objects; and
- have each $\mathscr{A}(A_1, A_2) \to \mathscr{B}(FA_1, FA_2)$ a surjective equivalence (a trivial fibration in **Cat**).

Note that the trivial fibrations don't use the 2-category structure; you don't need anything about the composition to know what these things are, only the "2-graph" structure. So they're much simpler to work with.

There's an ω-categorical analogue to these things, which permeates Makkai's work on ω-categories. You don't need the ω-category structure, only a globular set, to say what this means. The corresponding notion of "cofibrant object" is then what he calls a "computad."

It's a bit less trivial than with the other model structures to prove that this all works, but it's not really hard. Everything is directly a lifting property (once you use the version with adjoint equivalences), so finding generating cofibrations and trivial cofibrations is easy.

All objects are fibrant, but *not* all objects are cofibrant. We have a "special" cofibrant replacement $q : \mathscr{A}' \to \mathscr{A}$ with the property that pseudofunctors out of $\mathscr{A}$ are the same as 2-functors out of $\mathscr{A}'$:

$$\frac{\mathscr{A} \rightsquigarrow \mathscr{B}}{\mathscr{A}' \to \mathscr{B}}$$

and $\mathscr{A}$ is cofibrant (flexible) if and only if the trivial fibration q has a section in **2-Cat**. This happens exactly when the underlying category $\mathscr{A}_0$ is free on some graph (you haven't imposed any equations on 1-cells, but you may have introduced isomorphisms between them). In principle, a cofibrant $\mathscr{A}_0$ could be a retract of something free, but it turns out that this already implies that it is free.

There are three main things of interest to me in relation to the model structure on **2-Cat**. The first is the equation "cofibrant = flexible." The second involves the monoidal structures. The model structure is *not* compatible with the cartesian product $\times$. The thing to have in mind is that the locally discrete 2-category $\mathbf{2} = (0 \to 1)$ is cofibrant, but $\mathbf{2} \times \mathbf{2}$ is not, since the commutative square

$$
\begin{array}{ccc}
(0,0) & \longrightarrow & (0,1) \\
\downarrow & & \downarrow \\
(1,0) & \longrightarrow & (1,1)
\end{array}
$$

is not free. There are various tensor products you can put on **2-Cat**. The cartesian product is also called the *ordinary product* (since it is also a special case of the tensor product of $\mathscr{V}$-categories), but I like to call it the *black product* since the square is "filled in," in the sense that the square commutes. (Think of nerves of categories: 2-simplices, or solid triangles, represent commutative triangles in a category, whereas the boundary of a 2-simplex represents a not necessarily commutative triangle.)

There's also the *white* or *funny* product, in which the square does not commute (think of it as just the boundary, not filled in). It's a theorem that on **Cat** there are exactly 2 symmetric monoidal closed structures: the ordinary one and the "funny" one. The closed structure corresponding to the funny product is the not-necessarily-natural transformations (just components). Enriching over this structure gives you a "sesquicategory" (perhaps an unfortunate name, but you can see how it came about), which has hom-categories and whiskering, but no middle-four interchange, hence no well-defined horizontal composition of 2-cells. This funny tensor product can also be defined on **2-Cat**, or indeed on $\mathscr{V}$-**Cat** for any $\mathscr{V}$.

In the case of **2-Cat**, there's an intermediate possibility: the *Gray* or *grey* tensor product, due to John Gray, in which you put an isomorphism in the square, so it's "partially filled in."

$$
\begin{array}{ccc}
(0,0) & \longrightarrow & (0,1) \\
\downarrow & \cong & \downarrow \\
(1,0) & \longrightarrow & (1,1)
\end{array}
$$

(This is the "pseudo" version of the Gray tensor product; there's also a "lax" version: different shade of grey!)

The black and white tensor product make sense for any $\mathscr{V}$ at all, but the grey one doesn't. There's a canonical comparison from the funny/white product to the ordinary/black one, and the Gray/grey tensor product is a sort of "cofibrant replacement" in between.

The Gray tensor product $2 \otimes 2$ is cofibrant, and more generally, the model structure is compatible with the Gray tensor product.

The third thing of interest is the connection between 2-categories and bicategories. There is a model structure on the category **Bicat** of bicategories and strict homomorphisms. The notion of biequivalence still makes perfectly good sense, and these are the weak equivalences. The fibrations are once again the maps which are isofibrations on the hom-categories, and have the equivalence lifting property. Thus the full inclusion **2-Cat** $\hookrightarrow$ **Bicat** preserves and reflects weak equivalences and fibrations. This inclusion has a left adjoint "free strictification"

$$\textbf{2-Cat} \quad \perp \quad \textbf{Bicat}$$

given by universally making the associativity and identity isomorphisms into identities. This left adjoint is not a pseudomorphism classifier, since we are using strict morphisms of bicategories and of 2-categories, and it's not the usual strictification functor "st" either: in general the unit is *not* a biequivalence. But the component of the unit at a *cofibrant* bicategory *is* an equivalence. This fits well into the model category picture: it's part of what makes this adjunction a Quillen equivalence (one that induces an equivalence of homotopy categories). In fact the usual strictification can be seen as a derived version of the free strictification.

There exist bicategories (even monoidal categories) for which there does not exist a strict map into a 2-category which is a biequivalence, although we know that any $\mathscr{B}$ has a pseudofunctor $\mathscr{B} \rightsquigarrow \mathscr{B}_{st}$ which is a biequivalence. The point about cofibrant bicategories $\mathscr{B}$ is that "there aren't any equations between 1-cells," and this is what makes the unit at such a $\mathscr{B}$ an equivalence.

Just as for **2-Cat**, we have pseudomorphism classifiers in **Bicat**, which serve as "special cofibrant replacements."

The model structure on **2-Cat** is proper: showing that it's left proper (biequivalences are stable under pushout along cofibrations) is harder than any of the other results mentioned above, but the fact that it is right proper is an immediate consequence of the fact that every object is fibrant.

7.6. Back to 2-monads. There's a connection between the model structure on $\mathbf{Mnd}_f(\mathscr{K})$ and that on **2-Cat**. There's a 2-functor

$$\mathsf{sem} \colon \mathbf{Mnd}_f(\mathbf{Cat})^{\mathrm{op}} \to \mathbf{2\text{-}CAT}/\mathbf{Cat}$$

which you might call *semantics*, defined by:

$$T \mapsto (T\text{-Alg} \xrightarrow{U} \mathbf{Cat}).$$

and

$$(S \xrightarrow{f} T) \mapsto (T\text{-Alg} \xrightarrow{f^*} S\text{-Alg})$$

since if $a : TA \to A$ is a T-algebra, then its composite with $fA : SA \to TA$ makes A into an S-algebra.

In the ordinary unenriched case or the $\mathscr{V}$-enriched case, or even here, if we used $T \mapsto T\text{-Alg}_s$ rather than $T \mapsto T\text{-Alg}$, the semantics functor would be fully faithful. But the semantics functor defined above, using T-Alg, is not: to give a map $\mathsf{sem}(T) \to \mathsf{sem}(S)$ in **2-CAT**/**Cat** corresponds to giving a weak morphism from S to T, but not in the sense of pseudomorphisms of monads, considered above; rather in a still broader sense, in which the $fA : SA \to TA$ need not even be natural.

Now the definitions of fibration, weak equivalence, and trivial fibration in **2-Cat** have nothing to do with smallness, and make perfectly good sense in the category **2-CAT** of not-necessarily-small 2-categories. We can therefore define a morphism in **2-CAT**/**Cat** to be a fibration, weak equivalence, or trivial fibration if the underlying 2-functor in **2-CAT** is one.

Under these definitions, sem preserves limits, fibrations, and trivial fibrations, as one verifies using the 2-monads $\langle A, A \rangle$, $\{f, f\}_\ell$, and so on. Limits, fibrations, and trivial fibrations in $\mathbf{Mnd}_f(\mathbf{Cat})^{\mathrm{op}}$, correspond to colimits, cofibrations, and trivial cofibrations in $\mathbf{Mnd}_f(\mathbf{Cat})$. Thus, it should in principle be the right adjoint part of a Quillen adjunction. It's not, of course, because of size problems: **2-CAT**/**Cat** has large hom-categories, and sem lacks a left adjoint.

The assertion that sem preserves the weak equivalence $q : T' \to T$ is equivalent to the assertion that every pseudo T-algebra is equivalent to a strict one. More generally, sem preserves all weak equivalences if and only if pseudo algebras are equivalent to strict ones for every T. Whether or not this is the case is an open problem in the current generality, but it is true that sem preserves weak equivalences between cofibrant objects (flexible monads).

7.7. References to the literature. Model categories go back to [53]. There are now several modern treatments: [18] is one which emphasizes the compatibility between model structures and monoidal structures. The "categorical" model structure on **Cat** seems to be folklore; the first reference I know is [20]. The Thomason model structure on **Cat** comes from [61], which does have an error in the proof of properness, corrected in [9]. For the model structures on 2-Cat and Bicat see [35, 36]. There is also a "Thomason-style" model structure on 2-Cat [65]. The theory of model 2-categories, the model structure on the category of monads, and its relation to structure and semantics, all come from [38].

8. The formal theory of monads. In this section we return to formal category theory; in fact, to one of its high points: the formal theory of monads.

8.1. Generalized algebras. Let's start by thinking about ordinary monads. Let A be a category, $t = (t, \mu, \eta)$ a monad on A. Write A^t for the Eilenberg-Moore category (the category of algebras). The starting point is to think about the universal property of this construction. What is it to give a functor $C \xrightarrow{a} A^t$? We give an algebra ac for each $c \in C$, and use ac also for the name of the underlying object, with structure map $tac \xrightarrow{\alpha c} ac$. And for every $\gamma : c \to d$, we have an $a\gamma : ac \to ad$ with a commutative square

$$
\begin{array}{ccc}
tac & \xrightarrow{\alpha c} & ac \\
{\scriptstyle ta\gamma}\downarrow & & \downarrow{\scriptstyle a\gamma} \\
tad & \xrightarrow[\alpha d]{} & ad.
\end{array}
$$

This square awfully like a naturality square; it wants to say that α is natural with respect to γ, and indeed this is in fact the case. What we're actually doing is giving a functor $C \xrightarrow{a} A$ and a natural transformation

with equations of natural transformations

$$t^2a \xrightarrow{\mu a} ta \xleftarrow{\eta a} a$$

$$t\alpha \downarrow \qquad \downarrow \alpha \qquad \diagup 1$$

$$ta \xrightarrow{\alpha} a$$

which just says that on components, it makes each ac into a t-algebra.

You might call this a *generalized algebra*, or a *t-algebra with domain C*. Think of a usual algebra as an algebra with domain 1.

Similarly, you can look at natural transformations. To give a natural transformation

$$C \underset{b}{\overset{a}{\Downarrow}} A^t$$

amounts to giving

$$C \underset{b}{\overset{a}{\Downarrow\varphi}} A$$

which is suitably compatible, in the sense that

$$ta \xrightarrow{t\varphi} tb$$

$$\alpha \downarrow \qquad \downarrow \beta$$

$$a \xrightarrow{\varphi} b$$

This is the universal property of the Eilenberg-Moore construction, and the starting point of the theory.

I've been talking all along about categories, but once we've moved beyond algebras with domain 1, there's no reason to restrict in that way, so we can talk instead about a monad on an object A in any 2-category $\mathcal{K}$. (The notion of monad has not been weakened in any way. The 2-category $\mathcal{K}$ might be **Cat**, or **2-Cat**, or $\mathcal{V}$-**Cat**, but we use the same definition.)

We can't just *construct* A^t as we did before, but we can *ask* whether there exists an object A^t with the universal property. A slick way to do this is as follows. The hom-category $\mathcal{K}(C, A)$ has a monad $\mathcal{K}(C, t)$ on it (since 2-functors take monads to monads), and this is the ordinary type of monad in **Cat**. The endofunctor part of this monad sends $a \colon C \to A$ to $ta \colon C \to A$. This generalized notion of algebra is then nothing but the usual sort of algebra for the ordinary monad $\mathcal{K}(C, t)$. So what we want is an isomorphism

$$\mathcal{K}(C, A^t) \cong \mathcal{K}(C, A)^{\mathcal{K}(C,t)}$$

naturally in C (where the right hand side means the ordinary Eilenberg-Moore category of algebras for the ordinary monad $\mathscr{K}(C,t)$). We call A^t the Eilenberg-Moore object of t, or EM-object for short.

It turns out that in some 2-categories, such as **Cat**, it's enough to check the universal property for $C = 1$, since 1 generates **Cat**, in a suitable sense; but in an abstract 2-category there may not be a 1, and if there is one, it may not be enough to get the full universal property. Of course there are similar phenomena in ordinary category theory.

The universal property of A^t makes it look like a limit, and indeed it is one, but we'll look at some other points of view first.

8.2. Monads in $\mathscr{K}$. Let $\mathscr{K}$ be a 2-category. Previously we looked at the 2-category $\mathbf{Mnd}_f(\mathscr{K})$ of finitary 2-monads *on* $\mathscr{K}$ (as a fixed base object). We now consider the 2-category $\mathbf{mnd}(\mathscr{K})$ of all the (internal) monads *in* $\mathscr{K}$, with variable base object.

- Its objects are monads in $\mathscr{K}$.
- Its 1-cells correspond to morphisms which lift to the level of algebras:

$$
\begin{array}{ccc}
A^t & \xrightarrow{\overline{m}} & B^s \\
{\scriptstyle u^t}\downarrow & & \downarrow{\scriptstyle u^s} \\
A & \xrightarrow{m} & B
\end{array}
$$

where we assume temporarily that A^t and B^s exist, and we can think of this commutative diagram as an identity 2-cell and then take its mate, since the u's are right adjoints:

$$
\begin{array}{ccc}
A & \xrightarrow{m} & B \\
{\scriptstyle f^t}\downarrow & \nearrow & \downarrow{\scriptstyle f^s} \\
A^t & \xrightarrow{\overline{m}} & B^t \\
{\scriptstyle u^t}\downarrow & & \downarrow{\scriptstyle u^s} \\
A & \xrightarrow{m} & B
\end{array}
$$

which we then paste together to get a 2-cell, and the forgetful-free composite gives us the monads. Thus we should define a morphism of monads to be a morphism $m : A \to B$ with a 2-cell $\varphi : sm \to mt$ such that the diagrams

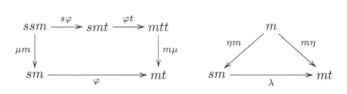

commute. We could *define* a morphism of monads to be a morphism $m : A \to B$ with a lifting $\overline{m} : A^t \to B^s$ as above, except that the EM-objects need not exist in a general 2-category; but when they do exist, the two descriptions are equivalent.

- The 2-cells in $\mathbf{mnd}(\mathcal{K})$ are 2-cells

$$A \underset{n}{\overset{m}{\Longrightarrow}}{\Downarrow\rho} \; B$$

in $\mathcal{K}$ with a compatibility condition, which you could express as saying that ρ lifts to a $\overline{\rho}$ between EM-objects, or you could equivalently express as saying that

$$
\begin{array}{ccc}
sm & \overset{\varphi}{\longrightarrow} & mt \\
{\scriptstyle s\rho}\downarrow & & \downarrow{\scriptstyle \rho t} \\
sn & \underset{\psi}{\longrightarrow} & nt
\end{array}
$$

commutes.

There's a full embedding $\mathrm{id} \colon \mathcal{K} \hookrightarrow \mathbf{mnd}(\mathcal{K})$ sending A to the identity monad $(A, 1)$ on A, and doing the obvious thing on 1-cells and 2-cells. This is particularly clear in the EM-objects picture, since if $t = 1$ then $A^t = A$, so obviously m will lift uniquely to an $\overline{m}$, which is what fully faithfulness of id says.

A trivial observation is that for any monad we can always choose to forget the monad and be left with the object, and this is left adjoint to id. The more interesting thing, however, is the existence of a right adjoint: this amounts exactly to a choice of an EM-object for each monad in $\mathcal{K}$. Why? Look at the universal property: If $A \mapsto A^t$ is the right adjoint, this says that

$$\mathcal{K}(C, A^t) \cong \mathbf{mnd}(\mathcal{K})((C, 1), (A, t))$$

The key point is that the right hand side is equal to $\mathcal{K}(C, A)^{\mathcal{K}(C,t)}$, since an object (that is, a morphism $(C, 1) \to (A, t)$) involves a and α as in

$$
\begin{array}{ccc}
C & \overset{a}{\longrightarrow} & A \\
{\scriptstyle 1}\downarrow & {\Swarrow}{\scriptstyle \alpha} & \downarrow{\scriptstyle t} \\
C & \underset{a}{\longrightarrow} & A
\end{array}
$$

subject to exactly the conditions which make (a, α) into a generalized algebra.

Now the really beautiful thing happens: we can start looking at duals of $\mathcal{K}$ and see what happens. Consider first $\mathcal{K}^{\mathrm{co}}$, where we reverse the

2-cells but not the 1-cells. A monad in $\mathscr{K}^{\mathrm{co}}$ is then a *comonad* in $\mathscr{K}$. And an EM-object in $\mathscr{K}^{\mathrm{co}}$ is the obvious analogue for comonads. If $\mathscr{K} = \mathbf{Cat}$, we get ordinary comonads, and the EM-object is the usual category of coalgebras for the comonad.

That's nice, but not incredibly surprising. What's more interesting is what happens in $\mathscr{K}^{\mathrm{op}}$. When we consider monads not in $\mathscr{K}$ but in $\mathscr{K}^{\mathrm{op}}$ we have to reverse the direction of the 1-cells, as in

but this is nothing but a monad in $\mathscr{K}$!

But what about the EM-object? The arrows are reversed, so we get a different universal property. An algebra for this monad consists of

The wonderful thing is that in the case $\mathscr{K} = \mathbf{Cat}$ this is the same thing as a map $A_t \to C$ where A_t is the Kleisli category. Recall that the Kleisli category can be defined as the full subcategory of the Eilenberg-Moore category consisting of the *free* T-algebras (the algebras of the form $(tx, \mu x)$ for some $x \in A$), or equivalently as the category with the same objects as A, but with morphisms from x to y given by the morphisms in A from x to ty (the monad structure is then used to make this into a category). The latter description is more convenient here: given $(a : A \to C, \alpha : at \to a)$ as above, the induced functor $A_t \to C$ sends an object $x \in A$ to ax, and a morphism $\varphi : x \to ty$ to the composite

$$ax \xrightarrow{\ a\varphi\ } aty \xrightarrow{\ \alpha y\ } ay$$

in C.

For a general 2-category $\mathscr{K}$, the Eilenberg-Moore objects in $\mathscr{K}^{\mathrm{co}}$ are called *Kleisli objects* (in $\mathscr{K}$). It's true in any 2-category that the EM-object is the terminal adjunction giving rise to the monad, and the Kleisli object is the initial one, but the universal property given above is richer in that it refers to maps with arbitrary domains.

Using $\mathscr{K}^{\mathrm{coop}}$, of course, gives you Kleisli objects for comonads.

8.3. The monad structure of mnd . Now, where does the construction $\mathbf{mnd}(\mathscr{K})$ really live? Consider the category $\mathbf{2\text{-}Cat}$ of 2-categories and 2-functors. Completely banish from your mind all concerns about size, which doesn't have any role here. So far we've constructed a 2-category $\mathbf{mnd}(\mathscr{K})$ for any 2-category $\mathscr{K}$. and this is clearly completely functorial, so we get a functor

$$\mathbf{mnd} \colon \textbf{2-Cat} \to \textbf{2-Cat}$$

and the inclusion id is clearly natural in $\mathscr{K}$, so we get a natural transformation

$$\textbf{2-Cat} \quad \overset{1}{\underset{\mathbf{mnd}}{\Downarrow \mathsf{id}}} \quad \textbf{2-Cat}$$

A certain sort of person is tempted to wonder whether this is part of the structure of a monad on 2-Cat! We do have a composition map

$$\textbf{2-Cat} \quad \overset{\mathbf{mnd}^2}{\underset{\mathbf{mnd}}{\Downarrow \mathsf{comp}}} \quad \textbf{2-Cat}$$

and what it does is one of the most striking aspects of the formal theory of monads.

This composition map sends a monad in $\mathbf{mnd}(\mathscr{K})$ to a monad in $\mathscr{K}$. What is a monad in $\mathbf{mnd}(\mathscr{K})$? It consists of

- a monad (A, t) in $\mathscr{K}$ (an object of $\mathbf{mnd}(\mathscr{K})$)
- an endo-1-cell, which consists of a morphism $A \xrightarrow{s} A$ in $\mathscr{K}$ with a 2-cell $ts \xrightarrow{\lambda} st$ (with conditions)
- A multiplication $(s, \lambda)(s, \lambda) \to (s, \lambda)$, corresponding to $s^2 \xrightarrow{\nu} s$ (with conditions)
- a unit $1 \to (s, \lambda)$ corresponding to $1 \to s$ (with conditions).

As well as the conditions for these to be 1-cells and 2-cells in $\mathbf{mnd}(\mathscr{K})$, we need the conditions for this to be a monad there. These make s itself into a monad on A in $\mathscr{K}$. The 2-cell λ is now what's called a *distributive law* between these two monads, which is exactly what you need to "compose" these two monads and get another monad.

Think about this as being like the tensor product of rings. $R \otimes S$ is the tensor product of the underlying abelian groups, with multiplication

$$R \otimes S \otimes R \otimes S \xrightarrow{1 \otimes \mathrm{tw} \otimes 1} R \otimes R \otimes S \otimes S \xrightarrow{m_R \otimes m_S} R \otimes S.$$

The point is we're trying to do something very similar, but here we're in a world where the tensor product is not commutative, so we don't have the twist. So λ plays the role of the twist; it's a "local" commutativity that only applies to these two objects. The conditions put on it are exactly what we need to make the composite st into a monad.

For example, the multiplication on st is then

$$stst \xrightarrow{s\lambda t} sstt \xrightarrow{ss\mu} sst \xrightarrow{\mu t} st.$$

The notion of distributive law, in the ordinary case of categories, is due to Jon Beck, and he proved that we have a bijection between distributive laws $ts \to st$ and "compatible" monad structures on st, and also to liftings of s to A^t (whenever A^t exists). It's not as well known as it should be and is frequently rediscovered.

You can also do this for $\mathscr{K}^{\mathrm{co}}$ or $\mathscr{K}^{\mathrm{op}}$ or $\mathscr{K}^{\mathrm{coop}}$, of course. A distributive law in $\mathscr{K}^{\mathrm{op}}$ is formally the same as a distributive law in $\mathscr{K}$, but now rather than liftings of s to A^t, it gives you extensions of t to A_s along the left adjoint $f_s : A \to A_s$.

REMARK 8.1. Operads are monoids in a monoidal category, so there is a corresponding notion of distributive law between operads. Furthermore, the passage from the monoidal category of collections to the monoidal category of endofunctors is strong monoidal, so distributive laws between operads induce distributive laws between the induced monads, and this process is compatible with the formation of the composite operad/monad. Just as not every monad arises from an operad, not every distributive law between monads arises from a distributive law between operads, even when the monads themselves do arise from operads.

EXAMPLE 16. *Groups are particular monoids in* **Set**, *so there is a corresponding notion of distributive law. If a group G acts on a group H, then there is a distributive law $G \times H \to H \times G$ sending (g, h) to $(g.h, g)$, and the induced "composite" is the semidirect product $H \rtimes G$. This generalizes to arbitrary monoids in a cartesian monoidal category.*

8.4. Eilenberg-Moore objects as limits. There are two ways to see Eilenberg-Moore objects as weighted limits. Remember that way back in Section 3.1, we saw that monads t in $\mathscr{K}$ correspond to lax functors $\tilde{t}: 1 \to \mathscr{K}$. Then the *lax limit* of $\tilde{t}$ is exactly the EM-object A^t.

I haven't explicitly discussed lax limits of lax functors, but it's not hard to extend the definition of lax limit to cover this case. Alternatively one can replace the lax functor by the corresponding 2-functor out of the "lax morphism classifier," and then just take the lax limit of the 2-functor. Let's see how this would work.

First recall how $\tilde{t}$ is defined. It sends $*$ to A, and 1_* to an endomorphism $t: A \to A$, the unit is the lax unit comparison, and the multiplication is the lax composition comparison. To understand the lax limit of these sorts of things, we should think about lax cones. A lax cone would involve a vertex C of $\mathscr{K}$, with just one component $C \xrightarrow{a} A$, and a lax naturality 2-cell for every 1-cell in 1:

$$
\begin{array}{ccc}
C & \xrightarrow{\ a\ } & A \\[2pt]
\| & \Swarrow_{\alpha} & \downarrow t \\[2pt]
C & \xrightarrow{\ a\ } & A
\end{array}
$$

and some conditions.

The lax morphism classifier on 1 is a 2-category **mnd** with a bijection

$$
\frac{\text{2-functors } \mathbf{mnd} \longrightarrow \mathscr{K}}{\text{lax functors } 1 \longrightarrow \mathscr{K}}
$$

but such lax functors are in turn the same as monads in $\mathscr{K}$. Thus **mnd** is the universal 2-category containing a monad. Remember that a monad in $\mathscr{K}$ is the same as a monoid in a hom-category, and we know the universal monoidal category containing a monoid is the "algebraic Δ," the category $\mathbf{Ord_f}$ of (possibly empty) finite ordinals. This is not the Δ of simplicial sets: an extra object has been added. Thus **mnd** has one object $*$ and $\mathbf{mnd}(*, *) = \mathbf{Ord_f}$.

Now we have a limit notion $(\tilde{t} \mapsto A^t)$, and we want to know the corresponding weight $J : \mathbf{mnd} \to \mathbf{Cat}$, so that $\{J, \tilde{t}\} = A^t$. We saw in Section 6.9 that the recipe for calculating J is to consider the Yoneda functor $\mathbf{mnd} \to [\mathbf{mnd}, \mathbf{Cat}]^{\mathrm{op}}$ and form the limit of it, or equivalently the colimit of $\mathbf{mnd}^{\mathrm{op}} \to [\mathbf{mnd}, \mathbf{Cat}]$. The colimit is the Kleisli object; since we are in a presheaf 2-category $[\mathbf{mnd}, \mathbf{Cat}]$ it is computed pointwise. The weight is called **alg**; it's now a straightforward exercise to calculate it.

Of course, in general, **alg**-weighted limits may or may not exist. Subject to the existence of the relevant limits, they can be built up from other limits we already know:

- First form the inserter of $A \underset{1}{\overset{t}{\rightrightarrows}} A$. This is an $A_1 \xrightarrow{\ k\ } A$ equipped with a 2-cell $tk \xrightarrow{\ \kappa\ } k$.
- Then take the equifier of $k(\eta k)$ and 1 to get an $A_2 \xrightarrow{\ k'\ } A_1$ such that the identity law holds.
- Finally take the equifier of something else to get the associativity.

In particular, this shows that EM-objects are PIE-limits, in fact *finite* PIE-limits.

8.5. Limits in $T\text{-Alg}_\ell$ and $T\text{-Alg}_c$. $T\text{-Alg}_\ell$ and $T\text{-Alg}_c$ are, recall, the 2-categories of strict T-algebras with lax and with colax morphisms. Recall also that we had nice pseudo-limits in $T\text{-Alg}$; here it's much harder.

In $T\text{-Alg}_\ell$, you have oplax limits, and in $T\text{-Alg}_c$ you have lax limits (it's a twisted world we live in!) These are much more restricted classes of limits, not including inserters, equifiers, comma objects and many of our other favourite limits.

I described how to construct inserters and equifiers in T-Alg: form the limit downstairs and show that the thing you get canonically becomes an algebra. This involves morphisms $(f, \overline{f})$ and $(g, \overline{g})$ between T-algebras (A, a) and (B, b), and 2-cells $fk \to gk$ for some k. If you look carefully at the construction, you'll see that $\overline{f}$ needs to be invertible, but $\overline{g}$ can be arbitrary. So you can form inserters and equifiers in T-Alg$_\ell$ provided that one of the 1-cells (the one that is, or tries to be, the domain of the 2-cells) is actually pseudo.

Dually, in T-Alg$_c$, it's the other 1-cell which needs to be pseudo. Now the Eilenberg-Moore object of a monad (A, t) can be constructed using the inserter $k : C \to A$ of t and 1_A, and then an equifier (see Section 8.4). Furthermore 1_A will always be strict, and it turns out that T-Alg$_c$ does have Eilenberg-Moore objects for monads. The most important case is where T-algebras are monoidal categories and so T-Alg$_c$ has opmonoidal functors. Then a monad in T-Alg$_s$ is an ordinary monad for which the category is monoidal, the endofunctor opmonoidal, and the natural transformations are opmonoidal natural transformations; this is sometimes called a Hopf monad.

8.6. The limit-completion approach. We can now see EM-objects as weighted limits in the strict sense, and there's a well-developed theory of free completions under classes of weighted limits. So we can form the free completion $\mathbf{EM}(\mathscr{K})$ of a 2-category $\mathscr{K}$ under EM-objects; or we can form the corresponding colimit completion $\mathbf{KL}(\mathscr{K})$, which freely adds Kleisli objects. These are related: $\mathbf{EM}(\mathscr{K}) = \mathbf{KL}(\mathscr{K}^{\mathrm{op}})^{\mathrm{op}}$.

The colimit side is more familiar to construct. To freely add all colimits to an ordinary category, we take the presheaf category; to add a restricted class, we take the closure in the presheaf category under the colimits we want to add. So here, to get $\mathbf{KL}(\mathscr{K})$, we take the closure of the representables in $[\mathscr{K}^{\mathrm{op}}, \mathbf{Cat}]$ under Kleisli objects. It's part of a general theorem that this works, at least when $\mathscr{K}$ is small.

Sometimes it can be tricky to calculate exactly which things appear in this completion process. You start with the representables and throw in the relevant colimits of representables. There will now be new diagrams, and we may have to add colimits for these. This can continue transfinitely. The nice thing about the case of Kleisli objects is that, as we shall see, it stops after one step.

Colimits in the functor category are constructed pointwise, so we construct Kleisli objects as in **Cat**. The key facts are:
- A left adjoint in **Cat** is of Kleisli type if and only if it is bijective on objects.
- These are closed under composition.

Now, given a monad t on A we throw in the Kleisli object A_t in $[\mathscr{K}^{\mathrm{op}}, \mathbf{Cat}]$, which may have a new monad s on it. We then throw in its Kleisli object for s to get $(A_t)_s$, but then the composite

$$A \longrightarrow A_t \longrightarrow (A_t)_s$$

is also a bijective-on-objects left adjoint, hence $(A_t)_s$ is also a Kleisli object for a monad on A. Thus this is a 1-step process.

Therefore, we can identify (up to equivalence) the objects of $\mathbf{KL}(\mathscr{K})$ with monads in $\mathscr{K}$, and then explicitly describe morphisms and 2-cells between them in terms of $\mathscr{K}$ itself.

In the dual case $\mathbf{EM}(\mathscr{K}) = (\mathbf{KL}(\mathscr{K}^{\mathrm{op}}))^{\mathrm{op}}$ we get

- The objects are the monads in $\mathscr{K}$.
- The morphisms are the monad morphisms (same as in $\mathbf{mnd}(\mathscr{K})$), and

- The 2-cells $(A,t)\ \underset{(n,\psi)}{\overset{(m,\varphi)}{\Downarrow}}\ (B,s)$ are 2-cells $m \to sn$ (which should look "Kleisli-like") with some compatibility with t.

Composition is like in the Kleisli category. Think of sn as the "free s-algebra on n," so using the universal property of free algebras, can express this as something $sm \to sn$, and express compatibility that way.

Why is this a good thing to do?

(i) We still have a fully faithful inclusion id : $\mathscr{K} \to \mathbf{EM}(\mathscr{K})$, and by general nonsense for limit-completions, a right adjoint to id is just a choice of EM-objects in $\mathscr{K}$.

(ii) It comes up in examples. If we start with **Span**, we've seen that categories are just monads in **Span**, and that functors can be seen as special morphisms between such monads; now we can also deal with natural transformations. There is a 2-functor

$$\mathbf{Cat} \to \mathbf{KL}(\mathbf{Span})$$

which is bijective-on-objects and locally fully faithful, so that the bicategory $\mathbf{KL}(\mathbf{Span})$ captures precisely the notion of natural transformation. This works equally well for $\mathbf{Cat}(\mathbb{E})$, for $\mathscr{V}$-$\mathbf{Cat}$, or for generalized multicategories.

(iii) Remember that a distributive law is a monad in $\mathbf{mnd}(\mathscr{K})$. The multiplication and unit are 2-cells in $\mathbf{mnd}\mathscr{K}$, so if we change the 2-cells, the notion of monad changes. A monad in $\mathbf{EM}(\mathscr{K})$ is more general: we call it a *wreath*, since the composition operation is a wreath product.

A wreath still lives on a monad (A,t) in $\mathscr{K}$. We have an endomorphism $s : A \to A$ as before, along with a 2-cell $\lambda : ts \to st$ with some conditions as before, but s is no longer a monad: the multiplication is now something $\nu : ss \to st$, and the unit $\sigma : 1 \to st$. You can still make sense of associativity and unit using λ, but everything ends up in st. Ultimately this gives a monad structure on st, which is called the *wreath product* or *composite* of s and t.

For example, consider the monoidal category **Set** under cartesian product. This can be regarded as a one-object bicategory, and so, after strictification, as a 2-category. Let G be a group acting on an abelian group A, and consider a normalized 2-cocycle $G \times G \xrightarrow{\rho} A$. We consider A as a monoid (in **Set**). G happens also to be a monoid (in fact a group), but the monoid structure isn't used directly. Rather we have the action

$$\lambda \colon G \times A \to A \times G \colon (g, a) \mapsto ({}^g a, g)$$

and the "$A \times G$-valued multiplication"

$$\nu \colon G \times G \longrightarrow A \times G \colon (g, h) \mapsto (\rho(g, h), gh)$$

which gives a wreath, and so induces a monoid structure $A \rtimes G$ (which is actually a group). The multiplication is the usual one coming from the cocycle.

There's a corresponding thing for Hopf algebras, giving a type of "twisted smash product."

8.7. The module-theoretic approach. Here's a different point of view, which is particularly suggestive if we take $\mathscr{K}$ to be the monoidal category (1-object bicategory) **Ab** of abelian groups. Then a monad (monoid) in **Ab** is a ring R: the objects of **EM(Ab)** are the rings.

We defined a morphism $(f, \varphi) \colon (A, t) \to (B, s)$ in **EM**$(\mathscr{K})$ to consist of a 1-cell $f \colon A \to B$ and a 2-cell $\varphi \colon sf \to ft$ subject to two equations. A 1-cell $R \to S$ in **EM(Ab)** consists of an abelian group M and a map $S \otimes M \to M \otimes R$. Think of this as being a bimodule structure on $M \otimes R$; the left action is

$$S \otimes M \otimes R \longrightarrow M \otimes R \otimes R \longrightarrow M \otimes R$$

and the right action is the free one, and the conditions on φ are equivalent to the bimodule axioms. Thus the 1-cells are the *right-free bimodules*. The 2-cells are the bimodule homomorphisms.

Composition of 1-cells is the ordinary module composition, but because of the freeness condition, don't need to use any coequalizers. If we were to look at **KL**$(\mathscr{K})$, we'd get the *left-free* modules.

One could also consider arbitrary modules. This is an important construction, but it requires the bicategory to have coequalizers in the hom-categories in order to define composition; and these coequalizers to be preserved by whiskering on either side in order for this composition to be associative (up to isomorphism), and so this has rather a different flavour.

8.8. References to the literature. The formal theory of monads goes back of course to [54]; for the account using limit-completions, and the notion of wreath see the much later sequel [40]. Distributive laws (for ordinary monads) are due to Beck [3]. The Eilenberg-Moore object

was described as a lax limit of a lax functor in "Two constructions on
lax functors" [55]: it was the first construction, the Kleisli object was the
second. This was done using weighted limits in [56]. For limits in T-Alg$_l$,
including Eilenberg-Moore objects for comonads, see [37]; for Hopf monads
see [48] and also [47].

9. Pseudomonads. These are formally very similar to monoidal categories. A pseudomonad involves a thing T, which plays the role of a
category, a multiplication $m : T^2 \to T$, a unit $i : 1 \to T$, an associativity
isomorphism

$$
\begin{array}{ccc}
T^3 & \xrightarrow{\ mT\ } & T^2 \\
{\scriptstyle Tm}\downarrow & \cong & \downarrow{\scriptstyle m} \\
T^2 & \xrightarrow{\ m\ } & T
\end{array}
$$

unit isomorphisms λ, ρ, and so on, all looking very like a monoidal category. Just as monads can be defined in any 2-category or bicategory,
pseudomonads can be defined in any Gray-category or tricategory.

The monoidal 2-category **Cat** (with cartesian structure) can be regarded as a one-object tricategory, and a pseudomonad in this tricategory
is precisely a monoidal category. The associativity pentagon becomes a
cube, relating ways to go from T^4 to T, involving a bunch of μ's and
a pseudonaturality isomorphism. In monoidal categories, one side of the
cube corresponds to

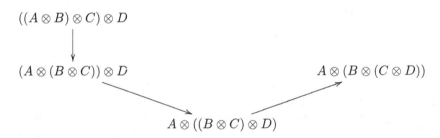

and the other side corresponds to

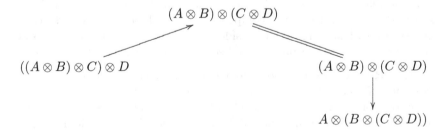

where in general, the equality will be replaced by an isomorphism, saying that it doesn't matter whether we tensor A and B first, then C and D, or vice versa.

Our unit isomorphisms have the form

If we were to consider lax monads, so that λ and ρ were not necessarily invertible, their direction would change, but for coherence problems it's useful to look at μ, λ, and ρ as rewriting rules, and then one wants them to go from the more complicated expression to the simpler.

It is convenient to work with Gray-categories rather than tricategories; by the coherence result that every tricategory is triequivalent to a Gray-category there is no loss of generality. Note, however, that the one-object tricategory corresponding to **Cat** is not a Gray-category, although it is a very special sort of tricategory.

One reason for working with Gray-categories is that we can then make use of the substantial machinery developed for enriched categories.

9.1. Coherence. The coherence result describes the fact that **there's a universal Gray-category with a pseudomonad in it**: there's a Gray-category Psm such that for any Gray-category $\mathbb{A}$, to give a Gray-functor Psm $\to \mathbb{A}$ is equivalent to giving a pseudomonad in $\mathbb{A}$. Corresponding to the identity Gray-functor Psm $\to$ Psm there is a pseudomonad in Psm, and this is the universal pseudomonad.

Psm is a sort of cofibrant replacement of **mnd**. More precisely, Psm like **mnd** has a single object $*$, and Psm$(*, *)$ is a cofibrant replacement of **mnd**$(*, *)$. It's not a pseudomorphism classifier: that would be too large; we need a smaller cofibrant replacement. Recall that **mnd**$(*, *) = \mathbf{Ord_f}$, the category of finite ordinals, or "algebraists' simplicial category." The underlying category of Psm$(*, *)$ (which is a 2-category, since Psm is a Gray-category) is freely generated by the face and degeneracy maps in $\mathbf{Ord_f}$ (forget the relations we expect to hold)

$$\longrightarrow \; \underset{\longrightarrow}{\overset{\longrightarrow}{\longleftarrow}} \; \cdots$$

Since this graph G generates $\mathbf{Ord_f}$, we have a map $FG \to \mathbf{Ord_f}$ which is bijective on objects and surjective on objects, so we can factor it as a bijective-on-objects-bijective-on-arrows 2-functor followed by an locally-fully-faithful one (throw in isomorphisms between the things that would become equal in $\mathbf{Ord_f}$), to get

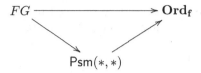

To construct the pseudofunctor classifier $\mathbf{Ord'_f}$, we would forgot all the way down to the underlying graph of $\mathbf{Ord_f}$, rather than a *generating* graph for it, and that would give a much larger cofibrant resolution, including, for example, a generating operation $T^n \to T$ for any n.

We also saw something like this for the Gray tensor product, which was obtained by factorizing the map from the funny tensor to the ordinary one.

You now have to define the composition in Psm

$$\mathbf{Psm}(*,*) \otimes \mathbf{Psm}(*,*) \longrightarrow \mathbf{Psm}(*,*)$$

to make it a Gray-category. You basically take the composition in $\mathbf{Ord_f}$, use that to define it on the generators, then build it up to deal with arbitrary 1-cells, but since the relations only hold up to isomorphism, that's why the Gray-tensor appears.

Now you prove that this has the universal property that I said it does, so it really does classify pseudo-monads in a Gray-category. I'm certainly not going to do that. Roughly, how does it go? Given a pseudomonad, we have

$$1 \xrightarrow{\ i\ } T \underset{\substack{\xrightarrow{iT} \\ \xleftarrow{m} \\ \xrightarrow{Ti}}}{} T^2 \qquad \cdots$$

and so on, which defines the putative Gray-functor $\mathsf{Psm} \to \mathbb{A}$ on objects, 1-cells, and 2-cells. The fun starts when we come to the 3-cells: we have μ, λ, and ρ, and we need to build up all the other required 3-cells. The idea is that for any 2-cell f in Psm (any 1-cell in the above picture, generated by m's and i's), there is a normal form $\overline{f}$ and a unique isomorphism $f \cong \overline{f}$ built up out of the 3-cells in Psm that one might expect to call μ, λ, and ρ. Thus any 3-cell $f \cong g$ in Psm can be written as a composite $f \cong \overline{f} = \overline{g} \cong g$, and this can be used to define the Gray-functor $\mathsf{Psm} \to \mathbb{A}$ on a 3-cells. The details of the rewrite system that these normal forms come from are a bit technical.

9.2. Algebras. The next step is to construct a particular weight $\mathsf{Psa} : \mathsf{Psm} \to \mathsf{Gray}$ such that for any Gray-functor $\mathbb{T} : \mathsf{Psm} \to \mathbb{A}$, the weighted limit $\{\mathsf{Psa}, \mathbb{T}\}$ is the object of pseudoalgebras, pseudomorphisms, and algebra 2-cells (all suitably defined) for the pseudomonad corresponding to $\mathbb{T}$. Again, this is sort of a "cofibrant replacement" for the corresponding one for 2-categories, although the domain has changed.

I won't do this, but I do want to make one point. It is the fact that we are working with Gray-categories rather than 3-categories which causes the pseudomorphisms to appear here. Recall that for ordinary monads, we talked about the fact that to give something $C \to A^t$ is the same as $a \colon C \to A$ with an action $\alpha : ta \to a$, where $c \mapsto (\alpha c : tac \to ac)$, and $\gamma \colon c \to d$ is sent to

$$
\begin{array}{ccc}
tac & \xrightarrow{\;\alpha c\;} & ac \\
{\scriptstyle ta\gamma}\downarrow & & \downarrow{\scriptstyle a\gamma} \\
tad & \xrightarrow[\;\alpha d\;]{} & ad
\end{array}
$$

and the fact that $a\gamma$ is a homomorphism can be seen as the naturality of α. There's an analogous fact for operads and Lawvere theories: the actions are natural with respect to homomorphisms.

When we come up to the Gray situation, we are thinking of pseudo-natural transformations, hence the square commutes up to isomorphism, so we get pseudo-morphisms, not strict ones. That's the "reason" for making the formal theory of pseudo-monads live in the Gray context. Even if you wanted only to consider 3-categories $\mathbb{A}$, the fact of working over **Gray** gives you the pseudomorphisms.

9.3. References to the literature. The basic definitions involving pseudomonads in Gray-categories were given in [46, 45], or in the equivalent language of pseudomonoids in [11]; for the universal pseudomonad, and the Gray-limit approach to pseudoalgebras, see [33]. For further work see also [8, 58, 41].

10. Nerves. In this section we use $\mathbf{\Delta}$ for the "topologists' delta," the category of *non-empty* finite ordinals. As usual, we write $[n]$ for the ordinal $\{0 < 1 < \ldots < n\}$. This section is particularly light on details; see [39] for more.

The nerve of an ordinary category $\mathscr{C}$ is the simplicial set $N\mathscr{C}$ in which
- a 0-simplex is an object
- a 1-simplex is a morphism
- a 2-simplex is a composable pair and its composite

and so on. This process gives a fully faithful embedding

$$\mathbf{Cat} \hookrightarrow [\mathbf{\Delta}^{\mathrm{op}}, \mathbf{Set}] = \mathbf{SSet}$$

into the category of simplicial sets.

The nerve of a *bicategory* $\mathscr{B}$ is the simplicial set $N\mathscr{B}$ in which
- a 0-simplex is an object
- a 1-simplex is a morphism
- a 2-simplex is a 2-cell living in a triangle

and so on. These 2-simplices are being overworked; they have to express both (something about) composition of 1-cells, as well as what the 2-cells are. Actually they don't ever encode what the composite of two 1-cells is, only what the morphisms out of the composite are. (This is like defining the tensor product of modules only in terms of bilinear maps, never actually specifying a choice of universal bilinear morphism.) Now this has its advantages, but it does make it hard to say when such composites are being preserved. In fact we get a fully faithful embedding

$$\mathbf{Bicat}_{\mathrm{nlax}} \hookrightarrow [\boldsymbol{\Delta}^{\mathrm{op}}, \mathbf{Set}]$$

where "nlax" indicates that we are taking the *normal lax functors* as morphisms: these are the lax functors which strictly preserve identities.

A lot of the time you want to talk about homomorphisms or strict homomorphisms rather than lax ones. If you want to get your hands on those there are various possibilities. One is to have a bit more structure than a simplicial set: specify as extra data a class of simplices called the *thin* simplices. Every degenerate simplex is thin, but there can also be non-degenerate thin simplices. The resulting structure is called a *stratified simplicial set*. The *stratified nerve* of a 2-category is the usual nerve, made equipped with a suitable stratification, in which a 2-simplex is thin when the 2-cell it contains is in fact an identity. (There is also a different stratification, often used for nerves of bicategories, in which a 2-simplex is then when the 2-cell is invertible.) One can now characterize the stratified simplicial sets which are stratified nerves of 2-categories (or bicategories); and indeed similarly for strict or weak ω-categories. The stratified simplicial sets which arise as nerves of ω-categories are called *complicial sets*.

A different way to specify this extra structure is to use simplicial objects not in **Set** but in **Cat**. For a bicategory $\mathscr{B}$, the 2-nerve $N_2\mathscr{B}$ of $\mathscr{B}$ (or just $N\mathscr{B}$ from now on) is a functor $N\mathscr{B} : \boldsymbol{\Delta}^{\mathrm{op}} \to \mathbf{Cat}$.

- The category $(N\mathscr{B})_0$ of 0-simplices is the discrete category consisting of the objects of $\mathscr{B}$.
- The category $(N\mathscr{B})_1$ of 1-simplices has morphisms of $\mathscr{B}$ as objects and 2-cells of $\mathscr{B}$ as morphisms. So far this looks like some kind of enriched nerve.
- The category $(N\mathscr{B})_2$ of 2-simplices doesn't need to include the 2-cells, since we already have them; we can therefore take 2-simplices to be *invertible* 2-cells

and morphisms of 2-simplices to consist of three 2-cells satisfying the evident compatibility conditions. (The domain and codomain 2-simplices will need to have the same three objects.)

- The category $(N\mathscr{B})_3$ of 3-simplices is the category of tetrahedra all of whose faces are isomorphisms

and so on. We'd like a functorial description of the 2-nerve. Consider **NHom**, the 2-category of bicategories, normal homomorphisms, and icons (recall that these are oplax natural transformations all of whose 1-cell components are identities). Now **Cat** $\hookrightarrow$ **NHom**, where **Cat** is the locally discrete 2-category consisting of categories, functors, and only identity natural transformations, embedding as a full sub-2-category consisting of the locally discrete bicategories. (An icon between functors can only be an identity.) And of course we have $\mathbf{\Delta} \hookrightarrow$ **Cat**, so the composite fully faithful $J\colon \mathbf{\Delta} \hookrightarrow$ **NHom** induces

$$\mathbf{NHom}(J,1)\colon \mathbf{NHom} \longrightarrow [\mathbf{\Delta}^{\mathrm{op}}, \mathbf{Cat}]$$

sending $\mathscr{B}$ to $\mathbf{NHom}(J-, \mathscr{B})$.

For instance, J sends $[0] \in \mathbf{\Delta}$ goes to the terminal bicategory, and a normal homomorphism from that into $\mathscr{B}$ is just an object of $\mathscr{B}$, with no room for icons, so $\mathbf{NHom}(J[0], \mathscr{B})$ is indeed the discrete category $(N\mathscr{B})_0$ of objects of $\mathscr{B}$. Similarly J sends $[1] \in \mathbf{\Delta}$ goes to the arrow category 2, and a normal homomorphism from this into $\mathscr{B}$ is an arrow in $\mathscr{B}$, and $\mathbf{NHom}(J[1], \mathscr{B})$ is just $(N\mathscr{B})_1$, and so on.

THEOREM 10.1. $\mathbf{NHom}(J,1) = N$ *is a fully faithful 2-functor (in a completely strict sense) and has a left biadjoint.*

The fact that $N = \mathbf{NHom}(J,1)$ is a straightforward direct calculation. The existence of the left biadjoint can be proved using techniques of 2-dimensional universal algebra.

How can we characterize the image of N? $X \in [\mathbf{\Delta}^{\mathrm{op}}, \mathbf{Cat}]$ is isomorphic to some $N\mathscr{B}$ if and only if

(a) X_0 is discrete;

(b) X is *3-coskeletal*; that is, isomorphic to the right Kan extension of its 3-truncation — the idea is that 4-simplices and higher are uniquely determined by their boundary;

(c) $X_2 \to \mathrm{cosk}_1(X)_2$ is a *discrete isofibration*. A functor $p\colon A \to B$ is a discrete isofibration if given $e \in E$ and $\beta\colon b \cong pe$, there exists a *unique* $\varepsilon\colon e' \cong e$ with $p\varepsilon = \beta$. This implies that if

$$X \underset{\Longrightarrow}{\overset{\varepsilon}{\rightrightarrows}} E$$

and $p\varepsilon = \mathrm{id}$, then $\varepsilon = \mathrm{id}$;

(d) $X_3 \to \mathrm{cosk}_1(X)_3$ (could also use the 2-coskeleton) is also a discrete isofibration;

(e) the Segal maps are equivalences.

A *Tamsamani weak 2-category*, or just *Tamsamani 2-category*, since no strict notion is considered, is a functor $X : \mathbf{\Delta}^{\mathrm{op}} \to \mathbf{Cat}$ satisfying

(a) and (d); thus the 2-nerve of a bicategory is a Tamsamani 2-category. Tamsamani suggests a way of getting from a bicategory to a Tamsamani 2-category, but it is not the 2-nerve construction given here.

The inclusion of **NHom** into Tamsamani 2-categories looks like it should be a biequivalence, but it's not quite. It would be if you broadened the definition of morphism of Tamsamani 2-category to include what might be called the "normal pseudonatural transformations."

Finally a warning: what you might guess for the nerve of a bicategory is to have

- $N\mathscr{B}_0$ the objects
- $N\mathscr{B}_1 = \sum_{x,y} \mathscr{B}(x,y)$
- $N\mathscr{B}_2 = \sum_{x,y,z} \mathscr{B}(y,z) \times \mathscr{B}(x,y)$
- $N\mathscr{B}_3 = \sum_{w,x,y,z} \mathscr{B}(y,z) \times \mathscr{B}(x,y) \times \mathscr{B}(w,x)$

in analogy with the case of nerves of ordinary categories. If you try to do this, the simplicial identities fail, due to the failure of associativity. Actually, what we do is to take the pseudo-limit of the composition functor

$$ \sum_{x,y,z} \mathscr{B}(y,z) \times \mathscr{B}(x,y) \longrightarrow \sum_{x,y} \mathscr{B}(x,y) $$

as $(N\mathscr{B})_2$. And this continues: for composable triples, we have

$$
\begin{array}{ccc}
\mathscr{B}^3 & \longrightarrow & \mathscr{B}^2 \\
\downarrow & \cong & \downarrow \\
\mathscr{B}^2 & \longrightarrow & \mathscr{B}
\end{array}
$$

and $N\mathscr{B}_3$ is the pseudo-limit of this whole diagram. Going on, we can construct each $(N\mathscr{B})_n$ as the pseudolimit of some higher cube.

10.1. References to the literature. The notion of nerve of a bicategory is due to Street. For nerves of ω-categories as stratified simplicial sets, see [62], and the references therein. The notion of 2-nerve of a bicategory is described in [39]. Tamsamani's definition of weak n-category is in [60].

Acknowledgements. It is a pleasure to acknowledge support and encouragement from a number of sources. I am grateful to the Institute for Mathematics and its Applications, Minneapolis for hosting and supporting the workshop on higher categories in 2004, and to John Baez and Peter May who organized the workshop and who encouraged me to publish these notes. The material here was based on lectures I gave at the University of Chicago in 2006, at the invitation of Peter May and Eugenia Cheng. I'm grateful to them for their hospitality, and the interest that they and the topology/categories group at Chicago took in these lectures. I'm particularly grateful to Mike Shulman, whose excellent TeXed notes of the lectures were the basis for the companion.

REFERENCES

[1] JIŘÍ ADÁMEK AND JIŘÍ ROSICKÝ. *Locally presentable and accessible categories*, Vol. 189 of *London Mathematical Society Lecture Note Series*. Cambridge University Press, Cambridge, 1994.

[2] MICHAEL BARR. Coequalizers and free triples. *Math. Z.*, **116**:307–322, 1970.

[3] JON BECK. Distributive laws. In *Sem. on Triples and Categorical Homology Theory (ETH, Zürich, 1966/67)*, pages 119–140. Springer, Berlin, 1969.

[4] JEAN BÉNABOU. Introduction to bicategories. In *Reports of the Midwest Category Seminar*, pages 1–77. Springer, Berlin, 1967.

[5] RENATO BETTI, AURELIO CARBONI, ROSS STREET, AND ROBERT WALTERS. Variation through enrichment. *J. Pure Appl. Algebra*, **29**(2):109–127, 1983.

[6] G.J. BIRD, G.M. KELLY, A.J. POWER, AND R.H. STREET. Flexible limits for 2-categories. *J. Pure Appl. Algebra*, **61**(1):1–27, 1989.

[7] R. BLACKWELL, G.M. KELLY, AND A.J. POWER. Two-dimensional monad theory. *J. Pure Appl. Algebra*, **59**(1):1–41, 1989.

[8] EUGENIA CHENG, MARTIN HYLAND, AND JOHN POWER. Pseudo-distributive laws. *Elec. Notes. Theoretical Comp. Sci*, **83**:1–3, 2004.

[9] DENIS-CHARLES CISINSKI. La classe des morphismes de Dwyer n'est pas stable par retractes. *Cahiers Topologie Géom. Différentielle Catég.*, **40**(3):227–231, 1999.

[10] JEAN-MARC CORDIER AND TIMOTHY PORTER. Homotopy coherent category theory. *Trans. Amer. Math. Soc.*, **349**(1):1–54, 1997.

[11] BRIAN DAY AND ROSS STREET. Monoidal bicategories and Hopf algebroids. *Adv. Math.*, **129**(1):99–157, 1997.

[12] EDUARDO J. DUBUC AND G.M. KELLY. A presentation of Topoi as algebraic relative to categories or graphs. *J. Algebra*, **81**(2):420–433, 1983.

[13] CHARLES EHRESMANN. *Catégories et structures*. Dunod, Paris, 1965.

[14] SAMUEL EILENBERG AND G. MAX KELLY. Closed categories. In *Proc. Conf. Categorical Algebra (La Jolla, Calif., 1965)*, pages 421–562. Springer, New York, 1966.

[15] THOMAS M. FIORE. Pseudo limits, biadjoints, and pseudo algebras: categorical foundations of conformal field theory. *Mem. Amer. Math. Soc.*, **182**(860):x+171, 2006.

[16] PETER GABRIEL AND FRIEDRICH ULMER. *Lokal präsentierbare Kategorien*. Springer-Verlag, Berlin, 1971.

[17] JOHN W. GRAY. *Formal category theory: adjointness for 2-categories*. Springer-Verlag, Berlin, 1974.

[18] MARK HOVEY. *Model categories*, Vol. 63 of *Mathematical Surveys and Monographs*. American Mathematical Society, Providence, RI, 1999.

[19] ANDRÉ JOYAL AND ROSS STREET. Pullbacks equivalent to pseudopullbacks. *Cahiers Topologie Géom. Différentielle Catég.*, **34**(2):153–156, 1993.

[20] ANDRÉ JOYAL AND MYLES TIERNEY. Strong stacks and classifying spaces. In *Category theory (Como, 1990)*, Vol. 1488 of *Lecture Notes in Math.*, pages 213–236. Springer, Berlin, 1991.

[21] G.M. KELLY. Coherence theorems for lax algebras and for distributive laws. In *Category Seminar (Proc. Sem., Sydney, 1972/1973)*, pages 281–375. Lecture Notes in Math., Vol. 420. Springer, Berlin, 1974.

[22] G.M. KELLY. Doctrinal adjunction. In *Category Seminar (Proc. Sem., Sydney, 1972/1973)*, pages 257–280. Lecture Notes in Math., Vol. 420. Springer, Berlin, 1974.

[23] G.M. KELLY. On clubs and doctrines. In *Category Seminar (Proc. Sem., Sydney, 1972/1973)*, pages 181–256. Lecture Notes in Math., Vol. 420. Springer, Berlin, 1974.

[24] G.M. KELLY. A unified treatment of transfinite constructions for free algebras, free monoids, colimits, associated sheaves, and so on. *Bull. Austral. Math. Soc.*, **22**(1):1–83, 1980.

[25] G.M. KELLY. Structures defined by finite limits in the enriched context. I. *Cahiers Topologie Géom. Différentielle*, **23**(1):3–42, 1982.

[26] G.M. KELLY. Elementary observations on 2-categorical limits. *Bull. Austral. Math. Soc.*, **39**(2):301–317, 1989.

[27] G.M. KELLY. Basic concepts of enriched category theory. *Repr. Theory Appl. Categ.*, (10):vi+137 pp. (electronic), 2005. Originally published as LMS Lecture Notes **64**, 1982.

[28] G.M. KELLY AND STEPHEN LACK. On property-like structures. *Theory Appl. Categ.*, **3**(9):213–250 (electronic), 1997.

[29] G.M. KELLY AND A.J. POWER. Adjunctions whose counits are coequalizers, and presentations of finitary enriched monads. *J. Pure Appl. Algebra*, **89**(1–2): 163–179, 1993.

[30] G.M. KELLY AND R. STREET. Review of the elements of 2-categories. In *Category Seminar (Proc. Sem., Sydney, 1972/1973)*, pages 75–103. Lecture Notes in Math., Vol. 420. Springer, Berlin, 1974.

[31] MAX KELLY, ANNA LABELLA, VINCENT SCHMITT, AND ROSS STREET. Categories enriched on two sides. *J. Pure Appl. Algebra*, **168**(1):53–98, 2002.

[32] STEPHEN LACK. On the monadicity of finitary monads. *J. Pure Appl. Algebra*, **140**(1):65–73, 1999.

[33] STEPHEN LACK. A coherent approach to pseudomonads. *Adv. Math.*, **152**(2): 179–202, 2000.

[34] STEPHEN LACK. Codescent objects and coherence. *J. Pure Appl. Algebra*, **175** (1–3):223–241, 2002.

[35] STEPHEN LACK. A Quillen model structure for 2-categories. *K-Theory*, **26**(2): 171–205, 2002.

[36] STEPHEN LACK. A Quillen model structure for bicategories. *K-Theory*, **33**(3): 185–197, 2004.

[37] STEPHEN LACK. Limits for lax morphisms. *Appl. Categ. Structures*, **13**(3): 189–203, 2005.

[38] STEPHEN LACK. Homotopy-theoretic aspects of 2-monads. *Journal of Homotopy and Related Structures*, to appear, available as arXiv.math.CT/0607646.

[39] STEPHEN LACK AND SIMONA PAOLI. 2-nerves of bicategories. *K-Theory*, to appear, available as arXiv.math.CT/0607271.

[40] STEPHEN LACK AND ROSS STREET. The formal theory of monads. II. *J. Pure Appl. Algebra*, **175**(1–3):243–265, 2002.

[41] AARON D. LAUDA. Frobenius algebras and ambidextrous adjunctions. *Theory Appl. Categ.*, **16**(4):84–122 (electronic), 2006.

[42] F. WILLIAM LAWVERE. Ordinal sums and equational doctrines. In *Sem. on Triples and Categorical Homology Theory (ETH, Zürich, 1966/67)*, pages 141–155. Springer, Berlin, 1969.

[43] F. WILLIAM LAWVERE. Metric spaces, generalized logic, and closed categories. *Rend. Sem. Mat. Fis. Milano*, **43**:135–166 (1974), 1973. Reprinted as Repr. Theory Appl. Categ. **1**:1–37, 2002.

[44] SAUNDERS MAC LANE AND ROBERT PARÉ. Coherence for bicategories and indexed categories. *J. Pure Appl. Algebra*, **37**(1):59–80, 1985.

[45] F. MARMOLEJO. Doctrines whose structure forms a fully faithful adjoint string. *Theory Appl. Categ.*, **3**(2):24–44 (electronic), 1997.

[46] F. MARMOLEJO. Distributive laws for pseudomonads. *Theory Appl. Categ.*, **5**(5):91–147 (electronic), 1999.

[47] PADDY MCCRUDDEN. Opmonoidal monads. *Theory Appl. Categ.*, **10**(19):469–485 (electronic), 2002.

[48] I. MOERDIJK. Monads on tensor categories. *J. Pure Appl. Algebra*, **168**(2–3): 189–208, 2002. Category theory 1999 (Coimbra).

[49] A.J. POWER. Coherence for bicategories with finite bilimits. I. In *Categories in computer science and logic (Boulder, CO, 1987)*, Vol. 92 of *Contemp. Math.*, pages 341–347. Amer. Math. Soc., Providence, RI, 1989.

[50] A.J. POWER. A general coherence result. *J. Pure Appl. Algebra*, **57**(2):165–173, 1989.

[51] JOHN POWER. Enriched Lawvere theories. *Theory Appl. Categ.*, **6**:83–93 (electronic), 1999. The Lambek Festschrift.

[52] JOHN POWER AND EDMUND ROBINSON. A characterization of pie limits. *Math. Proc. Cambridge Philos. Soc.*, **110**(1):33–47, 1991.

[53] DANIEL G. QUILLEN. *Homotopical algebra*. Lecture Notes in Mathematics, No. 43. Springer-Verlag, Berlin, 1967.

[54] ROSS STREET. The formal theory of monads. *J. Pure Appl. Algebra*, **2**(2): 149–168, 1972.

[55] ROSS STREET. Two constructions on lax functors. *Cahiers Topologie Géom. Différentielle*, **13**:217–264, 1972.

[56] ROSS STREET. Limits indexed by category-valued 2-functors. *J. Pure Appl. Algebra*, **8**(2):149–181, 1976.

[57] ROSS STREET. Fibrations in bicategories. *Cahiers Topologie Géom. Différentielle*, **21**(2):111–160, 1980.

[58] ROSS STREET. Frobenius monads and pseudomonoids. *J. Math. Phys.*, **45**(10): 3930–3948, 2004.

[59] ROSS STREET AND ROBERT WALTERS. Yoneda structures on 2-categories. *J. Algebra*, **50**(2):350–379, 1978.

[60] ZOUHAIR TAMSAMANI. Sur des notions de *n*-catégorie et *n*-groupoïde non strictes via des ensembles multi-simpliciaux. *K-Theory*, **16**(1):51–99, 1999.

[61] R.W. THOMASON. Cat as a closed model category. *Cahiers Topologie Géom. Différentielle*, **21**(3):305–324, 1980.

[62] DOMINIC VERITY. Complicial sets: Characterising the simplicial nerves of strict ω-categories. *Mem. Amer. Math. Soc.*, to appear.

[63] R.F.C. WALTERS. Sheaves and Cauchy-complete categories. *Cahiers Topologie Géom. Différentielle*, **22**(3):283–286, 1981.

[64] R.F.C. WALTERS. Sheaves on sites as Cauchy-complete categories. *J. Pure Appl. Algebra*, **24**(1):95–102, 1982.

[65] K. WORYTKIEWICZ, K. HESS, P.-E. PARENT, AND A. TONKS. A model structure à la Thomason on 2-cat. *J. Pure Appl. Algebra*, **208**(1):205–236, 2006.

NOTES ON 1- AND 2-GERBES

LAWRENCE BREEN*

The aim of these notes is to discuss in an informal manner the construction and some properties of 1- and 2-gerbes. They are for the most part based on the author's texts [1–4]. Our main goal is to describe the construction which associates to a gerbe or a 2-gerbe the corresponding non-abelian cohomology class.

We begin by reviewing the corresponding well-known theory for principal bundles and show how to extend this to biprincipal bundles (a.k.a bitorsors). After reviewing the definition of stacks and gerbes, we then construct the cohomology class associated to a gerbe. While the construction presented is equivalent to that in [4], it is clarified here by making use of diagrams (5.9) (a significant improvement on the corresponding diagram (2.4.7) of [4]) and (5.17). After a short discussion regarding the role of gerbes in algebraic topology, we pass from 1− to 2−gerbes. The construction of the associated cohomology classes follows the same lines as for 1-gerbes, but with the additional degree of complication entailed by passing from 1- to 2-categories, so that it now involves diagrams reminiscent of those in [5]. Our emphasis will be on explaining how the fairly elaborate equations which define cocycles and coboundaries may be reduced to terms which can be described in the traditional formalism of non-abelian cohomology.

Since the concepts discussed here are very general, we have at times not made explicit the mathematical objects to which they apply. For example, when we refer to "a space" this might mean a topological space, but also "a scheme" when one prefers to work in an algebro-geometric context, or even "a sheaf" and we place ourselves implicitly in the category of such spaces, schemes, or sheaves. Similarly, the standard notion of an X-group scheme G will correspond in a topological context to that of a bundle of groups on a space X. By this we mean a total space G above a space X that is a group in the cartesian monoidal category of spaces over X. In particular, the fibers G_x of G at points $x \in X$ are topological groups, whose group laws vary continuously with x.

Finally, in computing cocycles we will consider spaces X endowed with a covering $\mathcal{U} := (U_i)_{i \in I}$ by open sets, but the discussion will remain valid when the disjoint union $\coprod_{i \in I} U_i$ is replaced by an arbitrary covering morphism $Y \longrightarrow X$ of X for a given Grothendieck topology. The emphasis in vocabulary will be on spaces rather than schemes, and we have avoided any non-trivial result from algebraic geometry. In that sense, the text is implic-

*UMR CNRS 7539, Institut Galilée, Université Paris 13, 93430 Villetaneuse, France (breen@math.univ-paris13.fr).

J.C. Baez, J.P. May (eds.), *Towards Higher Categories*, The IMA Volumes
in Mathematics and its Applications 152, DOI 10.1007/978-1-4419-1524-5_5,
© Springer Science+Business Media, LLC 2010

itly directed towards topologists and category theorists rather than towards algebraic geometers, even though we have not sought to make precise the category of topological spaces in which we work.

It is a pleasure to thank Peter May, Bob Oliver and Jim Stasheff for their comments, and William Messing for his very careful reading of a preliminary version of this text.

1. Torsors and bitorsors.

1.1. Let G be a bundle of groups on a space X. The following definition of a principal space is standard, but note the occurence of a structural bundle of groups, rather than simply a constant one. We are in effect giving ourselves a family of groups G_x, parametrized by points $x \in X$, acting principally on the corresponding fibers P_x of P.

DEFINITION 1.1. *A left principal G-bundle (or left G-torsor) on a topological space X is a space $P \xrightarrow{\pi} X$ above X, together with a left group action $G \times_X P \longrightarrow P$ such that the induced morphism*

$$
\begin{array}{ccc}
G \times_X P & \simeq & P \times_X P \\
(g, p) & \mapsto & (gp, p)
\end{array}
\tag{1.1}
$$

is an isomorphism. We require in addition that there exists a family of local sections $s_i : U_i \longrightarrow P$, for some open cover $\mathcal{U} = (U_i)_{i \in I}$ of X. The groupoid of left G-torsors on X will be denoted $\mathrm{Tors}(X, G)$.

The choice of a family of local sections $s_i : U_i \longrightarrow P$, determines a G-valued 1-cochain $g_{ij} : U_{ij} \longrightarrow G$, defined above $U_{ij} := U_i \cap U_j$ by the equations

$$
s_i = g_{ij} s_j \qquad \forall\, i, j \in I.
\tag{1.2}
$$

The g_{ij} therefore satisfies the 1-cocycle equation

$$
g_{ik} = g_{ij} g_{jk}
\tag{1.3}
$$

above U_{ijk}. Two such families of local sections $(s_i)_{i \in I}$ and $(s'_i)_{i \in I}$ on the same open cover $\mathcal{U}$ differ by a G-valued 0-cochain $(g_i)_{i \in I}$ defined by

$$
s'_i = g_i s_i \qquad \forall i \in I
\tag{1.4}
$$

and for which the corresponding 1-cocycles g_{ij} and g'_{ij} are related to each other by the coboundary relations

$$
g'_{ij} = g_i\, g_{ij}\, g_j^{-1}
\tag{1.5}
$$

This equation determines an equivalence relation on the set $Z^1(\mathcal{U}, G)$ of 1-cocycles g_{ij}, and the induced set of equivalence classes for this equivalence relation is denoted $H^1(\mathcal{U}, G)$. Passing to the limit over open covers $\mathcal{U}$ of X yields the Čech non-abelian cohomology set $\check{H}^1(X, G)$, which classifies isomorphism classes of G-torsors on X. This set is endowed with a distinguished element, the class of the trivial left G-torsor $T_G := G \times X$.

DEFINITION 1.2. *Let X be a space, and G and H a pair of bundles of groups on X. A (G, H)-bitorsor on X is a space P over X, together with fiber-preserving left and right actions of G and H on P, which commute with each other and which define both a left G-torsor and a right H-torsor structure on P. For any bundle of groups G, a (G, G)-bitorsor is simply called a G-bitorsor.*

A family of local sections s_i of a (G, H)-bitorsor P above open sets U_i determines a local identification of P with both the trivial left G-torsor and the trivial right H-torsor. It therefore defines a family of local isomorphisms $u_i : H_{U_i} \longrightarrow G_{U_i}$ between the restrictions above U_i of the bundles H and G, which are explicitly given by the rule

$$s_i h = u_i(h) s_i \qquad (1.6)$$

for all $h \in H_{U_i}$. This however does not imply that the bundles of groups H and G are globally isomorphic.

EXAMPLE 1.

i) The trivial G-bitorsor on X: the right action of G on the left G-torsor T_G is the trivial one, given by fibrewise right translation. This bitorsor will also be denoted T_G.

ii) The group $P^{\mathrm{ad}} := \mathrm{Aut}_G(P)$ of G-equivariant fibre-preserving automorphisms of a left G-torsor P acts on the right on P by the rule

$$pu := u^{-1}(p)$$

so that any left G-torsor P is actually a (G, P^{ad})-bitorsor. The group P^{ad} is know as the gauge group of P (or more precisely the group of local gauge transformations). In particular, a left G-torsor P is a (G, H)-bitorsor if and only if the bundle of groups P^{ad} is isomorphic to H.

iii) Let

$$1 \longrightarrow G \xrightarrow{\ i\ } H \xrightarrow{\ j\ } K \longrightarrow 1 \qquad (1.7)$$

be a short exact sequence of bundles of groups on X. Then H is a G_K-bitorsor on K, where the left and right actions above K of the bundle of groups $G_K := G \times_X K$ are given by left and right multiplication in H:

$$(g, k) * h := f(g)\, h \qquad\qquad h * (g, k) := h\, f(g)$$

where $j(h) = k$.

1.2. Let P be a (G, H)-bitorsor and Q be an (H, K)-bitorsor on X. Let us define the contracted product of P and Q as follows:

$$P \wedge^H Q := \frac{P \times_X Q}{(ph, q) \sim (p, hq)}. \qquad (1.8)$$

It is a (G, K)-bitorsor on X via the action of G on P and the action of K on Q. To any (G, H)-bitorsor P on X is associated the opposite (H, G)-bitorsor P^o, with same underlying space as P, and for which the right action of G (resp. left action of H) is induced by the given left G-action (resp. right H-action) on P. For a given bundle of groups G on X, the category Bitors(X, G) of G-bitorsors on X is a group-like monoidal groupoid, in other words a monoidal category which is a groupoid, and in which every object has both a left and a right inverse. The tensor multiplication in Bitors(X, G) is the contracted product of G-bitorsors, the unit object is the trivial bitorsor T_G, and P^o is an inverse of the G-bitorsor P. Group-like monoidal groupoids are also known as *gr*-categories.

1.3. Twisted objects. Let P be a left G-torsor on X, and E a space over X on which G acts on the right. We say that the space $E^P := E \wedge^G P$ over X, defined as in (1.8), is the P-twisted form of E. The choice of a local section p of P above an open set U determines an isomorphism $\phi_p : E^P_{|U} \simeq E_{|U}$. Conversely, if E_1 is a space over X for which there exist a open cover $\mathcal{U}$ of X above which E_1 is locally isomorphic to E, then the space Isom$_X(E_1, E)$ is a left torsor on X under the action of the bundle of groups $G := \mathrm{Aut}_X E$.

PROPOSITION 1.1. *These two constructions are inverse to each other.*

EXAMPLE 2. Let G be a bundle of groups on X and H a bundle of groups locally isomorphic to G and let $P := \mathrm{Isom}_X(H, G)$ be the left Aut(G)-torsor of fiber-preserving isomorphisms from H to G. The map

$$
\begin{array}{ccc}
G \wedge^{\mathrm{Aut}(G)} P & \xrightarrow{\sim} & H \\
(g, u) & \mapsto & u^{-1}(g)
\end{array}
$$

identifies H with the P-twisted form of G, for the right action of Aut(G) on G induced by the standard left action. Conversely, given a fixed bundle of groups G on X, a G-torsor P determines a bundle of groups $H := G \wedge^{\mathrm{Aut}(G)} P$ on X locally isomorphic to G, and P is isomorphic to the left Aut(G)-torsor Isom(H, G).

The next example is very well-known, but deserves to be spelled out in some detail.

EXAMPLE 3. A rank n vector bundle $\mathcal{V}$ on X is locally isomorphic to the trivial bundle $\mathbb{R}^n_X := X \times \mathbb{R}^n$, whose group of automorphisms is the trivial bundle of groups

$$GL(n, \mathbb{R})_X := GL(n, \mathbb{R}) \times X$$

on X. The left principal $GL(n, \mathbb{R})_X$-bundle associated to $\mathcal{V}$ is its bundle of coframes $P_\mathcal{V} := \mathrm{Isom}(\mathcal{V}, \mathbb{R}^n_X)$. The vector bundle $\mathcal{V}$ may be recovered from $P_\mathcal{V}$ via the isomorphism

$$
\begin{array}{ccc}
\mathbb{R}^n_X \wedge^{GL(n, \mathbb{R})_X} P_\mathcal{V} & \xrightarrow{\sim} & \mathcal{V} \\
(y, p) & \mapsto & p^{-1}(y)
\end{array}
\tag{1.9}
$$

in other words as the $P_{\mathcal{V}}$-twist of the trivial vector bundle $\mathbb{R}_X^n$ on X. Conversely, for any principal $GL(n, \mathbb{R})_X$-bundle P on X, the twisted object $\mathcal{V} := \mathbb{R}_X^n \wedge^{GL(n,\mathbb{R})} P$ is known as the rank n vector bundle associated to P. Its coframe bundle $P_{\mathcal{V}}$ is canonically isomorphic to P.

REMARK 1.1. In (1.9), the right action on $\mathbb{R}_X^n$ of the linear group $GL(n, \mathbb{R})_X$ is given by the rule

$$
\begin{array}{ccc}
\mathbb{R}^n \times GL(n, \mathbb{R}) & \longrightarrow & \mathbb{R}^n \\
(Y, A) & \mapsto & A^{-1}Y
\end{array}
$$

where an element of $\mathbb{R}^n$ is viewed as a column matrix $Y = (\lambda_1, \ldots, \lambda_n)^T$. A local section p of $P_{\mathcal{V}}$ determines a local basis $\mathcal{B} = \{p^{-1}(e_i)\}$ of $\mathcal{V}$ and the arrow (1.9) then identifies the column vector Y with the element of $\mathcal{V}$ with coordinates (λ_i) in the chosen basis p. The fact that the arrow (1.9) factors through the contracted product is a global version of the familiar linear algebra rule which in an n-dimensional vector space V describes the effect of a change of basis matrix A on the coordinates Y of a given vector $v \in V$.

1.4. The cocyclic description of a bitorsor ([20, 1]). Consider a (G, H)-bitorsor P on X, with chosen local sections $s_i : U_i \longrightarrow P$ for some open cover $\mathcal{U} = (U_i)_{i \in I}$. Viewing P as a left G-torsor, we know by (1.2) that these sections define a family of G-valued 1-cochains g_{ij} satisfying the 1-cocycle condition (1.3). We have also seen that the right H-torsor structure on P is then described by the family of local isomorphisms $u_i : H_{U_i} \longrightarrow G_{U_i}$ defined by the equations (1.6) for all $h \in H_{U_i}$. It follows from (1.2) and (1.6) that the transition law for the restrictions of these isomorphisms above U_{ij} is

$$
u_i = i_{g_{ij}} u_j \tag{1.10}
$$

with i the inner conjugation homomorphism

$$
\begin{array}{ccc}
G & \xrightarrow{\; i \;} & \mathrm{Aut}(G) \\
g & \mapsto & i_g
\end{array}
\tag{1.11}
$$

defined by

$$
i_g(\gamma) = g\gamma g^{-1} . \tag{1.12}
$$

The pairs (g_{ij}, u_i) therefore satisfy the cocycle conditions

$$
\begin{cases}
g_{ik} & = \; g_{ij}\, g_{jk} \\
u_i & = \; i_{g_{ij}}\, u_j.
\end{cases}
\tag{1.13}
$$

A second family of local sections s_i' of P determines a corresponding cocycle pair (u_i', g_{ij}'), These new cocycles differ from the previous ones by the coboundary relations

$$\begin{cases} g'_{ij} &= g_i \, g_{ij} \, g_j^{-1} \\ u'_i &= i_{g_i} \, u_i \end{cases} \tag{1.14}$$

where the 0-cochains g_i are defined by (1.4). Isomorphism classes of (G, H)-bitorsors on X with given local trivialization on an open covering $\mathcal{U}$ are classified by the quotient of the set of cocycles (u_i, g_{ij}) (1.13) by the equivalence relation (1.14). Note that when P is a G-bitorsor, the terms of the second equation in both (1.13) and (1.14) live in the group $\mathrm{Aut}(G)$. In that case, the set of cocycle classes is the non-abelian hypercohomology set $H^0(\mathcal{U}, G \longrightarrow \mathrm{Aut}(G))$, with values in the length 1 complex of groups (1.11), where G is placed in degree -1. Passing to the limit over open covers, we obtain the Čech cohomology set $\check{H}^0(X, G \longrightarrow \mathrm{Aut}(G))$ which classifies isomorphism classes of G-bitorsors on X.

Let us see how the monoidal structure on the category of G-bitorsors is reflected at the cocyclic level. Let P and Q be a pair of G-bitorsors on X, with chosen local sections p_i and q_i. These determine corresponding cocycle pairs (g_{ij}, u_i) and (γ_{ij}, v_i) satisfying the corresponding equations (1.13). It is readily verified that the corresponding cocycle pair for the G-bitorsor $P \wedge^G Q$, locally trivialized by the family of local sections $p_i \wedge q_i$, is the pair

$$(g_{ij} \, u_i(\gamma_{ij}), \, u_i \, v_i) \tag{1.15}$$

so that the group law for cocycle pairs is simply the semi-direct product multiplication in the group $G \rtimes \mathrm{Aut}(G)$, for the standard left action of $\mathrm{Aut}(G)$ on G. The multiplication rule for cocycle pairs

$$(g_{ij}, u_i) * (\gamma_{ij}, v_i) = (g_{ij} \, u_i(\gamma_{ij}), \, u_i \, v_i)$$

passes to the set of equivalence classes, and therefore determines a group structure on the set $\check{H}^0(X, G \longrightarrow \mathrm{Aut}(G))$, which reflects the contracted product of bitorsors.

REMARK 1.2. Let us choose once more a family of local sections s_i of a (G, H)-bitorsor P. The local isomorphisms u_i provide an identification of the restrictions H_{U_i} of H with the restrictions G_{U_i} of G. Under these identifications, the significance of equations (1.10) is the following. By (1.10), we may think of an element of H as given by a family of local elements $\gamma_i \in G_i$, glued to each other above the open sets U_{ij} according to the rule

$$\gamma_i = i_{g_{ij}} \, \gamma_j \, .$$

For this reason, a bundle of groups H which stands in such a relation to a given group G may be called an *inner form* of G. This is the terminology used in the context of Galois cohomology, *i.e* when X is a scheme $\mathrm{Spec}(k)$ endowed with the étale topology defined by the covering morphism $\mathrm{Spec}(k') \longrightarrow \mathrm{Spec}(k)$ associated to a Galois field extension k'/k ([20] III § 1).

1.5.   The previous discussion remains valid in a wider context, in which the inner conjugation homomorphism i is replaced by an arbitrary homomorphism of groups $\delta : G \longrightarrow \Pi$. The cocycle and coboundary conditions (1.13) and (1.14) are now respectively replaced by the rules

$$\begin{cases} g_{ik} & = & g_{ij}\, g_{jk} \\ \pi_i & = & \delta(g_{ij})\, \pi_j \end{cases} \tag{1.16}$$

and by

$$\begin{cases} g'_{ij} = g_i\, g_{ij}\, g_j^{-1} \\ \pi'_i = \delta(g_i)\, \pi_i \end{cases} \tag{1.17}$$

and the induced Čech hypercohomology set with values in the complex of groups $G \longrightarrow \Pi$ is denoted $\check{H}^0(\mathcal{U}, G \longrightarrow \Pi)$. In order to extend to $\check{H}^0(\mathcal{U}, G \longrightarrow \Pi)$ the multiplication (1.15), we require additional structure:

DEFINITION 1.3. *A (left) crossed module is a group homomorphism $\delta : G \longrightarrow \Pi$, together with a left group action*

$$\begin{array}{ccc} \Pi \times G & \longrightarrow & G \\ (\pi,\, g) & \mapsto & {}^\pi g \end{array}$$

of Π on the group G, and such that the equations

$$\begin{cases} \delta({}^\pi g) = {}^\pi \delta(g) \\ {}^{\delta(\gamma)} g = {}^\gamma g \end{cases} \tag{1.18}$$

are satisfied, with G (resp. Π) acting on itself by the conjugation rule (1.12).

Crossed modules form a category, with a homomorphism of crossed modules

$$(G \xrightarrow{\;\delta\;} \pi) \longrightarrow (K \xrightarrow{\;\delta'\;} \Gamma)$$

defined by a pair of homomorphisms (u, v) such that the diagram of groups

$$\begin{array}{ccc} G & \xrightarrow{\;u\;} & K \\ {\scriptstyle \delta}\downarrow & & \downarrow{\scriptstyle \delta'} \\ \Pi & \xrightarrow{\;v\;} & \Gamma \end{array} \tag{1.19}$$

commutes, and such that $u({}^\pi g) = {}^{v(\pi)} u(g)$ (in other words such that u is v-equivariant).

A left crossed module $G \xrightarrow{\;\delta\;} \Pi$ defines a group-like monoidal category $\mathcal{C}$ with a strict multiplication on objects, by setting

$$\text{ob}\,\mathcal{C} := \Pi \qquad \text{ar}\,\mathcal{C} := G \times \Pi. \tag{1.20}$$

The source and target of an arrow (g, π) are as follows:

$$\pi \xrightarrow{\ (g,\pi)\ } \delta(g)\pi$$

and the composite of two composable arrows

$$\pi \xrightarrow{\ (g,\pi)\ } \delta(g)\pi \xrightarrow{\ (g',\,\delta(g)\pi)\ } \delta(g'g), \pi \qquad (1.21)$$

is the arrow $(g'g, \pi)$. The monoidal structure on this groupoid is given on the objects by the group multiplication in Π, and on the set $G \times \Pi$ of arrows by the semi-direct product group multiplication

$$(g, \pi) * (g'\, \pi') := (g\ {}^{\pi}g', \pi\, \pi') \qquad (1.22)$$

for the given left action of Π on G. In particular the identity element of the group Π is the unit object I of this monoidal groupoid.

Conversely, to a monoidal category $\mathcal{M}$ with strict multiplication on objects is associated a crossed module $G \xrightarrow{\ \delta\ } \Pi$, where $\Pi := \operatorname{ob}\mathcal{M}$ and G is the set $\operatorname{Ar}_I\mathcal{M}$ of arrows of $\mathcal{M}$ sourced at the identity object, with δ the restriction to G of the target map. The group law on G is the restriction to this set of the multiplication of arrows in the monoidal category $\mathcal{M}$. The action of an object $\pi \in \Pi$ on an arrow $g : I \longrightarrow \delta(g)$ in G has the following categorical interpretation: the composite arrow

$$I \xrightarrow{\ \sim\ } \pi\, I\, \pi^{-1} \xrightarrow{\ \pi\, g\, \pi^{-1}\ } \pi\, \delta(g)\, \pi^{-1}$$

corresponds to the element ${}^{\pi}g$ in G. Finally, given a pair elements $g, g' \in \operatorname{Ar}_I\mathcal{M}$, it follows from the composition rule (1.21) for a pair of arrows that the composite arrow

$$I \xrightarrow{\ (g,I)\ } \delta(g) \xrightarrow{\ (g',\delta(g))\ } \delta(g'g)$$

(constructed by taking advantage of the monoidal structure on the category $\mathcal{M}$ in order to transform the arrow g' into an arrow $(g', \delta(g))$ composable with g) is simply given by the element $g'g$ of the group $\Pi = \operatorname{Ar}_I\mathcal{M}$.

A stronger concept than that of a homomorphism of crossed module is what could be termed a "crossed module of crossed modules". This is the categorification of crossed modules and corresponds, when one extends the previous dictionary between strict monoidal categories and crossed modules, to strict monoidal bicategories. The most efficient description of such a concept is the notion of a crossed square, due to J.-L Loday. This consists of a homomorphism of crossed modules (1.19), together with a map

$$\begin{array}{ccc} K \times \Pi & \longrightarrow & G \\ (k, \pi) & \longmapsto & \{k, \pi\} \end{array} \qquad (1.23)$$

satisfying certain conditions for which we refer to [14] definition 5.1.

REMARK 1.3.

i) The definition (1.19) of a homomorphism of crossed modules is quite restrictive, and it is often preferable to relax in order that it correspond to a not necessarily strict monoidal functor between the associated (strict) monoidal groupoids. The definition of a weak homomorphism of crossed modules has been spelled out under various guise by B. Noohi in [16] (definition 8.4), [17] § 8.

ii) All these definitions obviously extend from groups to bundles of groups on X.

iii) The composition law (1.22) determines a multiplication

$$(g_{ij}, \pi_i) * (g'_{ij}, \pi'_i) := (g_{ij}\,{}^{\pi_i}g'_{ij}, \, \pi_i\,\pi'_i)$$

on $(G \longrightarrow \Pi)$-valued cocycle pairs, which generalizes (1.15), is compatible with the coboundary relations, and induces a group structure on the set $\check{H}^0(\mathcal{U}, G \longrightarrow \Pi)$ of degree zero cohomology classes with values in the crossed module $G \longrightarrow \Pi$ on X.

1.6. The following proposition is known as the Morita theorem, by analogy with the corresponding characterization in terms of bimodules of equivalences between certain categories of modules.

PROPOSITION 1.2. (Giraud [10])
i) A (G, H)-bitorsor Q on X determines an equivalence

$$\text{Tors}(H) \xrightarrow{\Phi_Q} \text{Tors}(G)$$

$$M \longmapsto Q \wedge^H M$$

between the corresponding categories of left torsors on X. In addition, if P is an (H, K)-bitorsor on X, then there is a natural equivalence

$$\Phi_{Q \wedge^H P} \simeq \Phi_Q \circ \Phi_P$$

between functors from $\text{Tors}(K)$ to $\text{Tors}(G)$. In particular, the equivalence Φ_{Q° in an inverse of Φ_Q.

ii) Any such equivalence Φ between two categories of torsors is equivalent to one associated in this manner to an (H, G)-bitorsor.

Proof of ii). To a given equivalence Φ is associated the left G-torsor $Q := \Phi(T_H)$. By functoriality of Φ, $H \simeq \text{Aut}_H(T_H) \xrightarrow{\Phi} \text{Aut}_G(Q)$, so that a section of H acts on the right on Q. $\square$

2. (1)-stacks.

2.1. The concept of a stack is the categorical analogue of a sheaf. Let us start by defining the analog of a presheaf.

DEFINITION 2.1.

i) A category fibered in groupoids above a space X consists in a family of groupoids $\mathcal{C}_U$, for each open set U in X, together with an inverse image functor

$$f^* : \mathcal{C}_U \longrightarrow \mathcal{C}_{U_1} \tag{2.1}$$

associated to every inclusion of open sets $f : U_1 \subset U$ (which is the identity whenever $f = 1_U$), and natural equivalences

$$\phi_{f,g} : (fg)^* \Longrightarrow g^* f^* \tag{2.2}$$

for every pair of composable inclusions

$$U_2 \overset{g}{\hookrightarrow} U_1 \overset{f}{\hookrightarrow} U . \tag{2.3}$$

For each triple of composable inclusions

$$U_3 \overset{h}{\hookrightarrow} U_2 \overset{g}{\hookrightarrow} U_1 \overset{f}{\hookrightarrow} U .$$

we also require that the composite natural transformations

$$\psi_{f,g,h} : (fgh)^* \Longrightarrow h^* (fg)^* \Longrightarrow h^* (g^* f^*)$$

and

$$\chi_{f,g,h} : (fgh)^* \Longrightarrow (gh)^* f^* \Longrightarrow (h^* g^*) f^*$$

coincide.

ii) A cartesian functor $F : \mathcal{C} \longrightarrow \mathcal{D}$ is a family of functors $F_U : \mathcal{C}_U \longrightarrow \mathcal{D}_U$ for all open sets $U \subset X$, together with natural transformations

$$
\begin{array}{ccc}
\mathcal{C}_U & \longrightarrow & \mathcal{C}_{U_1} \\
F_U \downarrow & \nearrow & \downarrow F_{U_1} \\
\mathcal{D}_U & \longrightarrow & \mathcal{D}_{U_1}
\end{array}
\tag{2.4}
$$

for all inclusion $f : U_1 \subset U$ compatible via the natural equivalences (2.2) for a pair of composable inclusions (2.3).

iii) A natural transformation $\Psi : F \Longrightarrow G$ between a pair of cartesian functors consists of a family of natural transformations $\Psi_U : F_U \Longrightarrow G_U$ compatible via the 2-arrows (2.4) under the inverse images functors (2.1).

The following is the analogue for fibered groupoids of the notion of a sheaf of sets, formulated here in a preliminary style:

DEFINITION 2.2. *A stack in groupoids above a space X is a fibered category in groupoids above X such that*
  • *("Arrows glue") For every pair of objects $x, y \in \mathcal{C}_U$, the presheaf $\operatorname{Ar}_{\mathcal{C}_U}(x, y)$ is a sheaf on U.*
  • *("Objects glue") Descent is effective for objects in $\mathcal{C}$.*

The gluing condition on arrows is not quite correct as stated. In order to be more precise, let us first observe that if x is any object in $\mathcal{C}_U$, and $(U_\alpha)_{\alpha \in I}$ an open cover of U, then x determines a family of inverse images x_α in $\mathcal{C}_{U_\alpha}$ which we will refer informally to as the restrictions of x above U_α, and sometimes denote by $x_{|U_\alpha}$. These are endowed with isomorphisms

$$ x_{\beta|U_{\alpha\beta}} \xrightarrow{\phi_{\alpha\beta}} x_{\alpha|U_{\alpha\beta}} \tag{2.5} $$

in $\mathcal{C}_{U_{\alpha\beta}}$ satisfying the cocycle equation

$$ \phi_{\alpha\beta}\, \phi_{\beta\gamma} = \phi_{\alpha\gamma} \tag{2.6} $$

when restricted to $\mathcal{C}_{U_{\alpha\beta\gamma}}$. An arrow $f : x \longrightarrow y$ in $\mathcal{C}_U$ determines arrows $f_\alpha : x_\alpha \longrightarrow y_\alpha$ in each of the categories $\mathcal{C}_{U_\alpha}$ such that the following diagram in $\mathcal{C}_{U_{\alpha\beta}}$ commutes

$$ \begin{array}{ccc} x_{\beta|U_{\alpha\beta}} & \xrightarrow{\ f_{\beta|U_{\alpha\beta}}\ } & y_{\beta|U_{\alpha\beta}} \\ {\scriptstyle\phi_{\alpha\beta}}\big\downarrow & & \big\downarrow{\scriptstyle\psi_{\alpha\beta}} \\ x_{\alpha|U_{\alpha\beta}} & \xrightarrow{\ f_{\alpha|U_{\alpha\beta}}\ } & y_{\alpha|U_{\alpha\beta}} \end{array} \tag{2.7} $$

The full gluing condition on arrows is the requirement that conversely, for any family of arrows $f_\alpha : x_\alpha \longrightarrow y_\alpha$ in $\mathcal{C}_{U_\alpha}$ for which the compatibility condition (2.7) is satisfied, there exists a unique arrow $f : x \longrightarrow y$ in $\mathcal{C}_U$ whose restriction above each open set U_α is the corresponding f_α. In particular, if we make the very non-categorical additional assumption that the $\phi_{\alpha\beta}$ are all identity arrows, then this gluing condition on arrows simply asserts that the presheaf of arrows from x to y is a sheaf on U. A fibered category for which the gluing property on arrows is satisfied is called a prestack.

Let us now pass to the gluing condition on objects. The term descent comes from algebraic geometry, where for a given family of objects $(x_\alpha) \in \mathcal{C}_{U_\alpha}$, a family of isomorphisms $\phi_{\alpha\beta}$ (2.5) satisfying the equation (2.6) is called descent data for the family of objects $(x_\alpha)_{\alpha \in I}$. The descent is said to be effective whenever any such descent data determines an object $x \in \mathcal{C}_U$, together with a family of arrows $x_{|U_\alpha} \longrightarrow x_\alpha$ in $\mathcal{C}_{U_\alpha}$ compatible with the given descent data on the given objects x_α, and the canonical descent data on the restrictions $x_{|U_\alpha}$ of x. A sheafification process, analogous to the one which transforms a presheaf into a sheaf, associates a stack to a given prestack. For a more detailed introduction to the theory of stacks in an algebro-geometric setting, see [24].

3. 1-gerbes.

3.1. We begin with the global description of the 2-category of gerbes, due to Giraud [10]. For another early discussion of gerbes, see [9].

DEFINITION 3.1.

*i) A (1)-gerbe on a space X is a stack in groupoids $\mathcal{G}$ on X which is locally **non-empty** and locally **connected**.*

ii) A morphism of gerbes is a cartesian functor between the underlying stacks.

iii) A natural transformation $\Phi : u \Longrightarrow v$ between a pair of such morphisms of gerbes $u, v : \mathcal{P} \longrightarrow \mathcal{Q}$ is a natural transformation between the corresponding pair of cartesian functors.

EXAMPLE 4. Let G be a bundle of groups on X. The stack $\mathcal{C} := \mathrm{Tors}(G)$ of left G-torsors on X is a gerbe on X: first of all, it is non-empty, since the category $\mathcal{C}_U$ always has at least one object, the trivial torsor T_{G_U}. In addition, every G-torsor on U is locally isomorphic to the trivial one, so the objects in the category $\mathcal{C}_U$ may be locally connected to each other by arrows.

A gerbe $\mathcal{P}$ on X is said to be *neutral* (or *trivial*) when the fiber category $\mathcal{P}_X$ is non empty. In particular, a gerbe $\mathrm{Tors}(G)$ is neutral with distinguished object the trivial G-torsor T_G on X. Conversely, the choice of a global object $x \in \mathcal{P}_X$ in a neutral gerbe $\mathcal{P}$ determines an equivalence of gerbes

$$
\begin{array}{ccc}
\mathcal{P} & \xrightarrow{\ \sim\ } & \mathrm{Tors}(G) \\
y & \mapsto & \mathrm{Isom}_{\mathcal{P}}(y, x)
\end{array}
\qquad (3.1)
$$

on X, where $G := \mathrm{Aut}_{\mathcal{P}}(x)$, acting on $\mathrm{Isom}_{\mathcal{P}}(x, y)$ by composition of arrows.

Let $\mathcal{P}$ be a gerbe on X and $\mathcal{U} = (U_i)_{i \in I}$ be an open cover of X. We now **choose** objects $x_i \in \mathrm{ob}\ \mathcal{P}_{U_i}$ for each $i \in I$. These objects determine corresponding bundles of groups $G_i := \mathrm{Aut}_{\mathcal{P}_{U_i}}(x_i)$ above U_i. When in addition there exists a bundle of groups G above X, together with U_i-isomorphisms $G_{|U_i} \simeq G_i$, for all $i \in I$, we say that $\mathcal{P}$ is a G-gerbe on X.

4. Semi-local description of a gerbe.

4.1. Let $\mathcal{P}$ be a G-gerbe on X, and let us choose a family of local objects $x_i \in \mathcal{P}_{U_i}$. These determine as in (3.1) equivalences

$$
\Phi_i : \mathcal{P}_{U_i} \longrightarrow \mathrm{Tors}(G)_{|U_i}
$$

above U_i. Choosing quasi-inverses for the Φ_i we get an induced family of equivalences

$$\Phi_{ij} := \Phi_{i\,|U_{ij}} \circ \Phi^{-1}_{j\,|U_{ij}} : \mathrm{Tors}(G)_{U_{ij}} \longrightarrow \mathcal{P}_{|U_{ij}} \longrightarrow \mathrm{Tors}(G)_{U_{ij}}$$

above U_{ij}, which corresponds by proposition 1.2 to a family of G-bitorsors P_{ij} above U_{ij}. By construction of the Φ_{ij}, there are also natural transformations

$$\Psi_{ijk} : \Phi_{ij}\,\Phi_{jk} \Longrightarrow \Phi_{ik}$$

above U_{ijk}, satisfying a coherence condition on U_{ijkl}. These define isomorphisms of G-bitorsors

$$\psi_{ijk} : P_{ij} \wedge^G P_{jk} \longrightarrow P_{ik} \tag{4.1}$$

above U_{ijk} for which this coherence condition is described by the commutativity of the diagram of bitorsors

$$
\begin{array}{ccc}
P_{ij} \wedge P_{jk} \wedge P_{kl} & \xrightarrow{\;\psi_{ijk}\wedge P_{kl}\;} & P_{ik} \wedge P_{kl} \\
{\scriptstyle P_{ij}\wedge\psi_{ijk}}\downarrow & & \downarrow{\scriptstyle \psi_{ikl}} \\
P_{ij} \wedge P_{jl} & \xrightarrow[\;\psi_{ijl}\;]{} & P_{il}
\end{array}
\tag{4.2}
$$

above U_{ijkl}.

4.2. Additional comments.

i) The isomorphism (4.1), satisfying the coherence condition (4.2), may be viewed as a 1-cocycle condition on X with values in the monoidal stack of G-bitorsors on X. We say that a family of such bitorsors P_{ij} constitutes a bitorsor cocycle on X.

ii) In the case of *abelian* G-gerbes[1] ([4] Definition 2.9), the monoidal stack of bitorsors on U_{ij} may be replaced by the symmetric monoidal stack of G-torsors on U_{ij}. In particular, for the multiplicative group $G = GL(1)$, the $GL(1)$-torsors P_{ij} correspond to line bundles L_{ij}. This point of view regarding abelian $GL(1)$ gerbes set forth by N. Hitchin in [11].

iii) The semi-local construction extends from G-gerbes to general gerbes. In that case a local group $G_i := \mathrm{Aut}_\mathcal{P}(x_i)$ above U_i is associated to each of the chosen objects x_i. The previous discussion remains valid, with the proviso that the P_{ij} are now (G_j, G_i)-bitorsors rather than simply G-bitorsors, and the ψ_{ijk} (4.1) are isomorphisms of (G_k, G_i)-bitorsors.

iv) If we replace the chosen trivializing open cover $\mathcal{U}$ of X by a single covering morphism $Y \longrightarrow X$ in some Grothendieck topology, the theory remains unchanged, but takes on a somewhat different flavor. An object

---

[1]which are not simply G-gerbes for which the structure group G is abelian!

$x \in \mathcal{P}_Y$ determines a bundle of groups $G := \text{Aut}_{\mathcal{P}_Y}(x)$ over Y, together with a (p_2^*G, p_1^*G)-bitorsor P above $Y \times_X Y$ satisfying the coherence condition analogous to (4.2) above $Y \times_X Y \times_X Y$. A bitorsor P on Y satisfying this coherence condition has been called a cocycle bitorsor by K.-H. Ulbrich [22], and a bundle gerbe by M.K. Murray [15]. It corresponds to a bouquet in Duskin's theory (see [21]). It is equivalent[2] to give oneself such a bundle gerbe P, or to consider a gerbe $\mathcal{P}$ on X, together with a trivialization of its pullback to Y, since to a trivializing object $x \in \text{ob}(\mathcal{P}_Y)$ we may associate the G-bitorsor $P := \text{Isom}(p_2^*x, p_1^*x)$ above $Y \times_X Y$.

5. Cocycles and coboundaries for gerbes.

5.1.   Let us keep the notations of section 4. In addition to choosing local objects $x_i \in \mathcal{P}_{U_i}$ in a gerbe $\mathcal{P}$ on X, we now **choose** arrows

$$x_j \xrightarrow{\phi_{ij}} x_i \tag{5.1}$$

in $\mathcal{P}_{U_{ij}}{}^3$. Since $G_i := \text{Aut}_{\mathcal{P}}(x_i)$, a chosen arrow ϕ_{ij} induces by conjugation a homomorphism of group bundles

$$G_j{}_{|U_{ij}} \xrightarrow{\lambda_{ij}} G_i{}_{|U_{ij}} \tag{5.2}$$

$$\gamma \longmapsto \phi_{ij}\,\gamma\,\phi_{ij}^{-1}$$

above the open sets U_{ij}. To state this slightly differently, such a homomorphism λ_{ij} is characterized by the commutativity of the diagrams

$$\begin{array}{ccc}
x_j & \xrightarrow{\gamma} & x_j \\
\phi_{ij}\downarrow & & \downarrow\phi_{ij} \\
x_i & \xrightarrow[\lambda_{ij}(\gamma)]{} & x_i
\end{array} \tag{5.3}$$

for every $\gamma \in G_j{}_{|U_{ij}}$. The choice of objects x_i and arrows ϕ_{ij} in $\mathcal{P}$ determines, in addition to the morphisms λ_{ij} (5.2), a family of elements $g_{ijk} \in G_i{}_{|U_{ijk}}$ for all (i, j, k), defined by the commutativity of the diagrams

---

[2]For a more detailed discussion of this when the covering morphism $Y \longrightarrow X$ is the morphism of schemes associated as in remark 1.2 to a Galois field extension k'/k, see [2] §5.

[3] Actually, this is a simplification, since the gerbe axioms only allow us to choose such an arrow locally, above each element U_{ij}^α of an open cover of U_{ij}. Such families of open sets (U_i, U_{ij}^α), and so on, form what is known as a hypercover of X. For simplicity, we assume from now on that our topological space X is paracompact. In that case, we may carry out the entire discussion without hypercovers.

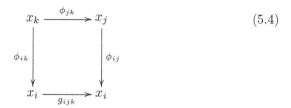

(5.4)

above U_{ijk}. These in turn induce by conjugation the following commutative diagrams of bundles of groups

$$
\begin{array}{ccc}
G_k & \xrightarrow{\lambda_{jk}} & G_j \\
\downarrow{\scriptstyle \lambda_{ik}} & & \downarrow{\scriptstyle \lambda_{ij}} \\
G_i & \xrightarrow[i_{g_{ijk}}]{} & G_i
\end{array}
$$

(5.5)

above U_{ijk}. The commutativity of diagram (5.5) may be stated algebraically as the cocycle equation

$$\lambda_{ij}\,\lambda_{jk} = i_{g_{ijk}}\,\lambda_{ik} \tag{5.6}$$

with i the inner conjugation arrow (1.11). The following equation is the second cocycle equation satisfied by the pair $(\lambda_{ij},\, g_{ijk})$. While the proof of lemma 5.1 given here is essentially the same as the one in [4], the present cubical diagram (5.9) is much more intelligible than diagram (2.4.7) of [4].

LEMMA 5.1. *The elements g_{ijk} satisfy the λ_{ij}-twisted 2-cocycle equation*

$$\lambda_{ij}(g_{jkl})\,g_{ijl} = g_{ijk}g_{ikl} \tag{5.7}$$

in $G_{i\,|U_{ijkl}}$.

Proof. Note that equation (5.7) is equivalent to the commutativity of the diagram

$$
\begin{array}{ccc}
x_i & \xrightarrow{g_{ijl}} & x_i \\
\downarrow{\scriptstyle g_{ikl}} & & \downarrow{\scriptstyle \lambda_{ij}(g_{jkl})} \\
x_i & \xrightarrow[g_{ijk}]{} & x_i
\end{array}
$$

(5.8)

above U_{ijkl}. Let us now consider the following cubical diagram:

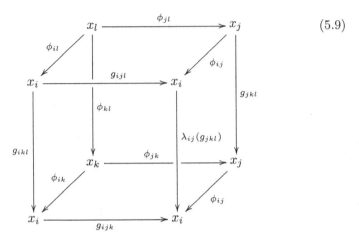

$$(5.9)$$

in which the left, back, top and bottom squares are of type (5.4), and the right-hand one of type (5.3). Since these five faces are commutative squares, and all the arrows in the diagram are invertible, the sixth (front) face is also commutative. Since the latter is simply the square (5.8), the lemma is proved. □

A pair (λ_{ij}, g_{ijk}) satisfying the equations (5.6) and (5.7):

$$\begin{cases} \lambda_{ij}\,\lambda_{jk} &= i_{g_{ijk}}\,\lambda_{ik} \\ \lambda_{ij}(g_{jkl})\,g_{ijl} &= g_{ijk}\,g_{ikl} \end{cases} \qquad (5.10)$$

is called a G_i-valued cocycle pair. It may be viewed as consisting of a 2-cocycle equation for the elements g_{ijk}, together with auxiliary data attached to the isomorphisms λ_{ij}. However, in contrast with the abelian case in which the inner conjugation term $i_{g_{ijk}}$ is trivial, these two equations cannot in general be uncoupled. When such a pair is associated to a G-gerbe $\mathcal{P}$ for a fixed bundle of groups G, the term λ_{ij} is a section above U_{ij} of the bundle of groups $\mathrm{Aut}_X(G)$, and g_{ijk} is a section of G above U_{ijk}. Such pairs (λ_{ij}, g_{ijk}) will be called G-valued cocycle pairs.

5.2. The corresponding coboundary relations will now be worked out by a similar diagrammatic process. Let us give ourselves a second family of local objects x'_i in $\mathcal{P}_{U_i}$, and of arrows

$$x'_j \xrightarrow{\phi'_{ij}} x'_i \qquad (5.11)$$

above U_{ij}. To these correspond by the constructions (5.3) and (5.4) a new cocycle pair $(\lambda'_{ij}, g'_{ijk})$ satisfying the cocycle relations (5.6) and (5.7). In order to compare the previous trivializing data (x_i, ϕ_{ij}) with the new one, we also choose (possibly after a harmless refinement of the given open cover $\mathcal{U}$ of X) a family of arrows

$$x_i \xrightarrow{\chi_i} x'_i \qquad (5.12)$$

in $\mathcal{P}_{U_i}$ for all i. The lack of compatibility between these arrows and the previously chosen arrows (5.1) and (5.11) is measured by the family of arrows $\vartheta_{ij} : x_i \longrightarrow x_i$ in $\mathcal{P}_{U_{ij}}$ determined by the commutativity of the following diagram:

$$
\begin{array}{ccc}
x_j & \xrightarrow{\ \phi_{ij}\ } & x_i \\
& & \Big\downarrow{\scriptstyle \chi_i} \\
{\scriptstyle \chi_j}\Big\downarrow & & x_i' \\
& & \Big\downarrow{\scriptstyle \vartheta_{ij}} \\
x_j' & \xrightarrow[\ \phi_{ij}'\ ]{} & x_i'
\end{array}
\tag{5.13}
$$

The arrow $\chi_i : x_i \longrightarrow x_i'$ induces by conjugation an isomorphism $r_i : G_i \longrightarrow G_i'$, characterized by the commutativity of the square

$$
\begin{array}{ccc}
x_i & \xrightarrow{\ u\ } & x_i \\
{\scriptstyle \chi_i}\Big\downarrow & & \Big\downarrow{\scriptstyle \chi_i} \\
x_i' & \xrightarrow[\ r_i(u)\ ]{} & x_i'
\end{array}
\tag{5.14}
$$

for all $u \in G_i$. The diagram (5.13) therefore conjugates to a diagram

$$
\begin{array}{ccc}
G_j & \xrightarrow{\ \lambda_{ij}\ } & G_i \\
& & \Big\downarrow{\scriptstyle r_i} \\
{\scriptstyle r_j}\Big\downarrow & & G_i' \\
& & \Big\downarrow{\scriptstyle i_{\vartheta_{ij}}} \\
G_j' & \xrightarrow[\ \lambda_{ij}'\ ]{} & G_i''
\end{array}
\tag{5.15}
$$

above U_{ij} whose commutativity is expressed by the equation

$$
\lambda_{ij}' = i_{\vartheta_{ij}}\, r_i\, \lambda_{ij}\, r_j^{-1}.
\tag{5.16}
$$

Consider now the diagram

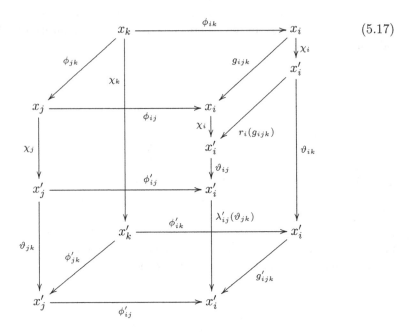

(5.17)

Both the top and the bottom squares commute, since these squares are of type (5.4). So do the back, the left and the top front vertical squares, since all three are of type (5.13). The same is true of the lower front square, and the upper right vertical square, since these two are respectively of the form (5.3) and (5.14). It follows that the remaining lower right square in the diagram is also commutative, since all the arrows in diagram (5.17) are invertible. The commutativity of this final square is expressed algebraically by the equation

$$g'_{ijk}\, \vartheta_{ik} = \lambda'_{ij}(\vartheta_{jk})\, \vartheta_{ij}\, r_i(g_{ijk}), \qquad (5.18)$$

an equation equivalent to [4] (2.4.17).

Let us say that two cocycle pairs $(\lambda_{ij},\, g_{ijk})$ and $(\lambda'_y,\, g'_{yk})$ are cohomologous if we are given a pair of elements $(r_i,\, \vartheta_{ij})$, with $r_i \in \mathrm{Isom}(G_i,\, G'_i)$ and $\vartheta_{ij} \in G'_{i\,|U_{ij}}$ satisfying the equations

$$\begin{cases} \lambda'_{ij} &= i_{\vartheta_{ij}}\, r_i\, \lambda_{ij}\, r_j^{-1} \\ g'_{ijk} &= \lambda'_{ij}(\vartheta_{jk})\, \vartheta_{ij}\, r_i(g_{ijk})\, \vartheta_{ik}^{-1}. \end{cases} \qquad (5.19)$$

Suppose now that $\mathcal{P}$ is a G-gerbe. All the terms in the first equations in both (5.10) and (5.19) are then elements of $\mathrm{Aut}(G)$, while the terms in the corresponding second equations live in G. The set of equivalence classes of cocycle pairs (5.10), for the equivalence relation defined by equations

(5.19), is then denoted $H^1(\mathcal{U}, G \longrightarrow \text{Aut}(G))$, a notation consistent with that introduced in §1.4 for $H^0(\mathcal{U}, G \longrightarrow \text{Aut}(G))$. The limit over the open covers $\mathcal{U}$ is the Čech hypercohomology set $\check{H}^1(X, G \longrightarrow \text{Aut}(G))$. We refer to [4] § 2.6 for the inverse construction, starting from a Čech cocycle pair, of the corresponding G-gerbe[4]. This hypercohomology set therefore classifies G-gerbes on X up to equivalence.

In geometric terms, this can be understood once we introduce the following definition, a categorification of the definition (1.1) of a G-torsor:

DEFINITION 5.1. *Let $\mathcal{G}$ be a monoidal stack on X. A left $\mathcal{G}$-torsor on X is a stack $\mathcal{Q}$ on X together with a left action functor*

$$\mathcal{G} \times \mathcal{Q} \longrightarrow \mathcal{Q}$$

which is coherently associative and satisfies the unit condition, and for which the induced functor

$$\mathcal{G} \times \mathcal{Q} \longrightarrow \mathcal{Q} \times \mathcal{Q}$$

defined as in (1.1) is an equivalence. In addition, we require that $\mathcal{Q}$ be locally non-empty.

The following three observations, when put together, explain in more global terms why G-gerbes are classified by the first cohomology set $H^1(X, G \longrightarrow \text{Aut}(G))$.

- To a G-gerbe $\mathcal{P}$ on X is associated its "bundle of coframes" $\mathcal{E}q(\mathcal{P}, \text{Tors}(G))$, and the latter is a left torsor under the monoidal stack $\mathcal{E}q(\text{Tors}(G), \text{Tors}(G))$.
- By the Morita theorem, this monoidal stack is equivalent to the monoidal stack $\text{Bitors}(G)$ of G-bitorsors on X.
- The cocycle computations leading up to (1.13) imply that the monoidal stack $\text{Bitors}(G)$ is the stack associated to the monoidal prestack defined by the crossed module $G \xrightarrow{i} \text{Aut}(G)$ (1.11).

REMARK 5.1. For a related discussion of non-abelian cocycles in a homotopy-theoretic context, see J.F. Jardine [12] Theorem 13 and [13] § 4, where a classification of gerbes equivalent to ours is given, including the case in which hypercovers are required.

5.3. A topological interpretation of a G-gerbe ([3] 4.2). The

context here is that of fibrewise topology, in which all constructions are done in the category of spaces above a fixed topological space X. Let G be a bundle of groups above X and $B_X G$ its classifying space, a space above X whose fiber at a point $x \in X$ is the classifying space BG_x of the group

---

[4]In [4] §2.7, we explain how this inverse construction extends to the more elaborate context of hypercovers, where a beautiful interplay between the Čech and the descent formalisms arises. This is also discussed, in more simplicial terms, in [1] §6.3–6.6.

G_x. By construction, $B_X G$ is the geometric realization of the simplicial space over X whose face and degeneracy operators above X are defined in the usual fashion (but now in the fibrewise context) starting from the multiplication and diagonal maps $G \times_X G \longrightarrow G$ and $G \longrightarrow G \times_X G$:

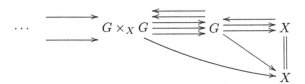

We attach to G the bundle $\mathrm{Eq}_X(BG)$ of group-like topological monoids of self-fiber-homotopy equivalences of $B_X G$ over X. The fibrewise homotopy fiber of the evaluation map

$$\mathrm{ev}_{X,*} : \mathrm{Eq}_X(BG) \longrightarrow BG,$$

which associates to such an equivalence its value at the distinguished section $*$ of $B_X G$ above X, is the submonoid $\mathrm{Eq}_{X,*}(B_X G)$ of pointed fibrewise homotopy self-equivalences of $B_X G$. The latter is fiber homotopy equivalent, by the fibrewise functor $\pi_1(-, *)$, to the bundle of groups $\mathrm{Aut}_X(G)$, whose fiber at a point $x \in X$ is the group $\mathrm{Aut}(G_x)$. This fibration sequence of spaces over X

$$\mathrm{Aut}_X(G) \longrightarrow \mathrm{Eq}_X(B_X G) \longrightarrow B_X G$$

is therefore equivalent to a fibration sequence of topological monoids over X, the first two of which are simply bundles of groups on X

$$G \overset{i}{\longrightarrow} \mathrm{Aut}(G) \longrightarrow \mathrm{Eq}_X(BG). \qquad (5.20)$$

This yields an identification of $\mathrm{Eq}_X(BG)$ with the fibrewise Borel construction[5] $E_X G \times_X^G \mathrm{Aut}_X(G)$. Our discussion in §1.4 asserts that this identification preserves the multiplications, so long as the multiplication on the Borel construction is given by an appropriate iterated semi-direct product construction, whose first non-trivial stage is defined as in (1.22). We refer to [3] for a somewhat more detailed discussion of this assertion, and to [8] § 4 for a related discussion, in the absolute rather than in the fibrewise context, of the corresponding fibration sequence

$$BG \longrightarrow B\,\mathrm{Aut}(G) \longrightarrow B\mathrm{Eq}(BG)$$

(or rather to its generalization in which the classifying space BG replaced by an arbitrary topological space Y). This proves:

---

[5] Our use here of the notation $\times^G$ is meant to be close to the topologists' $\times_G$. Algebraic geometers often denote such a G-equivariant product by $\wedge^G$, as we did in (1.8).

PROPOSITION 5.1. *The simplicial group over X associated to the crossed module of groups $G \longrightarrow \mathrm{Aut}(G)$ over X is a model for the group-like topological monoid $Eq_X(BG)$.*

For any group G, the set $H^1(X, G \longrightarrow \mathrm{Aut}(G))$ of 1-cocycle classes describes the classes of fibrations over X which are locally homotopy equivalent to the space BG, and the corresponding assertion when G is a bundle of groups on X is also true. We refer to the recent preprint of J. Wirth and J. Stasheff [23] for a related discussion of fiber homotopy equivalence classes of locally homotopy trivial fibrations, also from a cocyclic point of view.

EXAMPLE 5. Let us sketch here a modernized proof of O. Schreier's cocyclic classification (in 1926!) of (non-abelian, non-central) group extensions [19], which is much less well-known than the special case in which the extensions are central.

Consider a short exact sequence of groups (1.7). Applying the classifying space functor B, this induces a fibration

$$BG \longrightarrow BH \xrightarrow{\pi} BK$$

of pointed spaces, and all the fibers of π are homotopically equivalent to BG. It follows that this fibration determines an element in the pointed set $H^1(BK, G \longrightarrow \mathrm{Aut}(G))$. Conversely, such a cohomology class determines a fibration $E \xrightarrow{\phi} BK$ above BK, whose fibers are homotopy equivalent to the space BG. Since both BG and BK have distinguished points, so does E. Applying the fundamental group functor to this fibration of pointed spaces determines a short sequence of groups

$$1 \longrightarrow G \longrightarrow H \longrightarrow K \longrightarrow 1. \qquad \square$$

6. 2-stacks and 2-gerbes.

6.1. We will now extend the discussion of section 5 from 1- to 2-categories. A 2-groupoid is defined here as a 2-category whose 1-arrows are invertible up to a 2-arrow, and whose 2-arrows are strictly invertible.

DEFINITION 6.1. *A fibered 2-category in 2-groupoids above a space X consists in a family of 2-groupoids $\mathcal{C}_U$, for each open set U in X, together with an inverse image 2-functor*

$$f^* : \mathcal{C}_U \longrightarrow \mathcal{C}_{U_1}$$

associated to every inclusion of open sets $f : U_1 \subset U$ (which is the identity whenever $f = 1_U$), and a natural transfomation

$$\phi_{f,g} : (fg)^* \Longrightarrow g^* f^*$$

for every pair of composable inclusions

$$U_2 \xrightarrow{g} U_1 \xrightarrow{f} U.$$

For each triple of composable inclusions

$$U_3 \overset{h}{\hookrightarrow} U_2 \overset{g}{\hookrightarrow} U_1 \overset{f}{\hookrightarrow} U,$$

we require a modification

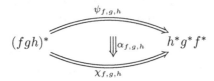

betweeen the composite natural transfomations

$$\psi_{f,g,h} : (fgh)^* \Longrightarrow h^* \, (fg)^* \Longrightarrow h^* \, (g^* f^*)$$

and

$$\chi_{f,g,h} : (fgh)^* \Longrightarrow (gh)^* \, f^* \Longrightarrow (h^* g^*) \, f^*.$$

Finally, for any $U_4 \overset{k}{\hookrightarrow} U_3$, the two methods by which the induced modifications α compare the composite 2-arrows

$$(fghk)^* \Longrightarrow (ghk)^* f^* \Longrightarrow ((hk)^* g^* f^* \Longrightarrow k^* h^* g^* f^*$$

and

$$(fghk)^* \Longrightarrow k^* (fgh)^* \Longrightarrow k^* (h^* (fg)^*) \Longrightarrow k^* h^* g^* f^*$$

must coincide.

DEFINITION 6.2. *A 2-stack in 2-groupoids above a space X is a fibered 2-category in 2-groupoids above X such that*

- *For every pair of objects $X, Y \in \mathcal{C}_U$, the fibered category $\mathrm{Ar}_{\mathcal{C}_U}(X, Y)$ is a stack on U.*
- *2-descent is effective for objects in $\mathcal{C}$.*

The 2-descent condition asserts that we are given, for an open covering $(U_\alpha)_{\alpha \in J}$ of an open set $U \subset X$, a family of objects $x_\alpha \in \mathcal{C}_{U_\alpha}$, of 1-arrows $\phi_{\alpha\beta} : x_\alpha \longrightarrow x_\beta$ between the restrictions to $\mathcal{C}_{U_{\alpha\beta}}$ of the objects x_α and x_β, and a family of 2-arrow

$$(6.1)$$

for which the tetrahedral diagram of 2-arrows whose four faces are the restrictions of the requisite 2-arrows ψ (6.1) to $\mathcal{C}_{U_{\alpha\beta\gamma\delta}}$ commutes:

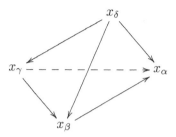

The 2-descent condition $(x_\alpha, \phi_{\alpha\beta}, \psi_{\alpha\beta\gamma})$ is effective if there exists an object $x \in \mathcal{C}_U$, together with 1-arrows $x_{|U_\alpha} \simeq x_\alpha$ in $\mathcal{C}_{U_\alpha}$ which are compatible with the given 1- and 2-arrows $\phi_{\alpha,\beta}$ and $\psi_{\alpha\beta\gamma}$.

DEFINITION 6.3. *A 2-gerbe $\mathcal{P}$ is a 2-stack in 2-groupoids on X which is locally non-empty and locally connected.*

To each object x in $\mathcal{P}_U$ is associated a group like monoidal stack (or *gr*-stack) $\mathcal{G}_x := \mathcal{A}r_U(x, x)$ above U.

DEFINITION 6.4. *Let $\mathcal{G}$ be a group-like monoidal stack on X. We say that a 2-gerbe $\mathcal{P}$ is a $\mathcal{G}$-2-gerbe if there exists an open covering $\mathcal{U} := (U_i)_{i \in I}$ of X, a family of objects $x_i \in \mathcal{P}_{U_i}$, and U_i-equivalences $\mathcal{G}_{U_i} \simeq \mathcal{G}_{x_i}$.*

6.2. Cocycles for 2-gerbes. In order to obtain a cocyclic description of a $\mathcal{G}$-2-gerbe $\mathcal{P}$, we will now categorify the constructions in §5. We choose paths

$$\phi_{ij} : x_j \longrightarrow x_i \qquad (6.2)$$

in the 2-groupoid $\mathcal{P}_{U_{ij}}$, together with quasi-inverses $x_i \longrightarrow x_j$ and pairs of 2-arrows

$$\phi_{ij}\,\phi_{ij}^{-1} \xRightarrow{r_{ij}} 1_{x_i} \qquad\qquad \phi_{ij}^{-1}\,\phi_{ij} \xRightarrow{s_{ij}} 1_{x_j} \ . \qquad (6.3)$$

These determine a monoidal equivalence

$$\lambda_{ij} : \mathcal{G}_{|U_{ij}} \longrightarrow \mathcal{G}_{|U_{ij}} \qquad (6.4)$$

as well as, functorially each object $\gamma \in \mathcal{G}_{|U_{ij}}$, a 2-arrow $M_{ij}(\gamma)$

$$
\begin{array}{ccc}
x_j & \xrightarrow{\ \gamma\ } & x_j \\
{\scriptstyle \phi_{ij}}\big\downarrow & \ \ \Nearrow & \big\downarrow{\scriptstyle \phi_{ij}} \\
x_i & \xrightarrow[\lambda_{ij}(\gamma)]{} & x_i
\end{array}
\qquad (6.5)
$$

which categorifies diagram (5.3). In fact, the 2-arrows r and s (6.3) can be chosen compatibly. For this reason, we will not label a 2-arrow such as $M_{ij}(\gamma)$ explicitly, and will treat diagrams such as (6.5) in appropriate contexts as if they were a commutative squares.

The paths ϕ_{ij} and their inverses also give us objects $g_{ijk} \in \mathcal{G}_{U_{ijk}}$ and 2-arrows m_{ijk}:

$$
\begin{array}{ccc}
x_k & \xrightarrow{\phi_{jk}} & x_j \\
{\scriptstyle \phi_{ik}}\big\downarrow & \overset{m_{ijk}}{\nearrow} & \big\downarrow{\scriptstyle \phi_{ij}} \\
x_i & \xrightarrow[g_{ijk}]{} & x_i
\end{array}
\tag{6.6}
$$

These in turn determine a 2-arrow ν_{ijkl} above U_{ijkl}

$$
\begin{array}{ccc}
x_i & \xrightarrow{g_{ijl}} & x_i \\
{\scriptstyle g_{ikl}}\big\downarrow & \overset{\nu_{ijkl}}{\swarrow} & \big\downarrow{\scriptstyle \lambda_{ij}(g_{jkl})} \\
x_i & \xrightarrow[g_{ijk}]{} & x_i
\end{array}
\tag{6.7}
$$

as the unique 2-arrow such that the following diagram of 2-arrows with right-hand face (6.5) and front face ν_{ijkl} commutes:

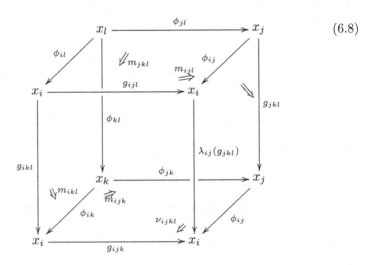

$$\tag{6.8}$$

This cube in $\mathcal{P}_{U_{ijkl}}$ will be denoted C_{ijkl}. Consider now the following diagram:

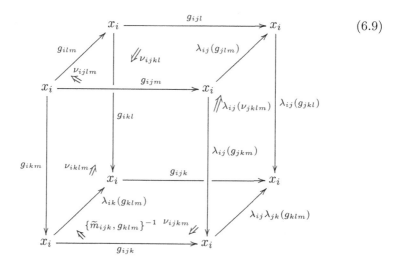

(6.9)

In order to avoid any possible ambiguity, we spell out in the following table the names of the faces of the cube (6.9):

TABLE 1
The faces of cube (6.9).

left	right	top	bottom	front	back
ν_{iklm}	$\lambda_{ij}(\nu_{jklm})$	ν_{ijlm}	$\{\widetilde{m}_{ijk},\, g_{klm}\}^{-1}$	ν_{ijkm}	ν_{ijkl}

As we see from this table, five of its faces are defined by arrows ν (6.7). The remaining bottom 2-arrow $\{\widetilde{m}_{ijk},\, g_{klm}\}^{-1}$ is essentially the inverse of the 2-arrow $\widetilde{m}_{ijk}(g_{klm})$ obtained by evaluating the natural transformation

$$\widetilde{m}_{ijk} : i_{g_{ijk}} \lambda_{ik} \Rightarrow \lambda_{ij}\,\lambda_{jk} \tag{6.10}$$

induced by conjugation from the 2-arrow m_{ijk} (6.6) on the object $g_{klm} \in G := \mathrm{Aut}_{\mathcal{P}}(x_k)$. More precisely, if we compose the latter 2-arrow as follows with the unlabelled 2-arrow $M_{g_{ijk}}(\lambda_{ik}(g_{klm}))$ associated to $i_{g_{ijk}}$:

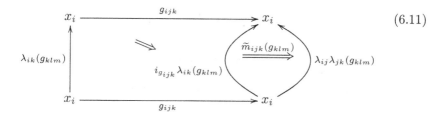

(6.11)

we obtain a 2-arrow

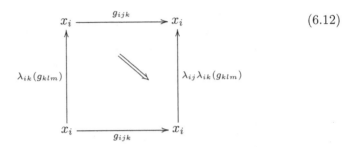

$$(6.12)$$

which we denote by $\{\tilde{m}_{ijk},\, g_{klm}\}$. It may be characterized as the unique 2-arrow such that the cube

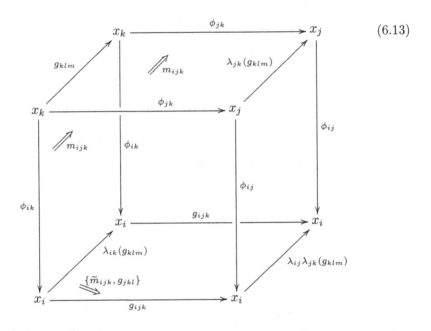

$$(6.13)$$

(with three unlabelled faces of type (6.5)) is commutative. For that reason, this cube will be denoted $\{\,,\,\}$. The following proposition provides a geometric interpretation for the cocycle equation which the 2-arrows ν_{ijkl} satisfy.

PROPOSITION 6.1. *The diagram of 2-arrows (6.9) is commutative.*

Proof. Consider the following hypercubic diagram, from which the 2-arrows have all been omitted for greater legibility.

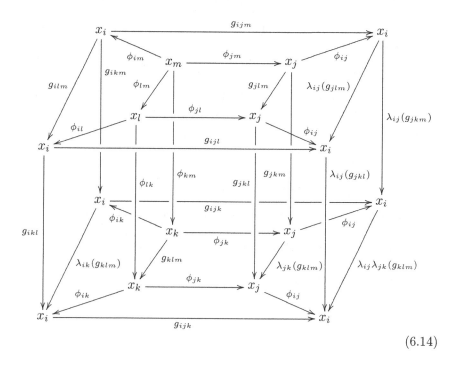

$$(6.14)$$

The following table is provided as a help in understanding diagram (6.14). The first line describes the position in the hypercube of each of the eight cubes from which it has been constructed, and the middle line gives each of these a name. Finally, the last line describes the face by which it is attached to the inner cube C_{jklm}.

TABLE 2
The constituent cubes of diagram (6.14).

inner	left	right	top	bottom	front	back	outer
C_{jklm}	C_{iklm}	Conj(ϕ_{ij})	C_{ijlm}	$\{\,,\,\}$	C_{ijkl}	C_{ijkm}	(6.9)
	m_{klm}	ν_{jklm}	m_{jlm}	$M_{jk}(m_{klm})$	m_{jkl}	m_{jkm}	

Only one cube in this table has not yet been described. It is the cube Conj(ϕ_{ij}) which appears on the right in diagram (6.14). It describes the construction of the 2-arrow $\lambda_{ij}(\nu_{jklm})$ starting from ν_{jklm}, by conjugation of its source and target arrows by the 1-arrows ϕ_{ij}.

Now that diagram (6.14) has been properly described, the proof of proposition 6.1 is immediate, and goes along the same lines as the proof of lemma 5.1. One simply observes that each of the first seven cubes in table 2 is a commutative diagram of 2-arrows. Since all their constituent 2-arrows are invertible, the remaining outer cube is also a commutative diagram of 2-arrows. The latter cube is simply (6.9), though with a different orientation, so the proof of the proposition is now complete. □

REMARK 6.1. When $i = j$, it is natural to choose as arrow ϕ_{ij} (6.2) the identity arrow 1_{x_i}. When $i = j$ or $j = k$, it is then possible to set $g_{ijk} = 1_{x_i}$ and to choose the identity 2-arrow for m_{ijk}. These choices yield the following normalization conditions for our cocycles:

$$\begin{cases} \lambda_{ij} = 1 \text{ whenever } i = j \\ g_{ijk} = 1 \text{ and } \tilde{m}_{ijk} = 1 \text{ whenever } i = j \text{ or } j = k \\ \nu_{ijkl} = 1 \text{ whenever } i = j, \ j = k, \text{ or } k = l \end{cases}$$

6.3. Algebraic description of the cocycle conditions. In order to obtain a genuinely cocyclic description of a $\mathcal{G}$-2-gerbe, it is necessary to translate proposition 6.1 into an algebraic statement. As a preliminary step, we implement such a translation for the cubical diagram C_{ijkl} (6.8) by which we defined the 2-arrow ν_{ijkl}. We reproduce this cube as

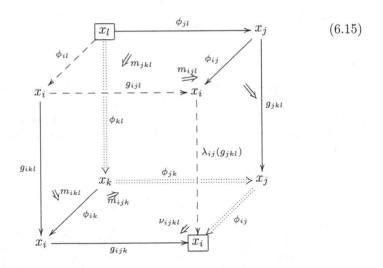

(6.15)

and consider the two composite paths of 1-arrows from the framed vertex x_l to the framed vertex x_i respectively displayed by arrows of type $--\!\!\succ$ and $\cdots\!\!\succ$. The commutativity of our cube is equivalent to the assertion that the two possible composite 2-arrows from the path $--\!\!\succ$ to the path $\cdots\!\!\succ$ coincide. This assertion translates, when taking into account the whiskerings which arise whenever one considers a face of the cube which does not contain the framed vertex x_i, to the equation

$$m_{ijk} \left(g_{ijk} * m_{ikl} \right) \nu_{ijkl} = \left(\phi_{ij} * m_{jkl} \right) \left(\lambda_{ij} (g_{jkl}) * m_{ijl} \right) \tag{6.16}$$

which algebraically defines the 2-arrow ν_{ijkl} in terms of the 2-arrows of type m_{ijk} (6.6). For reasons which will appear later on, we have neglected here the whiskerings by 1-arrows on the right, for faces of the cube which do not contain the framed vertex x_l from which all paths considered originate. With the left-hand side of this equality labelled "1" and the right-hand side "2", the two sides are compared according to the following scheme in the 2-category $\mathcal{P}_{U_{ijkl}}$:

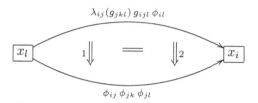

Consider now a 2-arrow

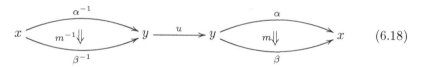

(6.17)

in $\mathcal{P}_U$, and denote by α_* and β_* the functors $\mathcal{G}_U \longrightarrow \mathcal{G}_U$ which conjugation by α and β respectively define. The conjugate of any 1-arrow $u \in \mathrm{ob}\, \mathcal{G}_U = \mathrm{Ar}_{\mathcal{P}_U}(y, y)$ by the 2-arrow m is the composite 2-arrow

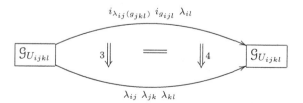

(6.18)

where m^{-1} is the horizontal inverse of the 2-arrow m. We denote by $\widetilde{m} : \alpha_* \Longrightarrow \beta_*$ the natural transformation which m defines in this way. It is therefore an arrow

$$\widetilde{m} : \alpha_* \longrightarrow \beta_*$$

in the monoidal category $\mathcal{E}q(\mathcal{G})_U$. With this notation, it follows that equation (6.16) conjugates according to the scheme

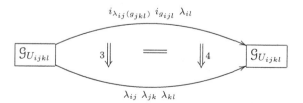

to the following equation between the arrows "3" and "4" in the category $\mathcal{E}q(\mathcal{G})_{U_{ijkl}}$:

$$\widetilde{m}_{ijk} \, {}^{g_{ijk}}\widetilde{m}_{ikl} \, j(\nu_{ijkl}) = \lambda_{ij}(\widetilde{m}_{jkl}) \, {}^{\lambda_{ij}(g_{jkl})}\widetilde{m}_{ijl} \qquad (6.19)$$

In such an equation, the term $j(\nu_{ijkl})$ is the image of the element $\nu_{ijkl} \in \mathrm{Ar}(\mathcal{G})$ under inner conjugation functor[6]

$$\mathcal{G} \xrightarrow{\;j\;} \mathcal{E}q(\mathcal{G}) \qquad (6.20)$$

associated to the group-like monoidal stack $\mathcal{G}$. By an expression such as ${}^{g_{ijk}}\widetilde{m}_{ikl}$, we mean the conjugate of the 1-arrow $\widetilde{m}_{ikl}$ by the object $j(g_{ijk})$ in the monoidal category $\mathcal{E}q(\mathcal{G})_{U_{ijkl}}$. We observe here that the right whiskerings of a 2-arrow m or ν (*i.e.* the composition a 2-arrow with a 1-arrow which precedes it) have no significant effect upon the conjugation operation which associates to a 2-arrow m (*resp.* ν) in $\mathcal{P}$ the corresponding natural transformation $\widetilde{m}$ (*resp.* $j(\nu)$), an arrow in $\mathcal{E}q(\mathcal{G})$. This is why it was harmless to ignore the right whiskerings in formula (6.16), and we will do so in similar contexts in the sequel.

Let us display once more the cube (6.9), but now decorated according to the same conventions as in (6.15):

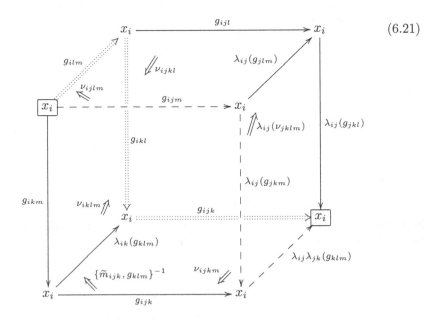

$$(6.21)$$

The commutativity of this diagram of 2-arrows translates (according to the recipe which produced the algebraic equation (6.16) from the cube (6.15)) to the following very twisted 3-cocycle condition for ν [7]:

$$\nu_{ijkl} \left(^{\lambda_{ij}(g_{jkl})}\nu_{ijlm}\right) \lambda_{ij}(\nu_{jklm}) =$$
$$= {}^{g_{ijk}}\nu_{iklm} \{\widetilde{m}_{ijk}, g_{klm}\}^{-1} \left(^{\lambda_{ij}\lambda_{jk}(g_{klm})}\nu_{ijkm}\right) \tag{6.22}$$

This is an equation satisfied by elements with values in $\mathrm{Ar}\,(\mathcal{G}_{U_{ijklm}})$. Note the occurrence here of the term $\{\widetilde{m}_{ijk}, g_{klm}\}^{-1}$, corresponding to the lower face of (6.21). While such a term does not exist in the standard definition of an abelian Čech 3-cocycle equation, non-abelian 3-cocycle relations of this type go back to the work of P. Dedecker [7]. They arise there in the context of group rather than Čech cohomology, with his cocycles taking their values in an unnecessarily restrictive precursor of a crossed square, which he calls a super-crossed group.

The following definition, which summarizes the previous discussion, may be also viewed as a categorification of the notion of a G-valued cocycle pair, as defined by equations (5.10):

DEFINITION 6.5. *Let $\mathcal{G}$ be a group-like monoidal stack on a space X, and $\mathcal{U}$ an open covering of X. A $\mathcal{G}$-valued Čech 1-cocycle quadruple is a quadruple of elements*

$$(\lambda_{ij}, \widetilde{m}_{ijk}, g_{ijk}, \nu_{ijkl}) \tag{6.23}$$

satisfying the following conditions. The term λ_{ij} is an object in the monoidal category $\mathcal{E}q_{U_{ij}}(\mathcal{G}_{|U_{ij}})$ and $\widetilde{m}_{ijk}$ is an arrow

$$\widetilde{m}_{ijk} : j(g_{ijk})\,\lambda_{ik} \longrightarrow \lambda_{ij}\,\lambda_{jk} \tag{6.24}$$

in the corresponding monoidal category $\mathcal{E}q_{U_{ijk}}(\mathcal{G}_{|U_{ijk}})$. Similarly, g_{ijk} is an object in the monoidal category $\mathcal{G}_{U_{ijk}}$ and

$$\nu_{ijkl} : \lambda_{ij}(g_{jkl})\,g_{ijl} \longrightarrow g_{ijk}\,g_{ikl}$$

an arrow (6.7) in the corresponding monoidal category $\mathcal{G}_{U_{ijkl}}$. Finally we require that the two equations (6.19) and (6.22), which we reproduce here for the reader's convenience, be satisfied:

---

[7]This is essentially the 3-cocycle equation (4.2.17) of [4], but with the terms in opposite order due to the fact that the somewhat imprecise definition of a 2-arrow ν given on page 71 of [4] yields the inverse of the 2-arrow ν defined here by equation (6.16).

$$\begin{cases} \widetilde{m}_{ijl} \,^{g_{ijk}}\widetilde{m}_{ikl} \, j(\nu_{ijkl}) = \lambda_{ij}(\widetilde{m}_{jkl}) \,^{\lambda_{ij}(g_{jkl})}\widetilde{m}_{ijl} \\[2mm] \nu_{ijkl} \,(^{\lambda_{ij}(g_{jkl})}\nu_{ijlm}) \, \lambda_{ij}(\nu_{jklm}) = \\[2mm] \qquad\qquad = \,^{g_{ijk}}\nu_{iklm} \, \{\widetilde{m}_{ijk}, g_{klm}\}^{-1}(^{\lambda_{ij}\lambda_{jk}(g_{klm})}\nu_{ijkm}). \end{cases} \qquad (6.25)$$

Returning to our discussion, let us consider such a $\mathcal{G}$-valued Čech 1-cocycle quadruple (6.23). In order to transform the categorical crossed module (6.20) into a weak analogue of a crossed square, it is expedient for us to restrict ourselves, in both the categories $\mathcal{G}$ and $\mathcal{E}q(\mathcal{G})$, to those arrows whose source is the identity object. Diagram (6.20) then becomes

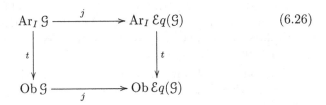

$$\qquad (6.26)$$

where t is the target map and the same symbol j describes the components on objects and on arrows of the inner conjugation functor (6.20). Recall that one can assign to any arrow $u : X \longrightarrow Y$ in a group-like monoidal category the arrow $uX^{-1} : I \longrightarrow YX^{-1}$ sourced at the identity, without loosing any significant information. In particular, the arrow $\widetilde{m}_{ijk}$ (6.24) may be replaced by an arrow

$$I \longrightarrow \lambda_{ij} \, \lambda_{jk} \, \lambda_{ik}^{-1} \, j(g_{ijk})^{-1}$$

in $(\mathrm{Ar}_I \, \mathcal{E}q(\mathcal{G}))_{U_{ijk}}$ and the arrow ν_{ijkl} (6.7) by an arrow

$$I \longrightarrow g_{ijk} \, g_{ikl} \, g_{ijl}^{-1} \, \lambda_{ij}(g_{jkl})^{-1},$$

in $(\mathrm{Ar}_I \, \mathcal{G})_{U_{ijkl}}$ which we again respectively denote by $\widetilde{m}_{ijk}$ and ν_{ijkl}. Our quadruple (6.23) then takes its values in the weak square

$$\begin{array}{ccc} (\mathrm{Ar}_I \, \mathcal{G})_{U_{ijkl}} & \xrightarrow{\ \ j\ \ } & (\mathrm{Ar}_I \, \mathcal{E}q(\mathcal{G}))_{U_{ijk}} \\[1mm] \Big\downarrow{\scriptstyle t} & & \Big\downarrow{\scriptstyle t} \\[3mm] (\mathrm{Ob}\,\mathcal{G})_{U_{ijk}} & \xrightarrow[\ \ j\ \ ]{} & (\mathrm{Ob}\,\mathcal{E}q(\mathcal{G}))_{U_{ij}} \end{array} \qquad (6.27)$$

in the positions

$$\begin{pmatrix} \nu_{ijkl} & \widetilde{m}_{ijk} \\ g_{ijk} & \lambda_{ij} \end{pmatrix}.$$ (6.28)

Since the evaluation action of $\mathcal{E}q(\mathcal{G})$ on $\mathcal{G}$ produces a map

$$\mathrm{Ar}_I \, \mathcal{E}q(\mathcal{G}) \times \mathrm{Ob}\,\mathcal{G} \longrightarrow \mathrm{Ar}_I \, \mathcal{G}$$

which is the analog of the morphism (1.23), the quadruple (6.23) may now be viewed as a cocycle with values in what might be termed the (total complex associated to the) weak crossed square (6.26). We will say that this modified quadruple (6.28) is a Čech 1-cocycle[8] for the covering $\mathcal{U}$ on X with values in the (weak) crossed square (6.26). Because of the position of the different terms of the quadruple (6.28) in the square (6.27), this terminology is consistent with the fact that the component ν_{ijkl} of such a 1-cocycle (6.23) satisfies a sort of 3-cocycle relation (6.22). The discussion in paragraph 6.3 will now be summarized as follows in purely algebraic terms:

PROPOSITION 6.2. *To a $\mathcal{G}$-2-gerbe $\mathcal{P}$ on X, locally trivialized by the choice of objects x_i in $\mathcal{P}_{U_i}$ and local paths ϕ_{ij} (6.2), is associated a 1-cocycle (6.23) with values in the weak crossed square (6.26).*

REMARK 6.2. When $\mathcal{G}$ is the *gr*-stack associated to a crossed module $\delta : G \longrightarrow \pi$, this coefficient crossed module of *gr*-stacks is a stackified version of the following crossed square associated by K.J.Norrie (see [18, 6] Theorem 3.5) to the crossed module $G \longrightarrow \pi$:

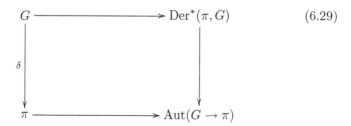

(6.29)

---

[8]This was called a 3-cocycle in [4], but the present terminology is more appropriate.

It is however less restrictive than Norrie's version, since the latter corresponds to the diagram of gr-stacks

$$\mathcal{G} \longrightarrow \mathrm{Isom}(\mathcal{G})$$

whereas we really need to consider, as in (6.20), self-equivalences of the monoidal stack $\mathcal{G}$, rather than automorphisms.

To phrase it differently, we need to replace the term $\mathrm{Aut}(G \longrightarrow \pi)$ in the square (6.29) by the weak automorphisms of the crossed module $G \longrightarrow \pi$, as discussed in remark 1.3, and modify the set of crossed homomorphisms $\mathrm{Der}^*(\pi, G)$ accordingly. Recently, B. Noohi has provided in [17] a very economical interpretation of weak morphisms of crossed modules and of the natural transformations between them, in terms of diagrams which he calls butterflies. This should yield a very concrete description of the terms λ_{ij} and $\tilde{m}_{ij}$ (6.28) in the 1-cocycle (6.23), as well as of the terms $\tilde{\zeta}_{ij}$ and r_i in the corresponding coboundary relations occuring in (6.44) below. Alternately, one could rely on his direct description in [16] Definition 8.4 of weak homomorphisms, as mentioned in Remark 1.10 above.

6.4. Coboundary relations. We now choose a second set of local objects $x'_i \in \mathcal{P}_{U_i}$, and of local arrows (6.2)

$$\phi'_{ij} : x'_j \longrightarrow x'_i.$$

By proposition 6.2, these determine a second crossed square valued 1-cocycle

$$\left(\lambda'_{ij}, \; \tilde{m}'_{ijk}, \; g'_{ijk}, \; \nu'_{ijkl} \right). \tag{6.30}$$

In order to compare it with the 1-cocycle (6.23), we proceed as we did in Section 5.2 above, but now in a 2-categorical setting. We choose once more an arrow χ_i (5.12). There now exist 1-arrows ϑ_{ij}, and 2-arrows ζ_{ij} in $\mathcal{P}_{U_{ij}}$.

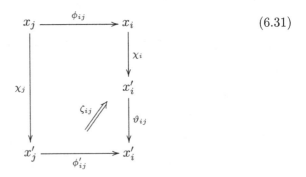

$$\tag{6.31}$$

The arrow χ_i induces by conjugation a self-equivalence $r_i : \mathcal{G} \longrightarrow \mathcal{G}$ and 2-arrows

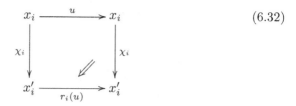

$$(6.32)$$

which are functorial in u. Furthermore, the diagram (6.31) induces by conjugation a diagram in $\mathcal{G}_{U_{ij}}$:

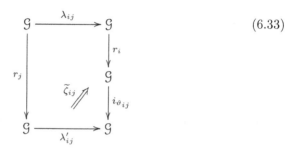

$$(6.33)$$

with $\widetilde{\zeta}_{ij}$ the natural transformation induced by ζ_{ij}. Consider now the diagram of 2-arrows

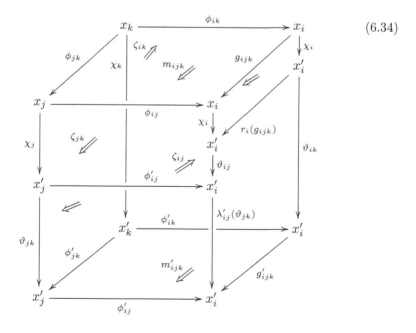

$$(6.34)$$

which extends (5.17). The three 2-arrows ζ_{ij}, ζ_{jk} and ζ_{ik} are of the form (6.31), the top and the bottom ones are both of the form m_{ijk} (6.6). The unlabelled lower front 2-arrow and the right-hand upper one are each part of the definition of the corresponding edges $\lambda'_{ij}(\vartheta_{jk})$ and of $r_i(g_{ijk})$. Since these seven 2-arrows are invertible, diagram (6.34) uniquely defines a 2-arrow b_{ijk} filling in the remaining lower right-hand square:

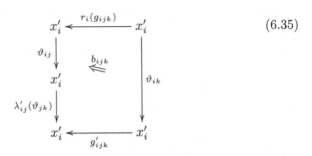

$$(6.35)$$

Diagram (6.34) becomes the following commutative diagram of 2-arrows, which we directly display in decorated form, according to the conventions of (6.15):

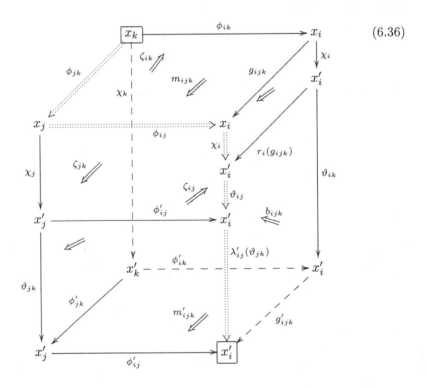

$$(6.36)$$

We derive from this diagram the algebraic equation

$$(\lambda'_{ij}(\vartheta_{kl}) * \zeta_{ij})\,(\phi'_{ij} * \zeta_{jk})\,m'_{ijk} = ((\lambda'_{ij}(\vartheta_{jk})\,\vartheta_{ij}\,\chi_i) * m_{ijk})\,b_{ijk}\,(g'_{ijk} * \zeta_{ik})$$

for the equality between the two corresponding 2-arrows between the decorated paths. With the same notations as for equation (6.19), the conjugated version of this equation is

$$\lambda'_{ij}(\vartheta_{jk})\widetilde{\zeta}_{ij}\ \lambda'_{ij}(\widetilde{\zeta}_{jk})\ \widetilde{m}'_{ijk} = {}^{\lambda'_{ij}(\vartheta_{jk})\,\vartheta_{ij}\,r_i}\widetilde{m}_{ijk}\ j(b_{ijk})\ ({}^{g'_{ijk}}\widetilde{\zeta}_{ik}) \qquad (6.37)$$

It is the analogue, with the present conventions, of equation [4] (4.4.12).

A second coboundary condition relates the cocycle quadruples (6.23) and (6.30). In geometric terms, it asserts the commutativity of the following diagram of 2-arrows, in which the unlabelled 2-arrow in the middle of the right vertical face is $\{\widetilde{\zeta}_{ij},\, g_{jkl}\}^{-1}$ defined in the same way as the 2-arrow which we denoted $\{\widetilde{m}_{ijk},\, g_{klm}\}$ (6.12):

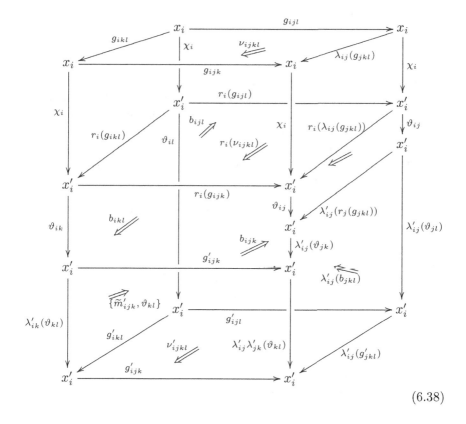

$$(6.38)$$

This cubic diagram compares the 2-arrows ν_{ijkl} and ν'_{ijkl}, which are respectively its top and bottom faces. It actually consists of two separate cubes. The upper one is trivially commutative, as it simply defines the 2-arrow $r_i(\nu_{ijkl})$, which is the common face between the two cubes considered.

LEMMA 6.1. *The cube of 2-arrows* (6.38) *is commutative*[9].

Proof. The proof that the full diagram (6.38) commutes is very similar to the proof of Proposition 6.1. We consider a hypercube analogous to diagram (6.14), and which therefore consists of eight cubes called left, right, top, bottom, front, back, inner and outer. The outer cube in this diagram is the cube (6.38). We will now describe the seven other cubes. Since these seven are commutative, this will imply that the outer one also is, so that the lemma will be proved. As this hypercubic diagram is somewhat more complicated than (6.14), we will describe it in words, instead of displaying it.

The top cube is a copy of cube (6.8), oriented so that its face ν_{ijkl} is on top, consistently with the top face of (6.38). The bottom cube is a cube of similar type which defines the 2-arrow ν'_{ijkl}. Since it is built from objects x', arrows ϕ' and g' and 2-arrows m' and ν', we will refer to it as the primed version of (6.8) . It is oriented so that ν'_{ijkl} is the bottom face.

We now describe the six other cubes. Four of these are of the type (6.36). If we denote the latter by the symbol P_{ijk} determined by its indices, these are respectively the left cube P_{ikl}, the back cube P_{ijl}, the inner cube P_{jkl} and the front cube P_{ijk}. Each of the first three rests on the corresponding face m'_{ikl}, m'_{ijl}, and m'_{jkl} of the bottom cube, and is attached at the top to the similar face m of the top cube. The cube P_{ijk} is attached to the corresponding face m_{ijk} of the top cube, but it does not constitute the full front cube. Below it is the following primed version of the cube (6.13), but now associated to the face $\{\tilde{m}'_{ijk}, \vartheta_{kl}\}$ (rather than as in (6.13) to the face $\{\tilde{m}_{ijk}, g_{klm}\}$).

<hr>

[9]In [4] § 4.9 the corresponding assertion is implicitly assumed to be true, although no proof is given there.

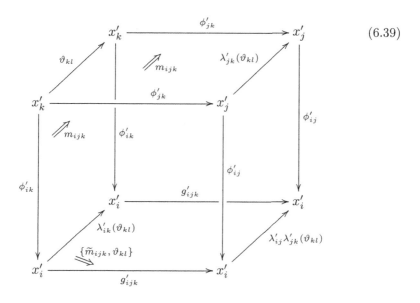

(6.39)

Finally, the right cube is itself constituted of two cubes. The lower one constructs the 2-arrow $\lambda'_{ij}(b_{jkl})$, starting from the 2-arrow b_{jkl} (6.35). The upper one is another commutative cube of type (6.13), but this time associated to the face $\{\widetilde{\zeta}_{ij},\, g_{jkl}\}$. This is the following commutative cube whose four unlabelled 2-arrows are the obvious ones.

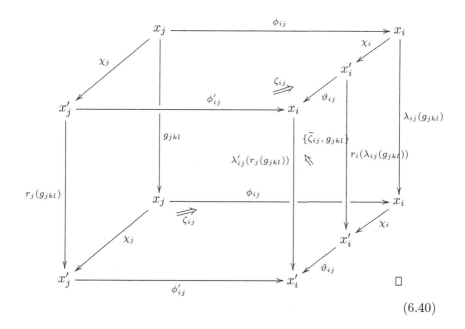

□

(6.40)

In order to translate the commutativity of the cube (6.38) into an algebraic expression, we now decorate it as follows, invoking once more the conventions of diagram (6.15):

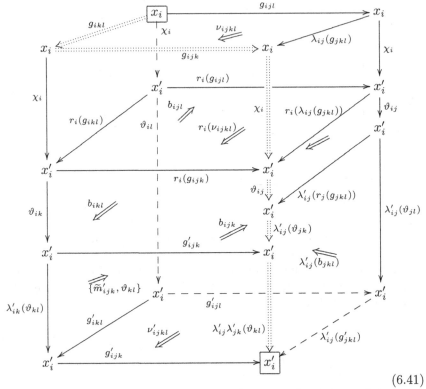

$$(6.41)$$

Reading off the two composite 2-arrows between the decorated 1-arrows (6.41), and taking into account the appropriate whiskerings, we see that the commutativity of diagram (6.41) is equivalent to the following algebraic equation

$$\left(^{\lambda'_{ij}\lambda'_{jk}(\vartheta_{kl})\lambda'_{ij}(\vartheta_{jk})\vartheta_{ij}r_i}\,\nu_{ijkl}\right)$$

$$\left(^{\lambda'_{ij}\lambda'_{jk}(\vartheta_{kl})\lambda'_{ij}(\vartheta_{jk})}\{\widetilde{\zeta}_{ij},\,g_{jkl}\}^{-1}\right)\,\lambda'_{ij}(b_{jkl})\,\left(^{\lambda'_{ij}(g'_{jkl})}b_{ijl}\right) = \qquad (6.42)$$

$$= \left(^{\lambda'_{ij}\lambda'_{jk}(\vartheta_{kl})}b_{ijk}\right)\{\widetilde{m}'_{ijk},\,\vartheta_{kl}\}\,\left(^{g'_{ijk}}b_{ikl}\right)\,\nu'_{ijkl}.$$

This equation is the analogue, under our present conventions, of equation (4.4.15) of [4]. It describes the manner in which the various terms of type b_{ijk} determine a coboundary relation between the non-abelian cocycle terms ν_{ijkl} and ν'_{ijkl}. A certain amount of twisting takes place, however, and the extra terms $\{\widetilde{\zeta}_{ij},\,g_{klm}\}^{-1}$ and $\{\widetilde{m}'_{ijk},\,\vartheta_{kl}\}$ need to be inserted in their proper locations, just as the factor $\{\widetilde{m}_{ijk},\,g_{klm}\}^{-1}$ was necessary in order to formulate equation (6.22). Once more, an equation such as (6.42) cannot be viewed in isolation from its companion equation (6.37). In addition, any arrow in either of the monoidal categories $\text{Ar}(\mathcal{G})$ or $\mathcal{G}$ must be replaced by the corresponding one which is sourced at the identity, without changing its name. The following definition summarizes the previous discussion.

DEFINITION 6.6. *Let* $(\lambda_{ij}, \widetilde{m}_{ijk}, g_{ijk}, \nu_{ijkl})$ *and* $(\lambda'_{ij}, \widetilde{m}'_{ijk}, g'_{ijk}, \nu'_{ijkl})$ *be a pair of 1-cocycle quadruples with values in the weak crossed square* (6.26). *These two cocycle quadruples are cohomologous if there exists a quadruple* $(r_i, \widetilde{\zeta}_{ij}, \vartheta_{ij}, b_{ijk})$ *with values in the weak crossed cube* (6.26). *More precisely, these elements take their values in the square*

$$
\begin{array}{ccc}
(\text{Ar}_I\,\mathcal{G})_{U_{ijk}} & \xrightarrow{\;\;j\;\;} & (\text{Ar}_I\,\mathcal{E}q(\mathcal{G}))_{U_{ij}} \\
\Big\downarrow{\scriptstyle t} & & \Big\downarrow{\scriptstyle t} \\
(\text{Ob}\,\mathcal{G})_{U_{ij}} & \xrightarrow{\;\;j\;\;} & (\text{Ob}\,\mathcal{E}q(\mathcal{G}))_{U_i}
\end{array}
\tag{6.43}
$$

in the positions

$$
\begin{pmatrix} b_{ijk} & \widetilde{\zeta}_{ij} \\ \vartheta_{ij} & r_i \end{pmatrix}.
\tag{6.44}
$$

The arrows b_{ijk} *and* $\widetilde{\zeta}_{ij}$ *are respectively of the form*

$$
I \xrightarrow{\;\;b_{ijk}\;\;} \lambda'_{ij}(\vartheta_{jk})\,\vartheta_{ij}\,r_i(g_{ijk})\,\vartheta_{ik}^{-1}\,(g'_{ijk})^{-1}
$$

and

$$
I \xrightarrow{\;\;\widetilde{\zeta}_{ij}\;\;} j(\vartheta_{ij})\,r_i\,\lambda_{ij}\,r_j^{-1}\,\lambda'^{\,-1}_{ij}
$$

and are required to satisfy the equations (6.37) *and* (6.42). *The set of equivalence classes of 1-cocycle quadruples* (6.28), *for the equivalence defined by these coboundary relations, will be called the Čech degree 1 cohomology set for the open covering* $\mathcal{U}$ *of* X *with values in the weak crossed square* (6.26). *Passing to the limit over the families of such open coverings of* X, *one obtains the Čech degree 1 cohomology set of* X *with values in this square.*

REMARK 6.3. If we consider as in Remark 6.6 a pair of normalized cocycle quadruples, the corresponding normalization conditions on the coboundary terms (6.44) are

$$\begin{cases} \vartheta_{ij} = 1 \text{ and } \widetilde{\zeta}_{ij} = 1 \text{ when } i = j \\ b_{ijk} = 1 \text{ whenever } i = j \text{ or } j = k. \end{cases}$$

The discussion in paragraphs 6.2-6.4 can now be entirely summarized as follows:

PROPOSITION 6.3. *The previous constructions associate to a $\mathcal{G}$-2-gerbe $\mathcal{P}$ on a space X an element of the Čech degree 1 cohomology set of X with values in the square (6.26), and this element is independent of the choice of local objects and arrows in $\mathcal{P}$.* We refer to chapter 5 of [4] for the converse to this proposition, which asserts that to each such 1-cohomology class corresponds a $\mathcal{G}$-2-gerbe, uniquely defined up to equivalence.

REMARK 6.4. As we observed in footnote 3, the proposition is only true as stated when the space X satisfies an additional assumption such as paracompactness. The general case is discussed in [4], where the open covering $\mathcal{U}$ of X is replaced by a hypercover.

REFERENCES

[1] L. BREEN, *Bitorseurs et cohomologie non abélienne* in *The Grothendieck Festschrift I*, Progress in Mathematics **86**:401–476 (1990), Birkhaüser.
[2] L. BREEN, *Tannakian Categories*, in *Motives*, Proc. Symp. Pure Math. **55**(1): 337–376 (1994).
[3] L. BREEN, *Théorie de Schreier supérieure*, Ann. Scient. École Norm. Sup. **25**(4):465–514 (1992).
[4] L. BREEN, *Classification of 2-gerbes and 2-stacks*, Astérisque **225** (1994), Société Mathématique de France.
[5] L. BREEN AND W. MESSING, *Differential Geometry of Gerbes*, Advances in Math. **198**:732–846 (2005).
[6] R. BROWN AND N.D. GILBERT, *Algebraic Models of 3-types and Automorphism structures for Crossed modules*, Proc. London Math. Soc. **59**:51–73 (1989).
[7] P. DEDECKER, *Three-dimensional Non-Abelian Cohomology for Groups*, in *Category theory, homology theory and their applications II*, Lecture Notes in Mathematics **92**:32–64 (1969).
[8] E. DROR AND A. ZABRODSKY, *Unipotency and Nilpotency in Homotopy Equivalences*, Topology **18**:187–197 (1979).
[9] J. DUSKIN, *An outline of non-abelian cohomology in a topos (1): the theory of bouquets and gerbes*, Cahiers de topologie et géométrie différentielle XXIII (1982).
[10] J. GIRAUD, *Cohomologie non abélienne*, Grundlehren **179**, Springer-Verlag (1971).
[11] N. HITCHIN, *Lectures on special Lagrangians on submanifolds*, Winter School on Mirror Symmetry, Vector Bundles and Lagrangian Submanifolds (Cambridge MA, 1999), AMS/IP Stud. Adv. Math **23**:151–182, American Math. Soc. (2001).
[12] J.F. JARDINE, *Cocycle categories*, arXiv:math.AT/0605198.
[13] J.F. JARDINE, *Homotopy classification of gerbes* arXiv:math.AT/0605200.
[14] J.-L. LODAY, *Spaces with finitely many homotopy groups*, J. Pure Appl. Alg. **24**:179–202 (1982).
[15] M.K. MURRAY, *Bundle gerbes*, J. of the London Math. Society **54**:403–416 (1996).
[16] B. NOOHI, *Notes on 2-groupoids, 2-groups and crossed modules*, Homology, Homotopy and Applications **9**(1):75–106 (2007).

[17] B. NOOHI, *On weak maps between 2-groups*, arXiv:math/0506313.

[18] K. NORRIE, *Actions and automorphisms*, Bull. Soc. Math. de France **118**:109–119 (1989).

[19] O. SCHREIER, *Uber die Erweiterung von Gruppen*, *I* Monatsh. Math. Phys. **34**: 165–180 (1926); *II* Abh. Math. Sem. Hamburg **4**:321–346 (1926).

[20] J.-P. SERRE, *Cohomologie galoisienne*, Lecture Notes in Math **5** (1964), fifth revised and completed edition (1994), Springer-Verlag.

[21] K.-H. ULBRICH, *On the correspondence between gerbes and bouquets*, Math. Proc. Cambridge Phil. Soc. **108**:1–5 (1990).

[22] K.-H. ULBRICH, *On cocycle bitorsors and gerbes over a Grothendieck topos*, Math. Proc. Cambridge Phil. Soc. **110**:49–55 (1991).

[23] J. WIRTH AND J. STASHEFF, *Homotopy Transition Cocycles*, J. Homotopy and Related Structure **1**:273–283 (2006).

[24] A. VISTOLI, *Grothendieck topologies, fibered categories and descent theory*, in *Fundamental Algebraic Geometry: Grothendieck's FGA explained*, Mathematical Monographs **123**:1–104, American Math. Soc. (2005).

AN AUSTRALIAN CONSPECTUS OF
HIGHER CATEGORIES*

ROSS STREET[†]

Much Australian work on categories is part of, or relevant to, the development of higher categories and their theory. In this note, I hope to describe some of the origins and achievements of our efforts that they might perchance serve as a guide to the development of aspects of higher-dimensional work.

I trust that the somewhat autobiographical style will add interest rather than be a distraction. For so long I have felt rather apologetic when describing how categories might be helpful to other mathematicians; I have often felt even worse when mentioning enriched and higher categories to category theorists. This is not to say that I have doubted the value of our work, rather that I have felt slowed down by the continual pressure to defend it. At last, at this meeting, I feel justified in speaking freely amongst motivated researchers who know the need for the subject is well established.

Australian Category Theory has its roots in homology theory: more precisely, in the treatment of the cohomology ring and the Künneth formulas in the book by Hilton and Wylie [71]. The first edition of the book had a mistake concerning the cohomology ring of a product. The Künneth formulas arise from splittings of the natural short exact sequences

$$0 \longrightarrow \mathrm{Ext}(HA, HB) \longrightarrow H[A, B] \xrightarrow{H} \mathrm{Hom}(HA, HB) \longrightarrow 0$$

$$0 \longrightarrow HA \otimes HB \xrightarrow{\otimes} H(A \otimes B) \longrightarrow \mathrm{Tor}(HA, HB) \longrightarrow 0$$

where A and B are chain complexes of free abelian groups; however, there are no choices of natural splittings. Wylie's former postgraduate student, Max Kelly, was intrigued by these matters and wanted to understand them conceptually.

So stimulated, in a series of papers [87–89, 91, 93] published in *Proc. Camb. Phil. Soc.*, Kelly progressed ever more deeply into category theory. He discussed equivalence of categories and proposed criteria for when a functor should provide "complete invariants" for objects of its domain category. Moreover, Kelly invented differential graded categories and used them to show homotopy nilpotence of the kernel of certain functors [93].

Around the same time, Sammy Eilenberg invented DG-categories probably for purposes similar to those that led Verdier to derived cate-

---

*Prepared for the Institute for Mathematics and its Applications Summer Program "n-Categories: Foundations and Applications" at the University of Minnesota (Minneapolis, 7–18 June 2004).

†Centre of Australian Category Theory, Macquarie University, New South Wales 2109, AUSTRALIA (street@math.mq.edu.au; http://www.math.mq.edu.au/~street).

J.C. Baez, J.P. May (eds.), *Towards Higher Categories*, The IMA Volumes in Mathematics and its Applications 152, DOI 10.1007/978-1-4419-1524-5_6, © Springer Science+Business Media, LLC 2010

gories. Thus began the collaboration of Eilenberg and Kelly on enriched categories. They realized that the definition of DG-category depended only on the fact that the category DGAb of chain complexes was what they called a *closed* or, alternatively, a *monoidal* category. They favoured the "closed" structure over "monoidal" since internal homs are usually more easily described than tensor products; good examples such as DGAb have both anyway.

The groundwork for the correct definition of monoidal category $\mathcal{V}$ had been prepared by Saunders Mac Lane with his *coherence theorem* for associativity and unit constraints. Kelly had reduced the number of axioms by a couple so that only the Mac Lane–Stasheff pentagon and the unit triangle remained. Enriched categories were also defined by Fred Linton; however, he had conditions on the base $\mathcal{V}$ that ruled out the examples $\mathcal{V} = \text{DGAb}$ and $\mathcal{V} = \text{Cat}$ that proved so vital in later applications.

The long Eilenberg–Kelly paper [46] in the 1965 LaJolla Conference Proceedings was important for higher category theory in many ways; I shall mention only two.

One of these ways was the realization that 2-categories could be used to organize category theory just as category theory organizes the theory of sets with structure. The authors provided an explicit definition of (strict) 2-category early in the paper although they used the term "hypercategory" at that point (probably just as a size distinction since, as we shall see, "2-category" was used near the end). So that the paper became more than a list of definitions with implications between axioms, the higher-categorical concepts allowed the paper to be summarized with theorems such as:

$$\mathcal{V}\text{-Cat is a 2-category} \quad \text{and} \quad (-)_* \colon \text{MonCat} \longrightarrow \text{2-Cat is a 2-functor.}$$

The other way worth mentioning here is their efficient definition of (strict) n-category and (strict) n-functor using enrichment. If $\mathcal{V}$ is symmetric monoidal then $\mathcal{V}$-Cat is too and so the enrichment process can be iterated. In particular, starting with $\mathcal{V}_0 = \text{Set}$ using cartesian product, we obtain cartesian monoidal categories $\mathcal{V}_n$ defined by $\mathcal{V}_{n+1} = \mathcal{V}_n$-Cat. This $\mathcal{V}_n$ is the category n-Cat of n-categories and n-functors. In my opinion, processes like $\mathcal{V} \mapsto \mathcal{V}$-Cat are fundamental in dimension raising.

With his important emphasis on categories as mathematical structures of the ilk of groups, Charles Ehresmann [44] defined categories internal to a category $\mathcal{C}$ with pullbacks. The category $\text{Cat}(\mathcal{C})$ of internal categories and internal functors also has pullbacks, so this process too can be iterated. Starting with $\mathcal{C} = \text{Set}$, we obtain the category $\text{Cat}^n(\text{Set})$ of n-tuple categories. In particular, $\text{Cat}^2(\text{Set}) = \text{DblCat}$ is the category of double categories; it contains 2-Cat in various ways as does $\text{Cat}^n(\text{Set})$ contain n-Cat.

At least two other papers in the LaJolla Proceedings volume had a strong influence on Australian higher-dimensional category theory. One was the paper [115] of Bill Lawvere suggesting a categorical foundations for mathematics; concepts such as comma category appeared there. The other was the paper [57] of John Gray developing the subject of Grothendieck fibred categories as a formal theory in Cat so that it could be dualized. This meant that Gray was essentially treating Cat as an arbitrary 2-category; the duality was that of reversing morphisms (what we call Cat^{op}) not 2-cells (what we call Cat^{co}). In stark contrast with topology, Grothendieck had unfortunately used the term "cofibration" for the Cat^{co} case.

Kelly developed the theory of enriched categories describing enriched adjunction [94] and introducing the variety of limit he called *end*. I later pointed out that Yoneda had used this concept in the special case of additive categories using an *integral notation* which Brian Day and Max Kelly adopted [33]. Following this, Mac Lane [121] discussed ends for ordinary categories.

Meanwhile, as Kelly's graduate student, I began addressing his concerns with the Künneth formulas. The main result of my thesis [134] (also see [137, 149]) was a Künneth hom formula for finitely filtered complexes of free abelian groups. I found it convenient to express the general theory in terms of DG-categories and triangulated categories; my thesis involved the development of some of their theory. In particular, I recognized that completeness of a DG-category should involve the existence of a suspension functor. The idea was consistent with the work of Day and Kelly [33] who eventually defined completeness of a $\mathcal{V}$-category $\mathcal{A}$ to include *cotensoring* $\mathcal{A}$ with objects V of $\mathcal{V}$: the characterizing property is $\mathcal{A}(B, A^V) \cong \mathcal{V}(V, \mathcal{A}(B, A))$. The point is that, for ordinary categories where $\mathcal{V} = \text{Set}$, the cotensor A^V is the product of V copies of A and so is not needed as an extra kind of limit. Cotensoring with the suspension of the tensor unit in $\mathcal{V} = \text{DGAb}$ gives suspension in the DG-category $\mathcal{A}$. Experience with DG-categories would prove very helpful in developing the theory of 2-categories.

In 1968–9 I was a postdoctoral fellow at the University of Illinois (Champaign–Urbana) where John Gray worked on 2-categories. To construct higher-dimensional comprehension schema [58], Gray needed lax limits and even lax Kan extensions [61]. He also worked on a closed structure for the category 2-Cat for which the internal hom $[A, B]$ of two 2-categories A and B consisted of 2-functors from A to B, lax natural transformations, and modifications. (By "lax" we mean the insertion of compatible morphisms in places where there used to be equalities. We use "pseudo" when the inserted morphisms are all invertible.) The next year at Tulane University, Jack Duskin and I had one-year (1969–70) appointments where we heard for a second time Mac Lane's lectures that led to his book [121]; we had all been at Bowdoin College (Maine) over the Summer. Many category theorists visited Tulane that year. Duskin and Mac Lane convinced Gray that this closed category structure on 2-Cat should be monoidal. Thus appeared the (lax) Gray tensor product of 2-categories that Gray was able to prove satisfied the coherence pentagon using Artin's braid groups (see [62, 63]).

Meanwhile Jean Bénabou [12] had invented weak 2-categories, calling them *bicategories*. He also defined a weak notion of morphism that I like to call *lax functor*. His convincing example was the bicategory Span($\mathcal{C}$) of spans in a category $\mathcal{C}$ with pullbacks; the objects are those of the category $\mathcal{C}$ while it is the morphisms of Span($\mathcal{C}$) that are spans; composition of spans requires pullback and so is only associative up to isomorphism. He pointed out that a lax functor from the terminal category **1** to Cat was a category $\mathcal{A}$ equipped with a "standard construction" or "triple" (that is, a monoid in the monoidal category $[A, A]$ of endofunctors of $\mathcal{A}$ where the tensor product is composition); he introduced the term *monad* for this concept. Thus we could contemplate monads in any bicategory. In particular, Bénabou observed that a monad in Span($\mathcal{C}$) is a category internal to $\mathcal{C}$.

The theory of monads (or "triples" [47]) became popular as an approach to universal algebra. A monad T on the category Set of sets can be regarded as an algebraic theory and the category Set^T of "T-modules" regarded as the category of models of the theory. Michael Barr and Jon Beck had used monads on categories to define an abstract cohomology that included many known examples.

The category $\mathcal{C}^T$ of T-modules (also called "T-algebras") is called, after its inventors, the Eilenberg–Moore category for T. The underlying functor $U^T: \mathcal{C}^T \longrightarrow \mathcal{C}$ has a left adjoint which composes with U^T to give back T. There is another category $\mathcal{C}_T$, due to Kleisli, equivalent to the full subcategory of $\mathcal{C}^T$ consisting of the free T-modules; this gives back T in the same way. In fact, whenever we have a functor $U: \mathcal{A} \longrightarrow \mathcal{C}$ with left adjoint F, there is a "generated" monad $T = UF$ on $\mathcal{C}$. There are comparison functors $\mathcal{C}_T \longrightarrow \mathcal{A}$ and $\mathcal{A} \longrightarrow \mathcal{C}^T$; if the latter functor is an equivalence, the functor U is said to be *monadic*. See [121] for details. Beck [11] established necessary and sufficient conditions for a functor to be monadic. Erny Manes showed that compact Hausdorff spaces were the modules for the ultrafilter monad β on Set (see [121]). However, Bourbakifying the definition of topological space via Moore–Smith convergence, Mike Barr [7] showed that general topological spaces were the relational modules for the ultrafilter lax monad on the 2-category Rel whose objects are sets and morphisms are relations. (One of my early Honours students at Macquarie University baffled his proposed Queensland graduate studies supervisor who asked whether the student knew the definition of a topological space. The aspiring researcher on dynamical systems answered positively: "Yes, it is a relational β-module!" I received quite a bit of flak from colleagues concerning that one; but the student Peter Kloeden went on to become a full professor of mathematics in Australia then Germany.)

I took Bénabou's point that a lax functor $W: \mathcal{A} \longrightarrow$ Cat became a monad when $\mathcal{A} = \mathbf{1}$ and in [136] I defined generalizations of the Kleisli and Eilenberg–Moore constructions for a lax functor W with any category $\mathcal{A}$

as domain. These constructions gave two universal methods of assigning strict functors $\mathcal{A} \longrightarrow \mathrm{Cat}$ to a lax one; I pointed out the colimit- and limit-like nature of the constructions. I obtained a generalized Beck monadicity theorem that we have used recently in connection with natural Tannaka duality. The Kleisli-like construction was applied by Peter May to spectra under the recommendation of Robert Thomason.

When I was asked to give a series of lectures on universal algebra from the viewpoint of monads at a Summer Research Institute at the University of Sydney, I wanted to talk about the lax functor work. Since the audience consisted of mathematicians of diverse backgrounds, this seemed too ambitious so I set out to develop the theory of monads in an arbitrary 2-category $\mathcal{K}$, reducing to the usual theory when $\mathcal{K} = \mathrm{Cat}$. This "formal theory of monads" [135] (see [113] for new developments) provides a good example of how 2-dimensional category theory provides insight into category theory. Great use could be made of duality: comonad theory became rigorously dual to monad theory under 2-cell reversal while the Kleisli and Eilenberg–Moore constructions became dual under morphism reversal. Also, a distributive law between monads could be seen as a monad in the 2-category of monads.

In 1971 Bob Walters and I began work on *Yoneda structures* on 2-categories [108, 165]. The idea was to axiomatize the deeper aspects of categories beyond their merely being algebraic structures. This work centred on the Yoneda embedding $A \longrightarrow \mathcal{P}A$ of a category A into its presheaf category $\mathcal{P}A = [A^{\mathrm{op}}, \mathrm{Set}]$. We covered the more general example of categories enriched in a base $\mathcal{V}$ where $\mathcal{P}A = [A^{\mathrm{op}}, \mathcal{V}]$. Clearly size considerations needed to be taken seriously although a motivating size-free example was preordered sets with $\mathcal{P}A$ the inclusion-ordered set of right order ideals in A. Size was just an extra part of the structure. With the advent of elementary topos theory and the stimulation of the work of Anders Kock and Christian Mikkelsen, we showed that the preordered objects in a topos provided a good example. We were happy to realize [108] that an elementary topos was precisely a finitely complete category with a *power object* (that is, a relations classifier). This meant that my work with Walters could be viewed as a higher-dimensional version of topos theory. As usual when raising dimension, what we might mean by a 2-dimensional topos could be many things, several of which could be useful. I looked [139, 141] at those special Yoneda structures where $\mathcal{P}A$ classified *two-sided discrete fibrations*.

At the same time, having made significant progress with Mac Lane on the coherence problem for symmetric closed monoidal categories [106, 107], Kelly was developing a general approach to coherence questions for categories with structure. In fact, Max Kelly and Peter May were in the same place at the same time developing the theories of "clubs" and "operads;" there was some interaction. As I have mentioned, clubs [95–99] were designed to address coherence questions in categories with structure; however, operads were initially for the study of topological spaces bearing homotopy invariant structure. Kelly recognized that at the heart of both notions were monoidal categories such as the category **P** of finite sets and permutations. May was essentially dealing with the category $[\mathbf{P}, \mathrm{Top}]$ (also written $\mathrm{Top}^{\mathbf{P}}$) of functors from **P** to the category Top of topological spaces; there is a tensor product on $[\mathbf{P}, \mathrm{Top}]$, called "substitution," and a monoid for this tensor product is a *symmetric topological operad*. Kelly was dealing with the slice 2-category $\mathrm{Cat}/\mathbf{P}$ with its "substitution" tensor product; a monoid here Kelly called a *club* (a special kind of 2-dimensional theory). There is a canonical functor $[\mathbf{P}, \mathrm{Top}] \longrightarrow [\mathrm{Top}, \mathrm{Top}]$ and a canonical 2-functor $\mathrm{Cat}/\mathbf{P} \longrightarrow [\mathrm{Cat}, \mathrm{Cat}]$; each takes substitution to composition. Hence each operad gives a monad on Top and each club gives a 2-monad on Cat. The modules for the 2-monad on Cat are the categories with the structure specified by the club. Kelly recognized that complete knowledge of the club solved the coherence problem for the club's kind of structure on a category.

That was the beginning of a lot of work by Kelly and colleagues on "2-dimensional universal algebra" [17]. There is a lot that could be said about this with some nice results and I recommend looking at that work; homotopy theorists will recognize many analogues. One theme is the identification of structures that are essentially unique when they exist (such as "categories with finite products," "regular categories," and "elementary toposes") as against those where the structure is really extra (such as "monoidal categories"). A particular class of the essentially unique case is those structures that are modules for a Kock–Zöberlein monad [111, 172]. In this case, the action of the monad on a category is provided by an adjoint to the unit of the monad. It turns out that these monads have an interesting relationship with the simplicial category [142]. It is well known (going right back to the days when monads were called *standard constructions*) that the coherence problem for monads is solved by the (algebraic) simplicial category Δ_{alg}: the monoidal category of finite ordinals (including the empty ordinal) and order-preserving functions. A monad on a category $\mathcal{A}$ is the same as a strict monoidal functor $\Delta_{\mathrm{alg}} \longrightarrow [\mathcal{A}, \mathcal{A}]$. In point of fact, Δ_{alg} is the underlying category of a 2-category $\mathrm{Ord}_{\mathrm{fin}}$ where the 2-cells give the pointwise order to the order-preserving functions. There are nice strings of adjunctions between the face and degeneracy maps. A Kock–Zöberlein monad on a 2-category $\mathcal{K}$ is the same as a strict monoidal 2-functor $\mathrm{Ord}_{\mathrm{fin}} \longrightarrow [\mathcal{K}, \mathcal{K}]$; see [142, 112]. My main example of algebras for a Kock–Zöberlein monad in [138, 142] was fibrations in a 2-category. The monad for fibrations needed an idea of John Gray that I will describe.

In the early 1970s, Gray [60] was working on 2-categories that admitted the construction which in Cat forms the arrow category $A^{\rightarrow}$ from a category A. This rang a bell, harking me back to my work on DG-categories: Gray's construction was like suspension. I saw that its existence should be part of the condition of completeness of a 2-category. A 2-category is complete if and only if it admits products, equalizers and cotensoring with the arrow category →.

Walters and I had a general concept of limit for an object of a 2-category bearing a Yoneda structure. As a special case I looked at what this meant for limits in 2-categories. Several people and collaborators had come to the same conclusion about what limit should mean for enriched categories. Borceux and Kelly called the notion "mean cotensor product." I used the term "indexed limit" for the 2-category case and Kelly adopted that name in his book on enriched categories. When preparing a talk to physicists and engineers in Milan, I decided a better term was *weighted limit*: roughly, the "weighting" J should provide the number of copies JA of each object SA in the diagram S whose limit we seek. Precisely, for $\mathcal{V}$-categories, the limit $\lim(J, S)$ of a $\mathcal{V}$-functor $S: \mathcal{A} \longrightarrow \mathcal{X}$ weighted by a $\mathcal{V}$-functor $J: \mathcal{A} \longrightarrow \mathcal{V}$ is an object of $\mathcal{X}$ equipped with a $\mathcal{V}$-natural isomorphism

$$\mathcal{X}\big(X, \lim(J, S)\big) \cong [\mathcal{A}, \mathcal{V}]\big(J, \mathcal{X}(X, S)\big).$$

Products, equalizers and cotensors are all examples. Conversely, if $\mathcal{X}$ admits these three particular examples, it admits all weighted limits; despite this, individual weighted limits can occur without being thus constructible.

The $\mathcal{V} = \mathrm{Cat}$ case is very interesting. Recall that a $\mathcal{V}$-category in this case is a 2-category. As implied above, it turns out that all weighted limits can be constructed from products, equalizers and cotensoring with the arrow category. Yet there are many interesting constructions that are covered by the notion of weighted limit: good examples are the Eilenberg–Moore construction on a monad and Lawvere's "comma category" of two morphisms with the same codomain.

Gray had defined what we call lax and pseudo limits of 2-functors. Mac Lane says that a limit is a universal cone; a cone is a natural transformation from a constant functor. A lax limit is a universal lax cone. A pseudo limit is a universal pseudo cone. Although these concepts seemed idiosyncratic to 2-category theory, I showed that all lax and pseudo limits were weighted limits and so were covered by "standard" enriched category theory. For example, the lax limit of a 2-functor $F: \mathcal{A} \longrightarrow \mathcal{X}$ is precisely $\lim(L_\mathcal{A}, F)$ where $L_\mathcal{A}: \mathcal{A} \longrightarrow \mathrm{Cat}$ is the 2-functor defined by $L_\mathcal{A}A = \pi_{0*}(\mathcal{A}/A)$; here $\mathcal{A}/A$ is the obvious slice 2-category of objects over A and π_{0*} applies the set-of-path-components functor $\pi_0: \mathrm{Cat} \longrightarrow \mathrm{Set}$ on the hom categories of 2-categories. Gray then pointed out that, for $\mathcal{V} = [\Delta^{\mathrm{op}}, \mathrm{Set}]$ (the category of simplicial sets), *homotopy limits* of $\mathcal{V}$-functors could be obtained as limits weighted by the composite $\mathcal{A} \xrightarrow{L_\mathcal{A}} \mathrm{Cat} \xrightarrow{\mathrm{Nerve}} [\Delta^{\mathrm{op}}, \mathrm{Set}]$.

In examining the limits that exist in a 2-category admitting finite limits (that is, admitting finite products, equalizers, and cotensors with →) I was led to the notion of *computad*. This is a 2-dimensional kind of graph: it has vertices, edges and faces. Each edge has a source and target vertex;

however, each face has a source and target directed path of edges. More 2-categories can be presented by finite computads than by finite 2-graphs. Just as for 2-graphs, the forgetful functor from the category of 2-categories to the category of computads is monadic: the monad formalizes the notion of pasting diagram in a 2-category while the action of the monad on a 2-category encapsulates the operation of *pasting in a 2-category*. Later, Steve Schanuel and Bob Walters pointed out that these computads form a presheaf category.

The step across from limits in 2-categories to limits in bicategories is fairly obvious. For bicategories $\mathcal{A}$ and $\mathcal{X}$, the limit $\lim(J, S)$ of a pseudofunctor $S\colon \mathcal{A} \longrightarrow \mathcal{X}$ weighted by a pseudofunctor $J\colon \mathcal{A} \longrightarrow \mathrm{Cat}$ is an object of $\mathcal{X}$ equipped with a pseudonatural equivalence

$$\mathcal{X}\big(X, \lim(J, S)\big) \simeq \mathrm{Psd}(\mathcal{A}, \mathrm{Cat})\big(J, \mathcal{X}(X, S)\big).$$

It is true that every bicategorical weighted limit can be constructed in a bicategory that has products, iso-inserters (or "pseudoequalizers"), and cotensoring with the arrow category (where the universal properties here are expressed by equivalences rather than isomorphisms of categories); however, the proof is a little more subtle than in the 2-category case. It is also a little tricky to determine which 2-categorical limits give rise to bicategorical ones: for example pullbacks and equalizers are not bicategorical limits *per se*; the weight needs to be *flexible* in a technical sense that would be natural to homotopy theorists.

Now I would like to say more about 2-dimensional topos theory. We have mentioned that Yoneda structures can be seen as a 2-dimensional version of elementary topos theory. However, given that a topos is a category of sheaves, there is a fairly natural notion of "2-sheaf," called *stack*, and a 2-topos should presumably be a 2-category of stacks. After characterizing Grothedieck toposes as categories possessing certain limits and colimits with exactness properties, Giraud developed a theory of stacks in connection with his non-abelian 2-dimensional cohomology. He expressed this in terms of fibrations over categories. Grothendieck had pointed out that a fibration $P\colon \mathcal{E} \longrightarrow \mathcal{C}$ over the category $\mathcal{C}$ was the same as a pseudofunctor $F\colon \mathcal{C}^{\mathrm{op}} \longrightarrow \mathrm{Cat}$ where, for each object U of $\mathcal{C}$, the category FU is the fibre of P over U. If $\mathcal{C}$ is a site (that is, it is a category equipped with a Grothendieck topology) then the condition that F should be a stack is that, for each covering sieve $R \longrightarrow \mathcal{C}(-, U)$, the induced functor

$$FU \longrightarrow \mathrm{Psd}(\mathcal{C}^{\mathrm{op}}, \mathrm{Cat})(R, F)$$

should be an equivalence of categories. We write $\mathrm{Stack}(\mathcal{C}^{\mathrm{op}}, \mathrm{Cat})$ for the full sub-2-category of $\mathrm{Psd}(\mathcal{C}^{\mathrm{op}}, \mathrm{Cat})$ consisting of the stacks. I developed this direction a little by defining 2-dimensional sites and proved a Giraud-like characterization of bicategories of stacks on these sites. Perhaps one point is worth mentioning here. In sheaf theory there are various ways of approaching the associated sheaf. Grothendieck used a so-called "L" construction. Applying L to a presheaf gave a separated presheaf (some "unit" map became a monomorphism) then applying it again gave the associated sheaf (the map became an isomorphism). I found that essentially the same L works for stacks. This time one application of L makes the unit map *faithful*, two applications make it *fully faithful*, and the associated stack is obtained after three applications when the map becomes an *equivalence*.

Just as Kelly was completing his book [101] on enriched categories, a remarkable development was provided by Walters who linked enriched category theory with sheaf theory. First, he extended the theory of enriched categories to allow a bicategory $\mathcal{W}$ (my choice of letter!) as base: *a category $\mathcal{A}$ enriched in $\mathcal{W}$* (or *$\mathcal{W}$-category*) has a set $\mathrm{Ob}\,\mathcal{A}$ of objects where each object A is assigned an object $e(A)$ of $\mathcal{W}$; each pair of objects A and B is assigned a morphism $\mathcal{A}(A,B)\colon e(A) \longrightarrow e(B)$ in $\mathcal{W}$ thought of as a "hom" of $\mathcal{A}$; and "composition" in $\mathcal{A}$ is a 2-cell $\mu_{A,C}^{B}\colon \mathcal{A}(B,C) \circ \mathcal{A}(A,B) \Rightarrow \mathcal{A}(A,C)$ which is required to be associative and unital. Walters regards each object A as a copy of "model pieces" $e(A)$ and $\mathcal{A}$ as a presentation of a structure that is made up of model pieces that are glued together according to "overlaps" provided by the homs. For example, each topological space T yields a bicategory $\mathcal{W} = \mathrm{Rel}(T)$ whose objects are the open subsets of T, whose morphisms $U \longrightarrow V$ are open subsets $R \subseteq U \cap V$, and whose 2-cells are inclusions. Each presheaf P on the space T yields a $\mathcal{W}$-category $el(P)$ whose objects are pairs (U,s) where U is an open subset of T and s is an element of PU; of course, $e(U,S) = U$. The hom $el(P)\big((U,s),(V,t)\big)$ is the largest open subset $R \subseteq U \cap V$ such that the "restrictions" of s and t to R are equal. As another example, for any monoidal category $\mathcal{V}$, let $\Sigma\mathcal{V}$ denote the bicategory with one object and with the endohom category of that single object being $\mathcal{V}$; then a $\mathcal{V}$-category in the Eilenberg–Kelly sense is exactly a $\Sigma\mathcal{V}$-category in Walters' sense.

For each Grothendieck site $(\mathcal{C}, J)$, Walters constructed a bicategory $\mathrm{Rel}_J\mathcal{C}$ such that the category of symmetric Cauchy-complete $\mathrm{Rel}_J\mathcal{C}$-categories became equivalent to the category of set-valued sheaves on $(\mathcal{C}, J)$. This stimulated the development of the generalization of enriched category theory to allow a bicategory as base. We established a higher-dimensional version of Walters' result to obtain stacks as enriched categories. Walters had been able to ignore many coherence questions because the base bicategories he needed were locally ordered (no more than one 2 cell between two parallel morphisms). However the base for stacks is not locally ordered.

I have mentioned the 2-category $\mathcal{V}$-Cat of $\mathcal{V}$-categories; the morphisms are $\mathcal{V}$-functors. However, there is another kind of "morphism" between $\mathcal{V}$-categories. Keep in mind that a category is a "monoid with several objects;" monoids can act on objects making the object into a module. There is a "several objects" version of module. Given $\mathcal{V}$-categories $\mathcal{A}$ and $\mathcal{B}$, we can speak of left $\mathcal{A}$-, right $\mathcal{B}$-bimodules [117]; I call this a *module from $\mathcal{A}$ to $\mathcal{B}$* (although earlier names were "profunctor" and "distributor"

[13]).   Provided $\mathcal{V}$ is suitably cocomplete, there is a bicategory $\mathcal{V}$-Mod whose objects are $\mathcal{V}$-categories and whose morphisms are modules. This is not a 2-category (although it is biequivalent to a fairly natural one) since the composition of modules involves a colimit that is only unique up to isomorphism. The generalization $\mathcal{W}$-Mod for a base bicategory $\mathcal{W}$ was explained in [147] and, using some monad ideas, in [15].

Also in [15] we showed how to obtain prestacks as Cauchy complete $\mathcal{W}$-categories for an appropriate base bicategory $\mathcal{W}$. This has some relevance to algebraic topology since Alex Heller and Grothendieck argue that homotopy theories can be seen as suitably complete prestacks on the category cat of small categories. I showed in [146] (also see [147] and [162]) that stacks are precisely the prestacks possessing colimits weighted by torsors. In [145] (accessible as [164]), I show that stacks on a (bicategorical) site are Cauchy complete $\mathcal{W}$-categories for an appropriate base bicategory $\mathcal{W}$.

Earlier (see [139] and [142]) I had concocted a construction on a bicategory $\mathcal{K}$ to obtain a bicategory $\mathcal{M}$ such that, if $\mathcal{K}$ is $\mathcal{V}$-Cat, then $\mathcal{M}$ is $\mathcal{W}$-Mod; the morphisms of $\mathcal{M}$ were *codiscrete two-sided cofibrations* in $\mathcal{K}$. I had used this as an excuse in [142] to develop quite a bit of bicategory theory: the bicategorical Yoneda Lemma, weighted bicategorical limits, and so on. The need for tricategories was also implicit.

The mathematical physicist John Roberts had asked Peter Freyd whether he knew how to recapture a compact group from its monoidal category of finite-dimensional unitary representations. While visiting the University of New South Wales in 1971, Freyd lectured on his solution of the finite group case. A decade and a half later Roberts with Doplicher did the general case using an idea of Cuntz: this is an analytic version of Tannaka duality. In 1977–8, Roberts visited Sydney. He spoke in the Australian Category Seminar (ACS) about non-abelian cohomology. It came out that he had worked on (strict) n-categories because he thought they were what he needed as coefficient structures in non-abelian cohomology. In the tea room at the University of Sydney, Roberts explained to me what the nerve of a 2-category should be: the dimension 2 elements should be triangles of 1-cells with 2-cells in them and the dimension 3 elements should be commutative tetrahedra. Furthermore, he had defined structures he called *complicial sets*: these were simplicial sets with distinguished elements (he originally called them "neutral" then later suggested "hollow," but I am quite happy to use Dakin's term [31] "thin" for these elements) satisfying some conditions, most notably, unique "thin horn filler" conditions. The important point was which horns need to have such fillers. Roberts believed that the category of complicial sets was equivalent to the category of n-categories.

I soon managed to prove that complicial sets, in which all elements of dimension greater than 2 were thin, were equivalent to 2-categories. I also obtained some nice constructions on complicial sets leading to new complicial sets. However the general equivalence seemed quite a difficult problem.

I decided to concentrate on one aspect of the problem. How do we rigorously define the nerve of an n-category? After unsuccessfully looking for an easy way out using multiple categories and multiply simplicial sets (I sent several letters to Roberts about this), I realized that the problem came down to defining the free n-category $\mathcal{O}_n$ on the n-simplex. Meaning had to be given to the term "free" in this context: free on what kind of structure? How was an n-simplex an example of the structure? The structure required was n-computad. The definition of n-computad and free n-category on an n-computad is done simultaneously by induction on n (see [150], [127, 154, 155, 162]). An element of dimension n of the nerve $N(A)$ of an ω-category A is an n-functor from $\mathcal{O}_n$ to A. Things began to click once I drew the following picture of the 4-simplex.

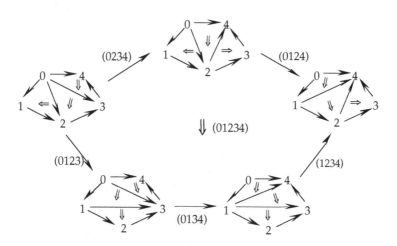

I was surprised to find out that Roberts had not drawn this picture in his work on complicial sets! It was only in studying this and the pictures for the 5- and 6-simplex that I understood the horn filler conditions for the nerve of an n-category. The resemblence to Stasheff's associahedra was only pointed out much later (I think by Jim Stasheff himself).

I think of the n-category $\mathcal{O}_n$ as a simplex with oriented faces; I call it the nth *oriental*. The problem in constructing it inductively starting with small n is *where to put that highest dimensional cell*. What are that cell's source and target? Even in the case of $\mathcal{O}_4$ above, the description of the 3-source and 3-target of the cell (01234), in terms of composites of lower dimensional cells, takes some work to write explicitly. To say a 4-functor out of $\mathcal{O}_4$ takes (01234) to the identity is the *non-abelian 3-cocycle condition*.

In mid-1982 I circulated a conjectural description of the free ω-category $\mathcal{O}_\omega$ on the infinite-dimensional simplex; the objects were to be the natural numbers and $\mathcal{O}_n$ would be obtained by restricting to the objects no greater than n. The description is very simple: however, it turns out to be hard even to prove $\mathcal{O}_\omega$ is an ω-category, let alone prove it free.

The starting point for my description is the fact that a path in a circuit-free (directed) graph is determined by the finite **set** of edges in the path: the edges order themselves using source and target. The set must be "well formed": there should be no two edges with the same source and no two with the same target. Moreover, the source of the path is the unique vertex which is a source of some edge but not the target of any edge in the set. What a **miracle** that this should work in higher dimensions.

Meanwhile, on the enriched category front, Walters had pointed out that in order for $\mathcal{W}$-Mod to be monoidal, the base bicategory $\mathcal{W}$ should be monoidal. You will recall that, in order to define tensor products and duals for $\mathcal{V}$-categories, Eilenberg–Kelly [46] had assumed $\mathcal{V}$ to be symmetric. In a talk in the ACS, Bob Walters reported on a discussion Carboni, Lawvere and he had had about the possibility of using an Eckmann–Hilton argument to show that a monoidal bicategory with one object was a symmetric monoidal category in the same way that a monoidal category with one object is a commutative monoid. It is perhaps not surprising that they did not pursue the calculation to completion at that time since monoidal bicategories had not appeared in print except for the locally ordered case. I was so taken by how much I could do *without* a monoidal structure on $\mathcal{W}$-Mod that I did not follow up the idea then either.

Duskin returned to Australia at the end of 1983 and challenged me to draw $\mathcal{O}_6$; this took me a weekend. The odd-faces-source and even-faces-target convention forces the whole deal!

By the end of 1984 I had prepared the oriented simplexes paper [150]. My conjectured description is correct. (Actually, Verity pointed out an error in the proof written in [150] which I corrected in [156].) The heart of proving things about $\mathcal{O}_\omega$ is the algorithm I call *excision of extremals* for deriving the non-abelian n-cocycle condition "from the top down."

The paper [150] has several other important features. Perhaps the most obvious are the diagrams of the orientals; they resemble Stasheff associahedra with some oriented faces and some commuting faces. I give the 1-sorted definition of ω-category and show the relationship between the 1-sorted definition of n-category and the inductive one in terms of enrichment. I make precise some facts about the category ω-Cat of ω-categories such as its cartesian closedness. I say what it means for a morphism in an n-category to be a weak equivalence.

The paper [150] defines what it means for an n-category to be free. I define the nerve of an n-category and make a conjecture about characterizing those nerves as "stratified" (or filtered) simplicial sets satisfying horn-filler conditions. The horns I suggested should be filled were a wider class than those of Roberts' complicial sets; I called my horns "admissible" and Roberts' "complicial." However, I really believed the admissible horns would still lead to complicial sets.

That there is a weaker notion of n-category than the strict ones was an obvious consequence of the introduction of weak 2-categories (bicategories) by Bénabou [12]. I later was reminded that Mac Lane, in 1969, had suggested tricategories as a possible area of study [120]. As a kind of afterthought in [150], I suggest a characterization of weak n-categories as stratified simplicial sets with horn filler conditions. My intuition was that, even in a strict n-category, the same horns should be fillable by only insisting that our thin elements be simplexes whose highest dimensional cell is a weak equivalence rather than a strict identity. So the same horns should have fillers even in a weak n-category. Of course, the fillers now would not be unique.

While travelling in North America, I submitted the preprint of [150] to expatriate Australian Graeme Segal as editor of *Topology*. I thought Graeme might have some interest in higher nerves as a continuation of his work in [132]. He rejected the paper without refereeing on grounds that it would not be of sufficient interest to topologists. I think this IMA Summer Program proves he was wrong. To make things worse, his rejection letter went to the institution I was visiting when I submitted and it was not forwarded to me at Macquarie University. I waited a year or so before asking Segal what happened! He sent me a copy of his short letter.

In April 1985, all excited about higher nerves, I began a trip to North America that would trigger two wonderful collaborations: one with Sammy Eilenberg and one with André Joyal. The first stop was a conference organized by Freyd and Scedrov at the University of Pennsylvania. After my talk, Sammy told me of his work on rewriting systems and that, in my orientals, he could see higher rewriting ideas begging to be explained. I left Philadelphia near the end of April as spring was beginning to bloom only to arrive in Montréal during a blizzard. Michael Barr had invited me to McGill where Robin Cockett was also visiting.

During my talk in the McGill Category Seminar, André Joyal started quizzing me on various aspects of the higher nerves. We probably remember that discussion differently. My memory is that André was saying that the higher associativities were not the important things as they could be coherently ignored; the more important things were the higher commutativities. In arguing that commutativities were already present in the "middle-of-four interchange," I was harking back to Walters' talk about applying an Eckmann–Hilton-like argument to a one-object monoidal bicategory. That night I checked what I could find out about a monoidal object (or pseudomonoid) in the 2-category of monoidal categories and strong monoidal functors. It was pretty clear that some kind of commutativity was obtained that was not as strong as a symmetry.

When I reported my findings to André the next day, he already knew what was going on. He told me about his work with Myles Tierney on homotopy 3-types as groupoids enriched in 2-groupoids with the Gray tensor product. I told André that I was happy enough with weak 3-groupoids as homotopy types and that ordinary cartesian product works just as well as the Gray tensor product when dealing with bicategories rather than stricter things. In concentrating on this philosophy, I completely put out of my memory the claim that André recently reminded me he made at that time about Gray-categories being a good 3-dimensional notion of weak 3-category. I believed we should come to grips with the fully weak n-categories and this dominated my thinking.

There had been other weakenings of the notion of symmetry for monoidal categories but this kind had not been considered by category theorists. I announced a talk on joint work with Joyal for the Isle of Thornes (Sussex, England) conference in mid-1985: the title was "Slightly incoherent symmetries for monoidal categories." Before the actual talk, we had settled on the name *braiding* for this kind of commutativity. I talked about the a coherence theorem for braided monoidal categories based on the braid groups just as Mac Lane had for symmetries based on the symmetric groups.

After this talk, Sammy Eilenberg told me about his use of the *braid monoid with zero* to understand the equivalence of derivations in rewriting systems. This was the basis of our unpublished work some of which is documented in [48]. We were going to finish the work after he finished his books on cellular spaces with Eldon Dyer.

I returned to Australia where Peter Freyd was again visiting. He became very excited when I lectured on braided monoidal categories in the ACS. He knew about his ex-student David Yetter's monoidal category of tangles. Freyd and Yetter had already entered low-dimensional topology with their participation in the "homfly" polynomial invariant for links. By the next year (mid-1986) at the Cambridge category meeting, I heard that Freyd was announcing his result with Yetter about the freeness of Yetter's category of tangles as a compact braided monoidal category. Their idea was that duals turned braids into links.

In the mid-1980s the low-dimensional topologist Iain Aitchison (Masters student of Hyam Rubenstein and PhD student of Robion Kirby) was my first postdoctoral fellow. He reminded me in more detail of the string diagrams for tensor calculations used by Roger Penrose. Max Kelly had mentioned these at some point, having seen Penrose using them in Cambridge. Moreover, Aitchison [1] developed an algorithm for the non-abelian n-cocycle condition "from the bottom up," something Roberts and I had failed to obtain. He did the same for oriented cubes in place of oriented simplexes. The algorithm is a kind of "Pascal's triangle" where a given entry is derived from two earlier ones; the simplex case is less symmetric because of the different lengths of sources and targets in that case. The algorithm appeared in combinatorial form in a Macquarie Math. Preprint but was nicely represented in terms of string diagrams drawn by hand with coloured pens.

Aitchison and I satisfied ourselves that the arguments of [150] carried over to cubes in place of simplexes but this was not published. That work was subsumed by my parity complexes [153, 156] and Michael Johnson's pasting schemes [73] which I intend to discuss below.

Following my talks on orientals in the ACS, Bob Walters and his student Mike Johnson obtained [74] a variant of my construction of the nerve of a (strict) category. The cells in their version of $\mathcal{O}_n$ were actual subsimplicial sets of a simplicial set and the compositions were all unions; they thought of these cells as simplicial "pasting diagrams." The cells in my $\mathcal{O}_n$ were only generators for the Walters–Johnson simplicial sets and so, while smaller objects to deal with, required some deletions from the unions defining composition.

Around this time I set my student Michael Zaks the problem of proving the equivalence between complicial sets and ω-categories. To get him started I proved [152] that the nerve of an ω-category satisfies the unique thin filler condition for admissible (and hence complicial) horns. So nerves of ω-categories are complicial sets. Zaks fell in love with the simplicial identities and came up with a construction he believed to be the zero-composition needed to make an n-category from a complicial set. We showed that this composition was the main ingredient required by using an induction based on showing an equivalence

$$\mathrm{Cmpl}_n \simeq \mathrm{Cmpl}_{n-1}\text{-Cat}$$

where the left-hand side is the category of n-trivial complicial sets; a stratified simplicial set is n-trivial when all elements of dimension greater than n are thin. Zaks did not complete the proof that his formula worked and we still do not know whether it does. In 1990, Dominic Verity was motivated by my paper [150] to work on this problem. Unaware of [152], Dominic independently came up with the machinery Zaks and I had developed. By mid-1991 Dominic had proved, amongst other things, that the nerve was fully faithful; he completed the details of the proof of the equivalence

$$\omega\text{-Cat} \simeq \mathrm{Cmpl}$$

in 1993; the details are still to appear [169].

Knowing the nerve of an n-category, we now knew the non-abelian cocycle conditions. So I turned attention to understanding the full cohomology. The idea was that, given a simplicial object X and an ω-category object A, there should be an ω-category to be called the *cohomology of X with coefficients in A*. Jack Duskin pointed out that this should be part of a general *descent* construction which obtains an ω-category $\mathrm{Desc}\,\mathcal{C}$ from any cosimplicial ω-category $\mathcal{C}$. For the cohomology case, the cosimplicial

ω-category is $\mathcal{C} = \mathrm{Hom}(X, A)$ taken in the ambient category. Furthermore, Jack drew a few low-dimensional diagrams.

It took me some time to realize that the diagrams Jack had drawn were really just products of globes with simplexes. I then embarked on a program of abstracting the structure possessed by simplexes, cubes and globes, and to show the structure was closed under products. For his PhD, Mike Johnson was also working on abstracting the notion of pasting diagram. In an ACS, I explained my idea about descent and gave an overly-simplistic description of the product of parity complexes. By the next week's ACS Mike Johnson had corrected my definition of product based on the usual tensor product of chain complexes. The next step was to find the right axioms on a parity complex in order for it to be closed under product. For this I invented a new order that I denoted by a solid triangle: let me denote it now by $\prec$. I need to give more detail.

A *parity complex* is a graded set dim: $P \longrightarrow \mathrm{N}$ together with functions

$$(-)^-, (-)^+ : P \longrightarrow \mathcal{P}_{\mathrm{fin}} P,$$

where $\mathcal{P}_{\mathrm{fin}} S$ is the set of finite subsets of the set S, such that

$$x \in y^- \cup y^+ \text{ implies } \dim x + 1 = \dim y.$$

For x in the fibre P_n we think of x^- as the set of elements in the source of x and x^+ as the set of elements in the target of x. For a subset S of P_n, put $S^\varepsilon = \bigcup_{x \in S} x^\varepsilon$ for $\varepsilon \in \{+, -\}$. There are some further conditions such as

$$x^- \cap x^+ = \emptyset, \quad x^{-+} \cap x^{+-} = \emptyset, \quad x^{--} \cap x^{++} = \emptyset, \quad x^{-+} \cup x^{+-} = x^{--} \cup x^{++}.$$

These conditions imply that we obtain a positive chain complex $\mathbf{Z}P$ consisting of the free abelian groups $\mathbf{Z}P_n$ with differential defined on generators by

$$\mathrm{d}(x) = x^+ - x^-$$

where we write S for the formal sum of the elements of a finite subset S of P_n. The order on P is the smallest reflexive transitive relation $\prec$ such that

$$x \prec y \text{ if either } x \in y^- \text{ or } y \in x^+.$$

The amazing axiom we require is that this order should be **linear**.

If the functions $(-)^-, (-)^+ : P \longrightarrow \mathcal{P}_{\mathrm{fin}} P$ land in singleton subsets of P, the parity complex is a globular set which represents a *globular pasting diagram*. As later shown by Michael Batanin, these globular sets hold the key to free n-categories on all globular sets. A very special globular pasting diagram is the "free-living globular k-cell;" it is a parity complex $\mathbf{G}_k$ with $2k + 1$ elements.

The original example of a parity complex is the infinite simplex $\Delta[\omega]$ whose elements of dimension n are injective order-preserving functions x:

$[n] \longrightarrow \omega$; we write such an x as an ordered $(n+1)$-tuple $(x_0, x_1, \ldots x_n)$. Also $\partial_i \colon [n-1] \longrightarrow [n]$ is the usual order-preserving function whose image does not contain i in $[n] = \{0, 1, \ldots, n\}$. Then

$$x^- = \{x\partial_i \mid i \text{ odd}\} \text{ and } x^+ = \{x\partial_i \mid i \text{ even}\}.$$

We obtain a parity complex $\Delta[k]$, called the *parity k-simplex*, by restricting attention to those x that land in $[k]$. In particular, $\Delta[1]$ is the parity interval and also denoted by $\mathbf{I}$.

The product of two parity complexes P and Q is the cartesian product $P \times Q$ with

$$\dim(x, y) = \dim x + \dim y \text{ and } (x, y)^\varepsilon = x^\varepsilon \times \{y\} \cup \{x\} \times y^{\varepsilon(m)}$$

where $\dim x = m$ and $\varepsilon(m)$ is the sign ε when m is even and the opposite of ε when m is odd. It can be shown that $P \times Q$ is again a parity complex. In particular, there is a parity k-cube

$$\mathbf{I}^k = \overbrace{\mathbf{I} \times \cdots \times \mathbf{I}}^{k}.$$

There is a canonical isomorphism of chain complexes

$$\mathbf{Z}(P \times Q) \cong \mathbf{Z}P \otimes \mathbf{Z}Q.$$

A parity complex P generates a free ω-category $\mathcal{O}(P)$. The description is rather simple because the conditions on a parity complex ensure sufficient "circuit-freeness" for the order of composition to sort itself out. The detailed description can be found in [153] or [162].

We shall now describe a monoidal structure on ω-Cat that was considered by Richard Steiner and Sjoerd Crans. It turns out that the full subcategory of ω-Cat, consisting of the free ω-categories $\mathcal{O}(\mathbf{I}^k)$ on the parity cubes, is dense in ω-Cat. The tensor product of the free ω-categories $\mathcal{O}(\mathbf{I}^h)$ and $\mathcal{O}(\mathbf{I}^k)$ is defined by

$$\mathcal{O}(\mathbf{I}^h) \otimes \mathcal{O}(\mathbf{I}^k) = \mathcal{O}(\mathbf{I}^{h+k}).$$

This is extended to a tensor product on ω-Cat by Kan extension along the inclusion. A result of Brian Day applies to show this is a monoidal structure. We call it the *Gray monoidal structure* on ω-Cat although John Gray only defined it on 2-Cat by forcing all cells of dimension higher than 2 to be identities. Dominic Verity has shown that, for a wide class of parity complexes P and Q, we have an isomorphism of categories

$$\mathcal{O}(P) \otimes \mathcal{O}(Q) \cong \mathcal{O}(P \times Q).$$

These ingredients allow us to define the descent ω-category $\operatorname{Desc} \mathcal{E}$ of a cosimplicial category $\mathcal{E}$ as follows. The functor $\operatorname{Cell}_n \colon \omega\text{-Cat} \longrightarrow \operatorname{Set}$, which assigns the set of n-cells to each ω-category, is represented by the free n-category $\mathcal{O}(\mathbf{G}^n)$ on the n-globe; that is,

$$\mathrm{Cell}_n(A) = \omega\text{-Cat}(\mathcal{O}(\mathbf{G}^n), A).$$

From this we see that $\mathcal{O}(\mathbf{G}^n)$ is a *co-n-category in* ω-Cat. Since the functor $- \otimes A$ preserves colimits, it follows that $\mathcal{O}(\mathbf{G}^n) \otimes A$ is a co-n-category in ω-Cat for all categories A. In particular,

$$\mathcal{O}(\mathbf{G}^n) \otimes \mathcal{O}_m = \mathcal{O}(\mathbf{G}^n) \otimes \mathcal{O}(\Delta[m]) \cong \mathcal{O}(\mathbf{G}^n \times \Delta[m])$$

is a co-n-category in ω-Cat. Allowing m to vary, we obtain a co-n-category $\mathcal{O}(\mathbf{G}^n \times \Delta)$ in the category $[\Delta, \omega\text{-Cat}]$ of cosimplicial ω-categories; so we define

$$\mathrm{Desc}\,\mathcal{E} = [\Delta, \omega\text{-Cat}](\mathcal{O}(\mathbf{G}^n \times \Delta), \mathcal{E}).$$

As a special case, the *cohomology ω-category* of a simplicial object X with coefficients in an ω-category object A (in some fixed category) is defined by

$$\mathcal{H}(X, A) = \mathrm{Desc}\,\mathrm{Hom}(X, A).$$

During 1986–7, André Joyal and I started to hear about Yang–Baxter operators from the Russian School. Drinfeld lectured on quantum groups at the World Congress in 1986. We attended Yuri Manin's lectures on quantum groups at the University of Montréal. My opinion at first was that, as far as monoidal categories were concerned, braidings were the good notion and Yang–Baxter operators were only their mere shadow. André insisted that we also needed to take these operators seriously. The braid category is not only the free braided monoidal category on a single object, it is the free monoidal category on an object bearing a Yang–Baxter operator. While we were at the Louvain-la-nerve category conference in mid-1987, Iain Aitchison brought us a paper by Turaev that had been presented at an Isle of Thorns low-dimensional topology meeting the week before. Turaev knew about Yetter's monoidal category of tangles and gave a presentation of it using Yang–Baxter operators. I had the impression that Turaev did not know about braided monoidal categories at the time. All André and I had put out in print were a handwritten Macquarie Mathematics Report at the end of 1985 and a typed revision about a year later.

I set my student Mei Chee Shum on the project of "adapting" Kelly–Laplaza's coherence for compact symmetric monoidal categories [105] to the braided case. She soon detected a problem with our understanding of the Freyd–Yetter result. Meanwhile, Joyal and I continued working on braided monoidal categories; there was a variant we called *balanced* monoidal categories based on braids of ribbons (not just strings). We started developing the appropriate string diagrams for calculating in the various monoidal categories with extra structure [76]; this could be seen as a formalization of the Penrose notation for calculating with tensors but now deepened the connection with low-dimensional topology. The notion of *tortile* monoidal category was established; Shum's thesis became a proof (based on Reidermeister calculus) that the free tortile monoidal category was the category of tangles on ribbons. Joyal and I proved in [77], just using universal properties, that this category was also freely generated as a monoidal category by a tortile Yang–Baxter operator. In doing this we introduced the notion of *centre* of a monoidal category $\mathcal{C}$: it is a braided monoidal category $\mathcal{ZC}$. This construction can be understood from the point of view of higher categories. For any bicategory $\mathcal{D}$, the braided monoidal category $\mathrm{Hom}(\mathcal{D}, \mathcal{D})(1_{\mathcal{D}}, 1_{\mathcal{D}})$, whose objects are pseudo-natural transformations of the identity of $\mathcal{D}$, whose morphisms are modifications, and whose tensor product is either of the two compositions, might be called the *centre* of the bicategory $\mathcal{D}$. If $\mathcal{D}$ is the one-object bicategory $\Sigma\mathcal{C}$ with hom monoidal category $\mathcal{C}$ then $\mathrm{Hom}(\mathcal{D}, \mathcal{D})(1_{\mathcal{D}}, 1_{\mathcal{D}})$ is the centre $\mathcal{ZC}$ of $\mathcal{C}$ in the sense of [77].

In statistical mechanics there are higher dimensional versions of the Yang–Baxter equations. The next one in the list is the Zamolodchikov equation. I began to hear about this from various sources; I think first from Aitchison who showed me the string diagrams. I talked a little about this at the category meeting in Montréal in 1991. This is where I was given a copy of Dominic Verity's handwritten notes on complicial sets. Moreover, Bob Gordon and John Power asked me whether I realized that my bicategorical Yoneda lemma in [142] could be used to give a one-line proof that every bicategory is biequivalent to a 2-category. I remembered that I had thought about using that lemma for some kind of coherence but it was probably along the lines of the Giraud result that every fibration was equivalent to a strict one (in the form that every pseudofunctor into Cat is equivalent to a strict 2-functor). Gordon and Power had been looking at categories on which a monoidal category acts and (I imagine) examined the "Cayley theorem" in that context, and then realized the connection with the bicategorical Yoneda lemma. Since this coherence theorem for bicategories was so easy, we decided we would use it as a model for a coherence theorem for trictegories. Tricategories had not been defined in full generality at that point. Our theorem was that every tricategory was triequivalent to a Gray-category; the latter is a little more general than a 3-category (there is an isomorphism instead of equality for the middle four law). In fact, Gray-categories are categories enriched in 2-Cat with a Gray-type tensor product. John Power has briefly described at this conference the rest of the story behind [54] so I shall say no more about that here.

Of course a tricategory is a "several object version" of a monoidal bicategory. The need for this had already come up in the Australian School: a monoidal structure was needed on the base bicategory $\mathcal{W}$ to obtain a tensor product of enriched $\mathcal{W}$-categories. Kapranov had also sent us rough notes on his work with Voevodsky (see [83–85]). Their monoidal bicategories were not quite as general as our one-object tricategories but they had ideas about braided monoidal bicategories and the relationship with the Zamolodchikov equation. Larry Breen and Martin Neuchl independently realized that Kapranov–Voevodsky needed an extra condition on their higher braiding. Kapranov–Voevodsky called Gray-categories *semi-strict 3-categories* and were advising us that they were writing a proof of coherence; I do not think that ever appeared.

By 1993, with Dominic Verity and Todd Trimble at Macquarie University, many interesting ideas were developed about monoidal and higher-order categories. Amongst other things Verity contributed vitally to the completion of work I had begun with other collaborators: modulated bicategories [23] and traced monoidal categories [81]. Todd was interested in operads and was establishing a use of Stasheff's associahedra to define weak n-categories. He seemed to know what was going on but could not write the general definition formally. I challenged him to write down a definition of weak 4-category which he did [167] against his better judgement: it is horrendous. Moreover, at Macquarie, Todd and Margaret McIntyre worked out the surface diagrams for monoidal bicategories; the paper was submitted to *Advances* and is still in revision limbo. I should point out that Todd was married just before taking the postdoctoral fellowship at Macquarie University. His wife stayed in the U.S. with her good job. So it was natural that, after two years (and only a couple of visits each way), he had to return to the U.S. This left one year of the fellowship to fill. Tim Porter mentioned a chap from Novosibirsk (Siberia). So Michael Batanin was appointed to Macquarie and began working on higher categories.

This brings me to the point of the letter John Baez and James Dolan sent me concerning their wonderful definition of weak ω-category. I think Michael Makkai caught on to their idea much quicker than me and I shall skip over the history in that direction.

I have learnt that when Michael Batanin comes to me starting a new topic with: "Oh Ross, have you ... ?," that something serious is about to come. If it is mathematics, it is something he has thought deeply about already. A few months after he arrived at Macquarie University, after returning to the Macquarie carpark from an ACS at Sydney University, Michael popped me one of these questions:

> "Oh Ross, have you ever thought of the free strict
> n-category on the terminal globular set?"

My response was that the terminal globular set is full of loops, so my approach to free n-categories using parity complexes did not apply. The loops frankly scared me! Soon after, Michael described the monad for ω-categories on globular sets. The clue was his answer to the carpark question: it involved plane trees which he used to codify globular pasting diagrams. Then the solution is like using what I tell undergraduates is my favourite mathematical object, the geometric series, to obtain free monoids.

Batanin's full fledged theory of higher (globular) operads quickly followed, including the operad for weak ω-categories and the natural monoidal environment for the operads; see [8, 158]. He also developed a theory of computads for the algebras of any globular operad [9]: the computads for weak n-categories differ from the ones for the strict case since you need to choose a pasting order for the source and target before placing your generating cell. (Verity's PhD thesis had a coherence theorem for bicategories that pointed out the need for this kind of thing.)

Let me finish with one further development I see as a highlight and a reference which contains many precise details of topics of interest to this conference. The highlight, arising from the development of the theory of monoidal bicategories jointly with Brian Day, is the realization of the connection among the concepts of quantum groupoids, $*$-autonomy in the sense of Michael Barr, and Frobenius algebras (see [38, 163]). The reference for further reading is [162] which I prepared for the Proceedings of the Workshop on "Categorical Structures for Descent and Galois Theory, Hopf Algebras and Semiabelian Categories" at the Fields Institute, Toronto 2002; it represents an improved and updated version of notes of three lectures presented at Oberwolfach in September 1995.

REFERENCES

[1] I. AITCHISON, String diagrams for non-abelian cocycle conditions, handwritten notes, talk presented at Louvain-la-neuve, Belgium, 1987.

[2] J. BAEZ AND J. DOLAN, Higher-dimensional algebra and topological quantum field theory, *J. Math. Phys.* **36** (1995), 6073–6105.

[3] J. BAEZ AND J. DOLAN, Higher-dimensional algebra III: n-categories and the algebra of opetopes, *Advances in Math.* **135** (1998), 145–206.

[4] J. BAEZ AND L. LANGFORD, Higher-dimensional algebra IV: 2-tangles, *Advances in Math.* **180** (2003), 705–764.

[5] J. BAEZ AND M. NEUCHL, Higher-dimensional algebra I: braided monoidal 2-categories, *Advances in Math.* **121** (1996), 196–244.

[6] C. BALTEANU, Z. FIERDEROWICZ, R. SCHWAENZL, AND R. VOGT, Iterated monoidal categories, *Advances in Math.* **176** (2003), 277–349.

[7] M. BARR, Relational algebras, Lecture Notes in Math. **137**, Springer, Berlin, 1970, 39–55.

[8] M. BATANIN, Monoidal globular categories as natural environment for the theory of weak n-categories, *Advances in Math.* **136** (1998), 39–103.

[9] M. BATANIN, Computads for finitary monads on globular sets, in *Higher Category Theory*, eds. E. Getzler and M. Kapranov, Contemp. Math. **230**, AMS, Providence, Rhode Island, 1998, pp. 1–36.

[10] M. BATANIN AND R. STREET, The universal property of the multitude of trees, *J. Pure Appl. Algebra* **154** (2000), 3–13.

[11] J. BECK, Triples, algebras and cohomology, *Reprints in Theory Appl. Cat.* **2** (2003), 1–59.

[12] J. BÉNABOU, Introduction to bicategories, Lecture Notes in Math. **47**, Springer, Berlin, 1967, pp. 1–77.

[13] J. BÉNABOU, Les distributeurs, Univ. Catholique de Louvain, Séminaires de Math. Pure, Rapport No. 33 (1973).

[14] C. BERGER, Double loop spaces, braided monoidal categories and algebraic 3-type of space, in *Higher Homotopy Structures in Topology and Mathematical Physics*, ed. J. McCleary, Contemp. Math. **227**, AMS, Providence, Rhode Island, 1999, pp. 49–66.

[15] R. BETTI, A. CARBONI, R. STREET, AND R. WALTERS, Variation through enrichment, *J. Pure Appl. Algebra* **29** (1983), 109–127.

[16] G.J. BIRD, G.M. KELLY, A.J. POWER, AND R. STREET, Flexible limits for 2-categories, *J. Pure Appl. Algebra* **61** (1989), 1–27.

[17] R. BLACKWELL, G.M. KELLY, AND A.J. POWER, Two-dimensional monad theory, *J. Pure Appl. Algebra* **59** (1989), 1–41.

[18] F. BORCEUX AND G.M. KELLY, A notion of limit for enriched categories, *Bull. Austral. Math. Soc.* **12** (1975), 49–72.

[19] R. BROWN, Higher dimensional group theory, in *Low Dimensional Topology*, London Math. Soc. Lecture Note Series **48** (1982), pp. 215–238.

[20] R. BROWN AND P.J. HIGGINS, The equivalence of crossed complexes and ∞-groupoids, *Cah. Top. Géom. Diff. Cat.* **22** (1981), 371–386.

[21] A. CARBONI, G.M. KELLY, AND R.J. WOOD, A 2-categorical approach to change of base and geometric morphisms. I., *Cah. Topologie Géom. Diff. Cat.* **32** (1991), 47–95.

[22] A. CARBONI, G.M. KELLY, D. VERITY, AND R.J. WOOD, A 2-categorical approach to change of base and geometric morphisms. II., *Theory Appl. Cat.* **4** (1998), 82–136.

[23] A. CARBONI, S. JOHNSON, R. STREET, AND D. VERITY, Modulated bicategories, *J. Pure Appl. Algebra* **94** (1994), 229–282.

[24] A. CARBONI, S. KASANGIAN, AND R. WALTER, An axiomatics for bicategories of modules, *J. Pure Appl. Algebra* **45** (1987), 127–141.

[25] A. CARBONI AND R. WALTERS, Cartesian bicategories I, *J. Pure Appl. Algebra* **49** (1987), 11–32.

[26] S.M. CARMODY, *Cobordism Categories*, Ph.D. Thesis, University of Cambridge, 1995.

[27] L. CRANE AND D. YETTER, A categorical construction of 4D topological quantum field theories, in *Quantum Topology*, eds. L. Kauffman and R. Baadhio, World Scientific Press, 1993, pp. 131–138.

[28] S. CRANS, Generalized centers of braided and sylleptic monoidal 2-categories, *Advances in Math.* **136** (1998), 183–223.

[29] S. CRANS, A tensor product for Gray-categories, *Theory Appl. Cat.* **5** (1999), 12–69.

[30] S. CRANS, On braidings, syllepses, and symmetries, *Cah. Top. Géom. Diff. Cat.* **41** (2000), 2–74 & 156.

[31] M. DAKIN, *Kan Complexes and Multiple Groupoid Structures*, Ph.D. Thesis, University of Wales, Bangor, 1977.

[32] B. DAY, On closed categories of functors, Lecture Notes in Math. **137**, Springer, Berlin, 1970, pp. 1–38.

[33] B. DAY AND G.M. KELLY, Enriched functor categories, *Reports of the Midwest Category Seminar, III*, Springer, Berlin, 1969, pp. 178–191.

[34] B. DAY, P. McCRUDDEN, AND R. STREET, Dualizations and antipodes, *Applied Categorical Structures* **11** (2003), 229–260.

[35] B. DAY AND R. STREET, Monoidal bicategories and Hopf algebroids, *Advances in Math.* **129** (1997), 99–157.

[36] B. DAY AND R. STREET, Lax monoids, pseudo-operads, and convolution, in *Diagrammatic Morphisms and Applications*, eds. D. Radford, F. Souza, and D. Yetter, Contemp. Math. **318**, AMS, Providence, Rhode Island, 2003, pp. 75–96.

[37] B. DAY AND R. STREET, Abstract substitution in enriched categories, *J. Pure Appl. Algebra* **179** (2003), 49–63.

[38] B. DAY AND R. STREET, Quantum categories, star autonomy, and quantum groupoids, in *Galois Theory, Hopf Algebras, and Semiabelian Categories*, Fields Institute Communications **43**, AMS, Providence, Rhode Island, 2004, pp. 193–231.

[39] V.G. DRINFEL'D, Quasi-Hopf algebras (Russian), *Algebra i Analiz* **1** (1989), 114–148; translation in *Leningrad Math. J.* **1** (1990), 1419–1457.

[40] J.W. DUSKIN, The Azumaya complex of a commutative ring, Lecture Notes in Math. **1348**, Springer, Berlin, 1988, pp. 107–117.

[41] J.W. DUSKIN, An outline of a theory of higher-dimensional descent, *Bull. Soc. Math. Belg. Sér. A* **41** (1989), 249–277.

[42] J.W. DUSKIN, A simplicial-matrix approach to higher dimensional category theory I: nerves of bicategories, *Theory Appl. Cat.* **9** (2002), 198–308.

[43] J.W. DUSKIN, A simplicial-matrix approach to higher dimensional category theory II: bicategory morphisms and simplicial maps (incomplete draft 2001).

[44] C. EHRESMANN, *Catégories et Structures*, Dunod, Paris, 1965.

[45] S. EILENBERG AND G.M. KELLY, A generalization of the functorial calculus, *J. Algebra* **3** (1966), 366–375.

[46] S. EILENBERG AND G.M. KELLY, Closed categories, *Proceedings of the Conference on Categorical Algebra at La Jolla*, Springer, Berlin, 1966, pp. 421–562.

[47] S. EILENBERG AND J.C. MOORE, Adjoint functors and triples, *Illinois J. Math.* **9** (1965), 381–398.

[48] S. EILENBERG AND R. STREET, Rewrite systems, algebraic structures, and higher-order categories (handwritten notes circa 1986, somewhat circulated).

[49] J. FISCHER, 2-categories and 2-knots, *Duke Math. Journal* **75** (1994), 493–526.

[50] P. FREYD AND R. STREET, On the size of categories, *Theory Appl. Cat.* **1** (1995), 174–178.

[51] P. GABRIEL AND M. ZISMAN, *Calculus of Fractions and Homotopy Theory*, Ergebnisse der Mathematik und ihrer Grenzgebiete, Band **35**, Springer, Berlin, 1967.

[52] J. GIRAUD, *Cohomologie Non Abélienne*, Springer, Berlin, 1971.

[53] R. GODEMENT, *Topologie Algébrique et Théorie des Faisceaux*, Hermann, Paris, 1964.

[54] R. GORDON, A.J. POWER, AND R. STREET, *Coherence for Tricategories*, Mem. Amer. Math. Soc. **117** (1995), No. 558.

[55] J.W. GRAY, Category-valued sheaves, *Bull. Amer. Math. Soc.* **68** (1962), 451–453.

[56] J.W. GRAY, Sheaves with values in a category, *Topology* **3** (1965), 1–18.

[57] J.W. GRAY, Fibred and cofibred categories, *Proceedings of the Conference on Categorical Algebra at La Jolla*, Springer, Berlin, 1966, pp. 21–83.

[58] J.W. GRAY, The categorical comprehension scheme, *Category Theory, Homology Theory and their Applications, III*, Springer, Berlin, 1969, pp. 242–312.

[59] J.W. GRAY, The 2-adjointness of the fibred category construction, *Symposia Mathematica IV (INDAM, Rome, 1968/69)*, Academic Press, London, 1970, pp. 457–492.

[60] J.W. GRAY, Report on the meeting of the Midwest Category Seminar in Zürich, Lecture Notes in Math. **195**, Springer, Berlin, 1971, 248–255.

[61] J.W. GRAY, Quasi-Kan extensions for 2-categories, *Bull. Amer. Math. Soc.* **80** (1974), 142–147.

[62] J.W. GRAY, *Formal Category Theory: Adjointness for 2-Categories*, Lecture Notes in Math. **391**, Springer, Berlin, 1974.

[63] J.W. GRAY, Coherence for the tensor product of 2-categories, and braid groups, in *Algebra, Topology, and Category Theory (a collection of papers in honour of Samuel Eilenberg)*, Academic Press, New York, 1976, pp. 63–76.

[64] J.W. GRAY, Fragments of the history of sheaf theory, Lecture Notes in Math. **753**, Springer, Berlin, 1979, pp. 1–79.

[65] J.W. GRAY, The existence and construction of lax limits, *Cah. Top. Géom. Diff. Cat.* **21** (1980), 277–304.

[66] J.W. GRAY, Closed categories, lax limits and homotopy limits, *J. Pure Appl. Algebra* **19** (1980), 127–158.

[67] J.W. GRAY, Enriched algebras, spectra and homotopy limits, Lecture Notes in Math. **962**, Springer, Berlin, 1982, pp. 82–99.

[68] J.W. GRAY, The representation of limits, lax limits and homotopy limits as sections, Contemp. Math. **30**, AMS, Providence, Rhode Island, 1984, pp. 63–83.

[69] M. HAKIM, *Topos annelés et schémas relatifs*, Ergebnisse der Mathematik und ihrer Grenzgebiete **64**, Springer, Berlin, 1972.

[70] C. HERMIDA, M. MAKKAI, AND J. POWER, On weak higher dimensional categories (preprint 1997 at http://www.math.mcgill.ca/makkai/).

[71] P.J. HILTON AND S. WYLIE, *Homology Theory: An Introduction to Algebraic Topology*, Cambridge University Press, Cambridge, 1960.

[72] M. JOHNSON, *Pasting Diagrams in n-Categories with Applications to Coherence Theorems and Categories of Paths*, Ph.D. Thesis, University of Sydney, Australia, 1987.

[73] M. JOHNSON, The combinatorics of n-categorical pasting, *J. Pure Appl. Algebra* **62** (1989), 211–225.

[74] M. JOHNSON AND R. WALTERS, On the nerve of an n-category, *Cah. Top. Géom. Diff. Cat.* **28** (1987), 257–282.

[75] A. JOYAL, Disks, duality and Θ-categories, preprint and talk at the AMS Meeting in Montréal (September 1997).

[76] A. JOYAL AND R. STREET, The geometry of tensor calculus I, *Advances in Math.* **88** (1991), 55–112.

[77] A. JOYAL AND R. STREET, Tortile Yang–Baxter operators in tensor categories, *J. Pure Appl. Algebra* **71** (1991), 43–51.

[78] A. JOYAL AND R. STREET, An introduction to Tannaka duality and quantum groups, Lecture Notes in Math. **1488**, Springer, Berlin, 1991, pp. 411–492.

[79] A. JOYAL AND R. STREET, Pullbacks equivalent to pseudopullbacks, *Cah. Top. Géom. Diff. Cat.* **34** (1993), 153–156.

[80] A. JOYAL AND R. STREET, Braided tensor categories, *Advances in Math.* **102** (1993), 20–78.

[81] A. JOYAL, R. STREET, AND D. VERITY, Traced monoidal categories, *Math. Proc. Cambridge Philos. Soc.* **119** (1996), 447–468.

[82] M.M. KAPRANOV AND V.A. VOEVODSKY, Combinatorial-geometric aspects of polycategory theory: pasting schemes and higher Bruhat orders (List of results), *Cah. Topologie et Géom. Diff. Cat.* **32** (1991), 11–27.

[83] M.M. KAPRANOV AND V.A. VOEVODSKY, ∞-Groupoids and homotopy types, *Cah. Top. Géom. Diff. Cat.* **32** (1991), 29–46.

[84] M.M. KAPRANOV AND V.A. VOEVODSKY, 2-Categories and Zamolodchikov tetrahedra equations, *Proc. Symp. Pure Math.* **56** (1994), 177–259.

[85] M.M. KAPRANOV AND V.A. VOEVODSKY, Braided monoidal 2-categories and Manin–Schechtman higher braid groups, *J. Pure Appl. Algebra* **92** (1994), 241–267.

[86] C. KASSEL, *Quantum Groups*, Springer, Berlin, 1995.

[87] G.M. KELLY, Observations on the Künneth theorem, *Proc. Cambridge Philos. Soc.* **59** (1963), 575–587.

[88] G.M. KELLY, Complete functors in homology. I. Chain maps and endomorphisms, *Proc. Cambridge Philos. Soc.* **60** (1964), 721–735.

[89] G.M. KELLY, Complete functors in homology. II. The exact homology sequence, *Proc. Cambridge Philos. Soc.* **60** (1964), 737–749.

[90] G.M. KELLY, On Mac Lane's conditions for coherence of natural associativities, commutativities, etc., *J. Algebra* **1** (1964), 397–402.

[91] G.M. KELLY, A lemma in homological algebra, *Proc. Cambridge Philos. Soc.* **61** (1965), 49–52.

[92] G.M. KELLY, Tensor products in categories, *J. Algebra* **2** (1965), 15–37.

[93] G.M. KELLY, Chain maps inducing zero homology maps, *Proc. Cambridge Philos. Soc.* **61** (1965), 847–854.

[94] G.M. KELLY, Adjunction for enriched categories, in *Reports of the Midwest Category Seminar, III*, Lecture Notes in Math. **106**, Springer, Berlin, 1969, pp. 166–177.

[95] G.M. KELLY, Many-variable functorial calculus. I, in *Coherence in Categories*, Lecture Notes in Math. **281**, Springer, Berlin, 1972, pp. 66–105.

[96] G.M. KELLY, An abstract approach to coherence, in *Coherence in Categories*, Lecture Notes in Math. **281**, Springer, Berlin, 1972, pp. 106–147.

[97] G.M. KELLY, A cut-elimination theorem, in *Coherence in Categories*, Lecture Notes in Math. **281**, Springer, Berlin, 1972, pp. 196–213.

[98] G.M. KELLY, Doctrinal adjunction, in *Category Seminar (Proc. Sem., Sydney, 1972/1973)*, Lecture Notes in Math. **420**, Springer, Berlin, 1974, pp. 257–280.

[99] G.M. KELLY, On clubs and doctrines, in *Category Seminar (Proc. Sem., Sydney, 1972/1973)*, Lecture Notes in Math. **420**, Springer, Berlin, 1974, pp. 181–256.

[100] G.M. KELLY, Coherence theorems for lax algebras and for distributive laws, in: *Category Seminar (Proc. Sem., Sydney, 1972/1973)*, Lecture Notes in Math. **420**, Springer, Berlin, 1974, pp. 281–375.

[101] G.M. KELLY, *Basic Concepts of Enriched Category Theory*, London Math. Soc. Lecture Notes Series **64**, Cambridge University Press, Cambridge, 1982. Also in *Reprints in Theory Appl. Cat.* **10** (2005), 1–136.

[102] G.M. KELLY, Elementary observations on 2-categorical limits, *Bull. Austral. Math. Soc.* **39** (1989), 301–317.

[103] G.M. KELLY, On clubs and data-type constructors, London Math. Soc. Lecture Note Ser. **177**, Cambridge Univ. Press, Cambridge, 1992, pp. 163–190.

[104] G.M. KELLY, A. LABELLA, V. SCHMITT, AND R. STREET, Categories enriched on two sides, *J. Pure Appl. Algebra* **168** (2002), 53–98.

[105] G.M. KELLY AND M.L. LAPLAZA, Coherence for compact closed categories, *J. Pure Appl. Algebra* **19** (1980), 193–213.

[106] G.M. KELLY AND S. MAC LANE, Coherence in closed categories, *J. Pure Appl. Algebra* **1** (1971), 97–140.

[107] G.M. KELLY AND S. MAC LANE, Erratum: Coherence in closed categories, *J. Pure Appl. Algebra* 1 (1971), p. 219.

[108] G.M. KELLY AND R. STREET, eds., *Abstracts of the Sydney Category Seminar 1972/3*. (First edition with brown cover, U. of New South Wales; second edition with green cover, Macquarie U.)

[109] G.M. KELLY AND R. STREET, Review of the elements of 2-categories, *Category Seminar (Proc. Sem., Sydney, 1972/1973)*, Lecture Notes in Math. **420**, Springer, Berlin, 1974, pp. 75–103.

[110] V. KHARLAMOV AND V. TURAEV, On the definition of the 2-category of 2-knots, *Amer. Math. Soc. Transl.* **174** (1996), 205–221.

[111] A. KOCK, Monads for which structures are adjoint to units, *rhus Univ. Preprint Series* **35** (1972–73), 1–15.

[112] A. KOCK, Monads for which structures are adjoint to units, *J. Pure Appl. Algebra* **104** (1995), 41–59.

[113] S. LACK AND R. STREET, The formal theory of monads II, *J. Pure Appl. Algebra* **175** (2002), 243–265.

[114] L. LANGFORD, 2-Tangles as a Free Braided Monoidal 2-Category with Duals, Ph.D. dissertation, U. of California at Riverside, 1997.

[115] F.W. LAWVERE, The category of categories as a foundation for mathematics, in *Proceedings of the Conference on Categorical Algebra at La Jolla*, Springer, Berlin, 1966, pp. 1–20.

[116] F.W. LAWVERE, Ordinal sums and equational doctrines, in *Seminar on Triples and Categorical Homology Theory*, Lecture Notes in Math. **80** (1969), 141–155.

[117] F.W. LAWVERE, Metric spaces, generalised logic, and closed categories, *Rend. Sem. Mat. Fis. Milano* **43** (1974), 135–166. Also in *Reprints in Theory Appl. Cat.* **1** (2002), pp. 1-37.

[118] J.-L. LODAY, Spaces with finitely many non-trivial homotopy groups, *J. Pure Appl. Algebra* **24** (1982), 179–202.

[119] M. MACKAAY, Spherical 2-categories and 4-manifold invariants, *Advances in Math.* **143** (1999), 288–348.

[120] S. MAC LANE, Possible programs for categorists, Lecture Notes in Math. **86**, Springer, Berlin, 1969, pp. 123–131.

[121] S. MAC LANE, *Categories for the Working Mathematician*, Springer, Berlin, 1971.

[122] S. MAC LANE AND R. PARÉ, Coherence for bicategories and indexed categories, J. Pure Appl. Algebra **37** (1985), 59–80.

[123] F. MARMOLEJO, Distributive laws for pseudomonads, *Theory Appl. Cat.* **5** (1999), 91–147.

[124] P. MAY, *The Geometry of Iterated Loop Spaces*, Lecture Notes in Math. **271**, Springer, Berlin, 1972.

[125] M. McINTYRE AND T. TRIMBLE, The geometry of Gray-categories, *Advances in Math.* (to appear).

[126] A. PITTS, Applications of sup-lattice enriched category theory to sheaf theory, *Proc. London Math. Soc. (3)* **57** (1988), 433–480.

[127] A.J. POWER, An n-categorical pasting theorem, in *Category Theory, Proceedings, Como 1990*, eds. A. Carboni, M. C. Pedicchio and G. Rosolini, Lecture Notes in Math. **1488** Springer, Berlin, 1991, pp. 326–358.

[128] A.J. POWER, Why tricategories?, *Inform. & Comput.* **120** (1995), 251–262.

[129] N. YU. RESHETIKHIN AND V.G. TURAEV, Ribbon graphs and their invariants derived from quantum groups, *Comm. Math. Phys.* **127** (1990), 1–26.

[130] J.E. ROBERTS, Mathematical aspects of local cohomology, *Algèbres d'Opérateurs et leurs Applications en Physique Mathématique (Proc. Colloq., Marseille, 1977)*, Colloq. Internat. CNRS **274**, CNRS, Paris, 1979, pp. 321–332.

[131] R.D. ROSEBRUGH AND R.J. WOOD, Proarrows and cofibrations, *J. Pure Appl. Algebra* **53** (1988), 271–296.

[132] G. SEGAL, Classifying spaces and spectral sequences, *Inst. Hautes Études Sci. Publ. Math.* **34** (1968), 105–112.

[133] M.C. SHUM, *Tortile Tensor Categories*, Ph.D. Thesis, Macquarie University, 1989; *J. Pure Appl. Algebra* **93** (1994), 57–110.

[134] R. STREET, *Homotopy Classification of Filtered Complexes*, Ph.D. Thesis, University of Sydney, 1968.

[135] R. STREET, The formal theory of monads, *J. Pure Appl. Algebra* **2** (1972), 149–168.

[136] R. STREET, Two constructions on lax functors, *Cah. Top. Géom. Diff. Cat.* **13** (1972), 217–264.

[137] R. STREET, Homotopy classification of filtered complexes, *J. Australian Math. Soc.* **15** (1973), 298–318.

[138] R. STREET, Fibrations and Yoneda's lemma in a 2-category, in *Category Seminar (Proc. Sem., Sydney, 1972/1973)*, Lecture Notes in Math. **420**, Springer, Berlin, 1974, pp. 104–133.

[139] R. STREET, Elementary cosmoi I, in *Category Seminar (Proc. Sem., Sydney, 1972/1973)*, Lecture Notes in Math. **420**, Springer, Berlin, 1974, pp. 134–180.

[140] R. STREET, Limits indexed by category-valued 2-functors, *J. Pure Appl. Algebra* **8** (1976), 149–181.

[141] R. STREET, Cosmoi of internal categories, *Trans. Amer. Math. Soc.* **258** (1980), 271–318.

[142] R. STREET, Fibrations in bicategories, *Cah. Top. Géom. Diff. Cat.* **21** (1980), 111–160.

[143] R. STREET, Conspectus of variable categories, *J. Pure Appl. Algebra* **21** (1981), 307–338.

[144] R. STREET, Cauchy characterization of enriched categories, *Rendiconti del Seminario Matematico e Fisico di Milano* **51** (1981), 217–233. (See [164].)

[145] R. STREET, Two dimensional sheaf theory, *J. Pure Appl. Algebra* **23** (1982), 251–270.

[146] R. STREET, Characterization of bicategories of stacks, Lecture Notes in Math. **962** (1982), 282–291.

[147] R. STREET, Enriched categories and cohomology, *Quaestiones Math.* **6** (1983), 265–283.

[148] R. STREET, Absolute colimits in enriched categories, *Cah. Top. Géom. Diff. Cat.* **24** (1983), 377–379.

[149] R. STREET, Homotopy classification by diagrams of interlocking sequences, *Math. Colloquium University of Cape Town* **13** (1984), 83–120.

[150] R. STREET, The algebra of oriented simplexes, *J. Pure Appl. Algebra* **49** (1987), 283–335.

[151] R. STREET, Correction to Fibrations in bicategories, *Cah. Top. Géom. Diff. Cat.* **28** (1987), 53–56.

[152] R. STREET, Fillers for nerves, Lecture Notes in Math. **1348**, Springer, Berlin, 1988, pp. 337–341.

[153] R. STREET, Parity complexes, *Cah. Top. Géom. Diff. Cat.* **32** (1991), 315–343.

[154] R. STREET, Categorical structures, in *Handbook of Algebra*, Volume 1, ed. M. Hazewinkel, Elsevier, Amsterdam, 1996, pp. 529–577.

[155] R. STREET, Higher categories, strings, cubes and simplex equations, *Applied Categorical Structures* **3** (1995), 29–77 & 303.

[156] R. STREET, Parity complexes: corrigenda, *Cah. Top. Géom. Diff. Cat.* **35** (1994), 359–361.

[157] R. STREET, Low-dimensional topology and higher-order categories, Proceedings of CT95, Halifax, July 9–15 1995; http://www.maths.mq.edu.au/~street/LowDTop.pdf.

[158] R. STREET, The role of Michael Batanin's monoidal globular categories, in *Higher Category Theory*, eds. E. Getzler and M. Kapranov, Contemp. Math. **230**, AMS, Providence, Rhode Island, 1998, pp. 99–116.

[159] R. STREET, The petit topos of globular sets, *J. Pure Appl. Algebra* **154** (2000), 299–315.

[160] R. STREET, Functorial calculus in monoidal bicategories, *Applied Categorical Structures* **11** (2003), 219–227.

[161] R. STREET, Weak omega-categories, in *Diagrammatic Morphisms and Applications*, eds. D. Radford, F. Souza and D. Yetter, Contemp. Math. **318**, AMS, Providence, Rhode Island, 2003, pp. 207–213.

[162] R. STREET, Categorical and combinatorial aspects of descent theory, *Applied Categorical Structures* **12** (2004), 537–576.

[163] R. STREET, Frobenius monads and pseudomonoids, *J. Math. Phys.* **45**(10) (October 2004), 3930–3948.

[164] R. STREET, Cauchy characterization of enriched categories, *Reprints in Theory and Applications of Categories* **4** (2004), 1–16. (See [144].)

[165] R. STREET AND R.F.C. WALTERS, Yoneda structures on 2-categories, *J. Algebra* **50** (1978), 350–379.

[166] Z. TAMSAMANI, Sur des notions de *n*-categorie et *n*-groupoide non-stricte via des ensembles multi-simpliciaux, *K-Theory* **16** (1999), 51–99.

[167] T. TRIMBLE, The definition of tetracategory (handwritten diagrams; August 1995).

[168] V.G. TURAEV, The Yang–Baxter equation and invariants of links, *Invent. Math.* **92** (1988), 527–553.

[169] D. VERITY, Complicial sets, *Mem. Amer. Math. Soc.* (to appear; arXiv:math/
       0410412v2).
[170] R.F.C. WALTERS, Sheaves on sites as Cauchy-complete categories, *J. Pure Appl.
       Algebra* **24** (1982), 95–102.
[171] H. WOLFF, Cat and graph, *J. Pure Appl. Algebra* **4** (1974), 123–135.
[172] V. ZÖBERLEIN, Doctrines on 2-categories, *Math. Z.* **148** (1976), 267–279.
       (Originally Doktrinen auf 2-Kategorien, manuscript, Math. Inst. Univ.
       Zürich, 1973.)

IMA SUMMER PROGRAM PARTICIPANTS

n-Categories: Foundations and Applications

- Scot Adams, Institute for Mathematics and its Applications, University of Minnesota
- Douglas N. Arnold, Institute for Mathematics and its Applications, University of Minnesota
- Donald G. Aronson, Institute for Mathematics and its Applications, University of Minnesota
- Steve Awodey, Department of Philosophy, Carnegie Mellon University
- Nils A. Baas, Department of Mathematical Sciences, Norewegian University of Science and Technology
- Bernard Badzioch, School of Mathematics, University of Minnesota
- John C. Baez, Department of Mathematics, University of California - Riverside
- Antar Bandyopadhyay, Institute for Mathematics and its Applications, University of Minnesota
- Michael Batanin, Department of Mathematics, Macquarie University
- Clemens Berger, Laboratoire J.A. Dieudonné, University of Nice
- Julia E. Bergner, Department of Mathematics, Kansas State University
- William Boshuck, Department of Mathematics, John Abbott College
- Lawrence Breen, UMR CNRS 7539 Institut Galilée, Université Paris 13
- Ronnie Brown, Department of Mathematics, University of Wales
- Manuel Bullejos, Departamento de Álgebra, University of Granada
- Jeffrey L. Caruso, Pageflex, Inc.
- Sunil Chebolu, Department of Mathematics, University of Washington
- Eugenia Cheng, Department of Pure Mathematics and Mathematical Statistics, Cambridge University
- David Neil Corfield, Department of Philosophy, Oxford University
- Alissa Crans, Department of Mathematics, University of California - Riverside
- Alexei Davydov, Department of Mathematics, Macquarie University

J.C. Baez, J.P. May (eds.), *Towards Higher Categories*, The IMA Volumes in Mathematics and its Applications 152, DOI 10.1007/978-1-4419-1524-5,
© Springer Science+Business Media, LLC 2010

- Aurora Ines Del Rio Cabeza, Algebra Department, University of Granada
- James Dolan, Department of Mathematics, University of California - Riverside
- Josep Elgueta, Departament Matematica Aplicada II, Universitat Politecnica de Catalunya
- Anthony D. Elmendorf, Department of Mathematics, Purdue University
- Ulrich Fahrenberg, Department of Mathematical Sciences, Aalborg University
- Lisbeth Fajstrup, Department of Mathematical Sciences, Aalborg University
- Mark Feshbach, Department of Mathematics, University of Minnesota
- Zbigniew Fiedorowicz, Department of Mathematics, Ohio State University
- Thomas M. Fiore, Department of Mathematics, University of Michigan
- Stefan Forcey, Department of Mathematics, Virginia Tech
- Carl A. Futia
- Philippe Gaucher, Université Paris 7, Denis-Diderot
- Eric Goubault, Commissariat l'Energie Atomique
- Nick Gurski, Department of Mathematics, University of Chicago
- Eric Harrelson, School of Mathematics, University of Minnesota
- Mehdi Hakim Hashemi, School of Mathematics, University of Minnesota
- Claudio Hermida, School of Computing, Queen's University
- Thomas Hunter, Department of Mathematics and Statistics, Swarthmore College
- Martin Hyland, Department of Pure Mathematics and Mathematical Statistics, Cambridge University
- Michael Johnson, Department of Mathematics and Computer Science, Macquarie University
- André Joyal, Shannon Laboratory Département de mathématiques, Universit du Québec à Montréal
- Hyeung-Joon Kim, School of Mathematics, University of Minnesota
- Joachim Kock, Department of Mathematics, Universite du Quebec a Montreal
- Wojciech Komornicki, Department of Mathematics, Hamline University

- Sanjeevi Krishnan, Department of Mathematics, University of Chicago
- Stephen Lack, School of Computing and Mathematics, University of Western Sydney
- Yves Lafont, Institut de Mathématiques de Luminy, Université de la Méditerranée (Marseille)
- Aaron Lauda, Department of Mathematics, University of California - Riverside
- Tom Leinster, Department of Mathematics, University of Glasgow
- Marco Mackaay, Area Departamental de Matematica, University do Algarve
- Michael Makkai, Department of Mathematics, McGill University
- Gianfranco Mascari, Consiglio Nazionale delle Ricerche (CNR)
- J. Peter May, Department of Mathematics, University of Chicago
- William Messing, School of Mathematics, University of Minnesota
- Francois Metayer, Department of Mathematics, Equipe PPS Universite Paris 7
- Jean-Pierre Meyer, Department of Mathematics, The Johns Hopkins University
- Gary Nan Tie, The St. Paul Companies
- Joshua Paul Nichols-Barrer, Department of Mathematics, Massachusetts Institute of Technology
- Simona Paoli, Department of Mathematics, Macquarie University
- Timothy Porter, School of Informatics, University of Wales
- John Power, School of Informatics, University of Edinburgh
- Victor Reiner, Department of Mathematics, University of Minnesota
- Tony Robbin, Yale University Press
- Jonathan Rogness, School of Mathematics, University of Minnesota
- Andrew Michael Salch, Department of Mathematics, University of Rochester
- Fadil Santosa, Institute for Mathematics and its Applications, University of Minnesota
- Dana Stewart Scott, Department of Computer Sciences, Carnegie Mellon University
- Michael Shulman, Department of Mathematics, University of Chicago
- Richard Steiner, Department of Mathematics, University of Glasgow
- Danny Stevenson, Department of Pure Mathematics, University of Adelaide

- Ross Street, Centre of Australian Category Theory, Macquarie University
- James Swenson, School of Mathematics, University of Minnesota
- Bertrand Toen, Laboratoire J.A. Dieudonne, University of Nice
- Gabriele Vezzosi, Dipartimento di Matematica Applicata 'G. Sansone', Universit degli Studi di Firenze
- Alexander A. (Sasha) Voronov, School of Mathematics, University of Minnesota
- Peter Webb, School of Mathematics, University of Minnesota
- Mark Weber, Department of Mathematics and Statistics, University of Ottawa
- Ittay Weiss, Department of Mathematics, Utrecht University
- Marek Zawadowski, Department of Mathematics, University of Warsaw
- Javier Zuniga, School of Mathematics, University of Minnesota

Continued from page ii

1997–1998 Emerging Applications of Dynamical Systems
1998–1999 Mathematics in Biology
1999–2000 Reactive Flows and Transport Phenomena
2000–2001 Mathematics in Multimedia
2001–2002 Mathematics in the Geosciences
2002–2003 Optimization
2003–2004 Probability and Statistics in Complex Systems: Genomics,
           Networks, and Financial Engineering
2004–2005 Mathematics of Materials and Macromolecules: Multiple Scales,
           Disorder, and Singularities
2005-2006 Imaging
2006-2007 Applications of Algebraic Geometry
2007-2008 Mathematics of Molecular and Cellular Biology
2008-2009 Mathematics and Chemistry
2009-2010 Complex Fluids and Complex Flows
2010-2011 Simulating Our Complex World: Modeling, Computation
           and Analysis

IMA SUMMER PROGRAMS

1987 Robotics
1988 Signal Processing
1989 Robust Statistics and Diagnostics
1990 Radar and Sonar (June 18–29)
      New Directions in Time Series Analysis (July 2–27)
1991 Semiconductors
1992 Environmental Studies: Mathematical, Computational, and
           Statistical Analysis
1993 Modeling, Mesh Generation, and Adaptive Numerical Methods
           for Partial Differential Equations
1994 Molecular Biology
1995 Large Scale Optimizations with Applications to Inverse Problems,
           Optimal Control and Design, and Molecular and Structural
           Optimization
1996 Emerging Applications of Number Theory (July 15–26)
      Theory of Random Sets (August 22–24)
1997 Statistics in the Health Sciences
1998 Coding and Cryptography (July 6–18)
      Mathematical Modeling in Industry (July 22–31)
1999 Codes, Systems, and Graphical Models (August 2–13, 1999)
2000 Mathematical Modeling in Industry: A Workshop for Graduate
           Students (July 19–28)
2001 Geometric Methods in Inverse Problems and PDE Control
           (July 16–27)
2002 Special Functions in the Digital Age (July 22–August 2)

2003	Probability and Partial Differential Equations in Modern Applied Mathematics (July 21–August 1)
2004	n-Categories: Foundations and Applications (June 7–18)
2005	Wireless Communications (June 22–July 1)
2006	Symmetries and Overdetermined Systems of Partial Differential Equations (July 17–August 4)
2007	Classical and Quantum Approaches in Molecular Modeling (July 23-August 3)
2008	Geometrical Singularities and Singular Geometries (July 14-25)
2009	Nonlinear Conservation Laws and Applications (July 13-31)

IMA "HOT TOPICS/SPECIAL" WORKSHOPS

- Challenges and Opportunities in Genomics: Production, Storage, Mining and Use, April 24–27, 1999
- Decision Making Under Uncertainty: Energy and Environmental Models, July 20–24, 1999
- Analysis and Modeling of Optical Devices, September 9–10, 1999
- Decision Making under Uncertainty: Assessment of the Reliability of Mathematical Models, September 16–17, 1999
- Scaling Phenomena in Communication Networks, October 22–24, 1999
- Text Mining, April 17–18, 2000
- Mathematical Challenges in Global Positioning Systems (GPS), August 16–18, 2000
- Modeling and Analysis of Noise in Integrated Circuits and Systems, August 29–30, 2000
- Mathematics of the Internet: E-Auction and Markets, December 3–5, 2000
- Analysis and Modeling of Industrial Jetting Processes, January 10–13, 2001
- Special Workshop: Mathematical Opportunities in Large-Scale Network Dynamics, August 6–7, 2001
- Wireless Networks, August 8–10 2001
- Numerical Relativity, June 24–29, 2002
- Operational Modeling and Biodefense: Problems, Techniques, and Opportunities, September 28, 2002
- Data-driven Control and Optimization, December 4–6, 2002
- Agent Based Modeling and Simulation, November 3–6, 2003
- Enhancing the Search of Mathematics, April 26–27, 2004
- Compatible Spatial Discretizations for Partial Differential Equations, May 11–15, 2004
- Adaptive Sensing and Multimode Data Inversion, June 27–30, 2004
- Mixed Integer Programming, July 25–29, 2005

- New Directions in Probability Theory, August 5–6, 2005
- Negative Index Materials, October 2–4, 2006
- The Evolution of Mathematical Communication in the Age of Digital Libraries, December 8–9, 2006
- Math is Cool! and Who Wants to Be a Mathematician?, November 3, 2006
- Special Workshop: Blackwell-Tapia Conference, November 3–4, 2006
- Stochastic Models for Intracellular Reaction Networks, May 11–13, 2008
- Multi-Manifold Data Modeling and Applications, October 27–30, 2008
- Mixed-Integer Nonlinear Optimization: Algorithmic Advances and Applications, November 17–21, 2008
- Higher Order Geometric Evolution Equations: Theory and Applications from Microfluidics to Image Understanding, March 23–26, 2009
- Career Options for Women in Mathematical Sciences, April 2-4, 2009
- MOLCAS, May 4-8, 2009
- IMA Interdisciplinary Research Experience for Undergraduates, June 29-July 31, 2009
- Research in Imaging Sciences, October 5-7, 2009

SPRINGER LECTURE NOTES FROM THE IMA:

The Mathematics and Physics of Disordered Media
    Editors: Barry Hughes and Barry Ninham
    (Lecture Notes in Math., Volume 1035, 1983)

Orienting Polymers
    Editor: J.L. Ericksen
    (Lecture Notes in Math., Volume 1063, 1984)

New Perspectives in Thermodynamics
    Editor: James Serrin
    (Springer-Verlag, 1986)

Models of Economic Dynamics
    Editor: Hugo Sonnenschein
    (Lecture Notes in Econ., Volume 264, 1986)

The IMA Volumes
in Mathematics and its Applications

Current Volumes: